MODERNIZATION AND CLEAVAGE IN DUTCH SOCIETY

To Yvonne and Allard

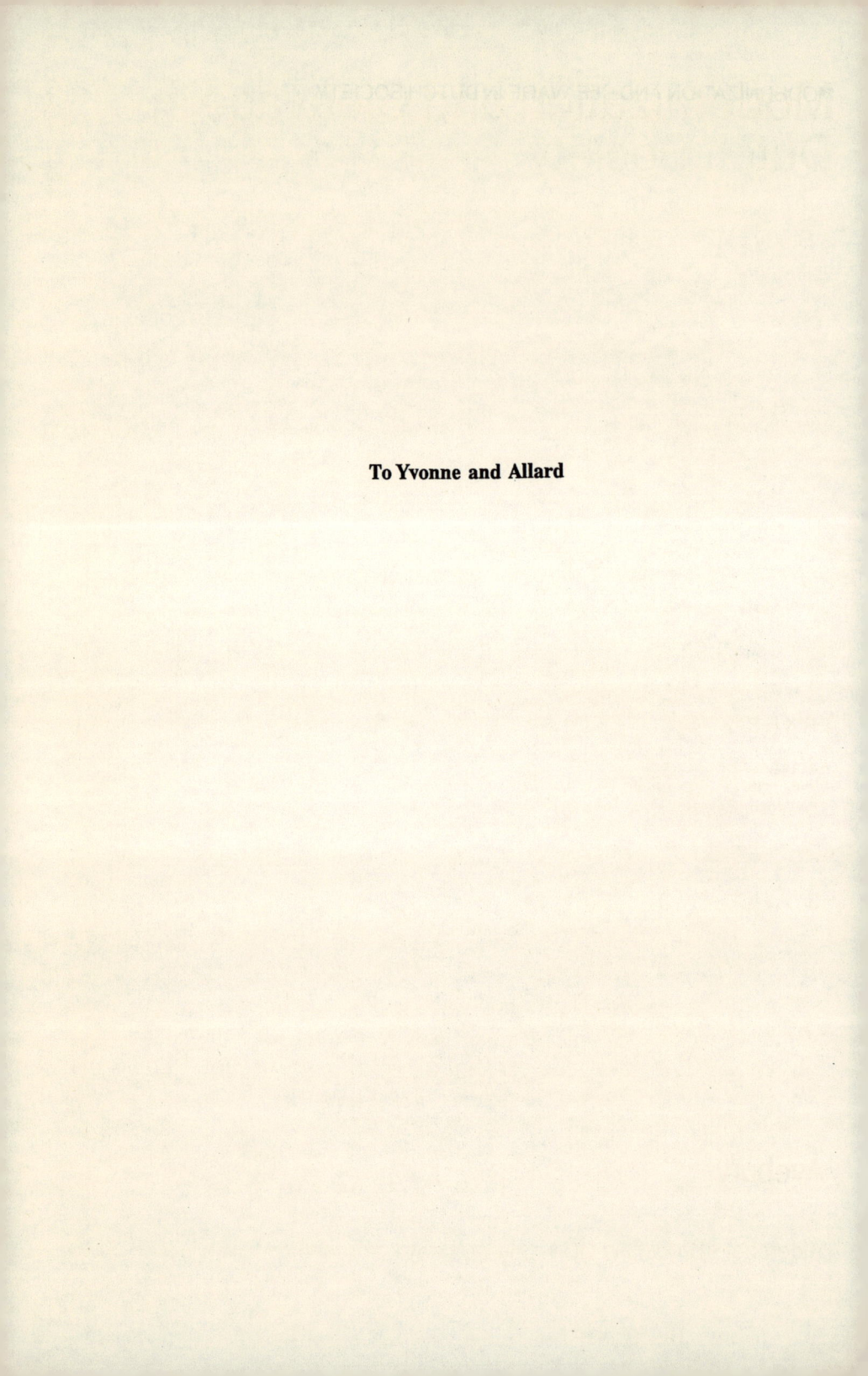

Modernization and Cleavage in Dutch Society

A Study of Long Term Economic and Social Change

ERIK H BAX

Faculty of Economics
University of Groningen
The Netherlands

Avebury

Aldershot · Brookfield USA · Hong Kong · Singapore · Sydney

© Erik H. Bax 1990

All rights reserved. No part of this publication may be reproduced, stored in a retrieval system, or transmitted in any form or by any means, electronic, mechanical photocopying, recording, or otherwise without the prior permission of Gower Publishing Company Limited.

Published by
Avebury
Gower Publishing Company Limited
Gower House
Croft Road
Aldershot
Hants GU11 3HR
England

Gower Publishing Company
Old Post Road
Brookfield
Vermont 05036
USA

HC
325
.B33
1990

Printed in Great Britain by
Athenaeum Press Ltd, Newcastle upon Tyne.

ISBN 0 566 07164 9

TABLE OF CONTENTS

List of Tables	viii
List of Figures	xi
Acknowledgements	xiii
Chapter I Introduction	1
Chapter II On Modernization	7
Introduction	7
The Idea of Modernization	7
The Social and Intellectual Origins of Modernization Theory	7
Defining Modernization	8
The Idea of Bipolarity: Tradition versus Modernity	13
The Problem of Homogeneity	15
The Problem of Chronology	15
The Problem of Finality	16
The Problem of Empirical Boundaries	16
The Problem of Universality	18
The Idea of Prerequisites and Consequences of Modernization	19
The Idea of Convergence	23
The Explanation of the Variance of Modernity	27
The Nature and Key Elements of Modernization	29
Chapter III Monitoring Dutch Modernization	35
Modern Economic Growth and Technological Development	35
Twentieth Century Technological Development and Economic Growth	41
Economic Growth and Technological Development Summarized	43
The Division of Labour, Inequality and Social Mobility	44
Twentieth Century Changes in the Division of Labour, Inequality and Social Mobility	47
Enlargement of Scale	51
Bureaucratization and the Formalization of Social Relations	58
Public Sector Growth	63
Standardization of Information	65
Monitoring Dutch Modernization: Conclusions	70
Chapter IV Long Term Modernization and Institutional Change in Dutch Society	73
Introduction	73
Segmented Pluralism in the Dutch Republic	74
The Origins of the Dutch Republic	74
The Power Structure	75
The Nature of Segmented Pluralism in the Republic	77
Mechanisms of Integration in the Republic	79
Structural Factors and Institutions	79

The Culture of "Living-Apart-Together"	81
Modernization and the Change of Dutch Pluralism	82
National Unification	82
The Rise of Liberalism and the Constitutional Change of 1848	83
Shifting Alliances After 1850	85
The Rise of Protestant Organization	87
The School Issue	89
The Rise of Roman-Catholic Organization	91
The Rise of a Segmented Working Class	93
Conclusions	96
Chapter V Dutch Pillarization: Theory and Process	**99**
Introduction	99
The Concepts of Pillar and Pillarization	99
Theories of Pillarization	105
The Emancipation Hypothesis	105
The Protection Hypothesis	106
The Social Control Approach	107
The Consociational Democracy Theory	107
Modernization and Pillarization	110
Monitoring Pillarization	114
Politics	114
Organizational Pillarization	117
Education	117
Health Service	120
Broadcasting	121
Economic Life and Industrial Relations	123
Individual Behaviour: Social Distance	130
Concluding Remarks	138
Chapter VI Dutch De-Pillarization: Theory and Process	**141**
Introduction	141
Monitoring De-pillarization	141
Politics	141
Organizational De-pillarization	147
Education	147
Broadcasting and Health Service	149
Economic Life and Industrial Relations	149
Individual Behaviour: Social Distance	153
Dissimilarities in the De-Pillarization of Dutch Society	154
De-pillarization Explained	156
De-pillarization and the Rise of the Welfare State	157
De-pillarization and Enlargement of Scale	159
De-pillarization and Secularization	163
De-pillarization and the Individualization of Dutch Society	167
Modernization and Pillarized Social Control: The Inverted U-Curve	170
Notes	**175**

Appendix A   **187**
   A.1. Per Capita National Income 1900-1986   187
   A.2. Employment by Sector of Industry   188
   A.3. Urbanization   188
   A.4. Size of Enterprises by Employed Persons   191
   A.5. Ministry Personnel as a Percentage of the Labour Force and of Total Population   192
   A.6. Civil Cases Brought to Courts of Justice   194
   A.7. Public Expenditures   198
   A.8. Education   200
   A.9. Newspapers Sent by Inland Mail   202

Appendix B   **203**
   B.1. Politics   203
   B.2. Education   204
   B.3. Health Service   205
   B.4. Broadcasting   207
   B.5. Membership Trade-Union Federations   209
   B.6. Strike Activity   210
   B.7. Connubium   214

Appendix C   **229**
   C.1. Marriages Consecrated   229

Summary   **233**

References   **237**

Index   **247**

LIST OF TABLES

3.1. Population Increase of Eleven European Countries, 1500-1820 (Percentages and Average Annual Compound Growth Rates) — 36
3.2. Average Annual Compound Growth Rates of Steampower Application (numbers of steam engines and horse power), 1853-1895 — 45
3.3. Numbers of Clerks as a Percentage of the Labour Force (1849-1909) — 46
3.4. Average Annual Compound Growth Rates of Employment by Sector of Industry, 1899-1981 — 48
3.5. Average Annual Compound Growth Rates of Employment by Sector of Industry as a Percentage of Total Employment, 1899-1981 — 49
3.6. Dutch Population by Degree of Urbanization as a Percentage of Total Dutch Population, 1899-1980 — 54
3.7. Average Annual Compound Growth Rates of Dutch Population by Degree of Urbanization as a Percentage of Total Population, 1899-1980 — 54
3.8. Balance of Dutch Inland Migration by Type of Municipality (per Thousand Inhabitants), 1957-1980 — 54
3.9. Number of Employed Persons in Medium and Large Size Industrial Enterprises as a Percentage of the Total Number of Employed Persons in Industry, 1859-1978 — 56
3.10. Average Annual Compound Growth Rates of the Number of Persons Employed in Medium and Large Size Industrial Enterprises as a Percentage of the Total Number of Persons Employed in Industry, 1859-1978 — 56
3.11. Average Number of Employed Workers per Enterprise Establishment by Sector of Industry, 1930, 1950, 1963 and 1978 — 56
3.12. Average Annual Compound Growth Rates of the Average Number of Employed Workers per Enterprise Establishment by Sector of Industry, 1930-1950, 1950-1963, 1963-1978 and 1930-1978 — 57
3.13. Average Annual Compound Growth Rates of Enterprises with Five Employees or Less and Enterprises with More than Five Employees as a Percentage of All Enterprises, 1930-1978 — 57
3.14. Average Annual Compound Growth Rates of Dutch Ministry Personnel as a Percentage of the Labour Force and of Total Population, 1849-1975 — 61
3.15. Average Annual Compound Growth Rates of Civil Cases Brought to All Courts of Justice and to All Courts of Justice Minus Districts Courts Per Thousand of the Population, 1852-1985 — 62
3.16. Public Expenditures (excl. Social Insurance) as a Percentage of Gross Domestic Product (current market prices), 1850-1982 — 64
3.17. Average Annual Compound Growth Rates of Public Expenditures as a Percentage of GDP, 1850-1982 — 64
3.18. Average Annual Compound Growth Rates of Participation in University Education and Higher Vocational Education per Thousand of the Population Aged between 18 and 24 Years of Age, 1930-1984 — 67
3.19. Number of Newspapers Sent by Inland Mail. Absolute Figures (x 1000) (1) and per Capita (2), 1860-1938 — 69
3.20. Average Annual Compound Growth Rates of the Per Capita Number of Newspapers Sent by Inland Mail, 1860-1938 — 69

4.1.	Distribution of Power in the Seventeenth Century Dutch Republic. Dutch Provinces by Percentage Share in the Republic's Budget and by Percentage Share of City Votes in the Provincial Staten	76
5.1.	Factual or Preferred Ratio of Friendship Relations within Ones Own Denominational Group as a Percentage of the Total Number of Friends	100
5.2.	Percentages of Votes in National Elections for Roman-Catholics, Protestants, Social-Democrats and Liberals as a Percentage of All Votes Cast, 1922-1986	115
5.3.	Number of Government Changes in Austria, Belgium, France, Germany and the Netherlands, 1918-1984	117
5.4.	The Number of Pupils of Elementary Schools by Denomination as a Percentage of All Pupils of Elementary Schools, 1850-1984	118
5.5.	Number of University Students by Denomination of the University as a Percentage of All University Students, 1920-1983	120
5.6.	Members of Home Nursing Services by Denomination of Home Nursing Service as a Percentage of All Members of Home Nursing Services, 1951-1972	121
5.7.	Number of Hospital Beds by Denomination, 1950-1985	121
5.8.	Members of Denominational and Non-Denominational Broadcasting Associations as a Percentage of All Members of Broadcasting Associations, 1950-1986	122
5.9.	Members of Social-Democratic (NVV), Roman-Catholic (NKV), Protestant (CNV) and Other Trade-Union Federations as a Percentage of the Total Number of Trade-Union Federations, 1914-1980	125
5.10.	Arithmetic Means and Standard Deviations of Strike Activity as a Percentage of the Demand of Labour for Twelve Industrialized Countries, 1950-1979	127
5.11.	Homogeneously and Mixed Married in the Netherlands per Denomination as a Percentage of All Married per Denomination in 1947 and 1960	131
5.12.	Probabilities to Be Homogeneously Married for Men and Women per Denomination in 1947 and 1960	132
5.13.	Homogeneously Married in Current Year by Denomination as a Percentage of All Married in Current Year by Denomination, 1936-1980	132
6.1.	Votes Cast for Denominational and Non-Denominational Parties in National Parliamentary Elections (2nd Chamber) as a Percentage of All Votes Cast, 1946-1986	144
6.2.	Public Opinion of All Respondents (a) and Church Members (b) on Preferences for Public Non-Denominational and Denominational Schools (percentages), 1966-1986	148
6.3.	Total Government Expenditure as a Percent of GDP at Current Prices, France, Germany, Japan, Netherlands, UK, USA, 1880-1981	151
6.4.	Public Opinion of Church Members on the Marriage of a Daughter to a Partner of a Different Denomination (percentages), 1966-1986	154
6.5.	Church Membership by Denomination as a Percentage of Total Population, 1899-1981	165
6.6.	Attitudes of Church Members on the Obligation to Conform to the Prescriptions of Their Church or Religious Group (Percentages of Total Number of Church Members in the Samples), 1966-1986	168

6.7. Public Opinion on Some Civil Liberties. Respondents in Favour as a Percentage of All Responding on the Questions Involved, 1966-1986   169

LIST OF FIGURES

3.1.	Alternative Proposed Growth Paths of Dutch National Income, 1688-1900	37
3.2.	Jan de Vries' Proposed Growth Path for the Netherlands, 1675-1900	39
3.3.	Dutch Per Capita National Income (Net Market Prices, in Guilders of 1900 (x 1,000)), 1900-1986	42
3.4.	Employment by Sector of Industry (agriculture (a), industry (b), services (c)) as a Percentage of Total Employment, 1899-1981	48
3.5.	Dispersion of the Inhabitants of 502 Dutch Municipalities, 1840-1970 (Coefficients of Variation)	53
3.6.	Enterprises with Five Employees or Less (a) and Enterprises with More than Five Employees (b) as a Percentage of All Enterprises, 1930-1978	58
3.7.	Dutch Ministry Personnel as a Percentage of the Labour Force (a) and of Total Population (b), 1849-1975	60
3.8.	The Number of Civil Cases Brought to All Courts of Justice (a) and to All Courts of Justice Minus Districts Courts (b) Per Thousand of the Population, 1852-1985	62
3.9.	Public Expenditures as a Percentage of GDP (current market prices), 1850-1982	65
3.10.	Numbers of Participants in Formal Education per Thousand of the Population Aged between 3 and 25 Years of Age, 1850-1960	66
3.11.	Participation in University Education (a) and Higher Vocational Education (b) per Thousand of the Population Aged between 18 and 24 Years of Age, 1930-1984	67
3.12.	Per Capita Number of Newspapers Sent by Inland Mail, 1860-1938	68
5.1.	Percentages of Votes in National Elections for Christian-Democrats (Roman-Catholics + Protestants) (a), Social-Democrats (b) and Liberals (c) as a Percentage of All Votes Cast, 1922-1986	116
5.2.	Denominational (a) and Non-Denominational (b) Quotum of Pupils of Elementary Schools as a Percentage of All Pupils of Elementary Schools, 1850-1984	119
5.3.	Members of Denominational (a) and Non-Denominational (b) Broadcasting Associations as a Percentage of All Members of Broadcasting Associations, 1949-1986	122
5.4.	Arithmetic Means and Standard Deviations of Strike Activity as a Percentage of the Demand for Labour for Switzerland (1), Austria (2), Netherlands (3), Germany (4), Sweden (5), Norway (6), Japan (7), Denmark (8), France (9), Belgium (10), UK (11) and Italy (12), 1950-1980	128
5.5.	Strike Activity as a Percentage of the Demand for Labour in the Period 1920-1979 and the Arithmetic Mean, The Netherlands	129
5.6.	Homogeneously Married in Current Year by Denomination as a Percentage of All Married in Current Year by Denomination, 1936-1980, (RC(a); Prot(b); All(c); No(d); Others(e))	133
5.7.	Differences between in Current Year Homogeneously Married Roman-Catholics (Men (a); Women (b)) as a Percentage of All in Current Year Married Roman-Catholics (Men resp. Women) and Their Probabilities of a Homogeneous Marriage, 1936-1980	134

5.8.	Differences between in Current Year Homogeneously Married Protestants (Men (a); Women (b)) as a Percentage of All in Current Year Married Protestants (Men resp. Women) and Their Probabilities of a Homogeneous Marriage, 1936-1980	135
5.9.	Differences between in Current Year Homogeneously Married "Others" (Men (a); Women (b)) as a Percentage of All in Current Year Married "Others" (Men resp. Women) and Their Probabilities of a Homogeneous Marriage, 1936-1980	136
5.10.	Differences between in Current Year Homogeneously Married "No Denomination" (Men (a); Women (b)) as a Percentage of All in Current Year Married "No Denomination" (Men resp. Women) and Their Probabilities of a Homogeneous Marriage, 1936-1980	137
6.1.	Votes Cast for Denominational (a) and Non-Denominational Parties (b) in National Parliamentary Elections (2nd Chamber) as a Percentage of All Votes Cast, 1946-1986	143
6.2.	Marriages Consecrated as a Percentage of All Marriages Contracted, 1951-1986	164
6.3.	Roman-Catholic (a), Re-Reformed (b) and Dutch Reformed (c) Homogeneous Marriages Consecrated by Denomination as a Percentage of the Number of Homogeneous Marriages Contracted by Denomination, 1951-1986	165
6.4.	The Inverted U-Curve Relation between Modernization and Pillarization	173

ACKNOWLEDGEMENTS

For financial assistance I am indebted to the Institute for Economic Research of Groningen University. The Netherlands Central Bureau of Statistics provided the data on marriages.

This book could not have been written without the support of many colleagues and friends. It is evident that I am responsible for the shortcomings of the end product.

For comments on earlier drafts I would like to thank Geurt Collenteur, Cees van der Meer, Henk van der Meulen, Graeme C. Moodie, Chris de Neubourg, Jan Pen, Wouter van Rossum, Jan de Vries and Hans Jürgen Wagener.

As an economic sociologist I needed a second opinion to check my ideas on the history of Dutch society. Discussions with Pim Kooij elicited fresh ideas and helped to broaden my knowledge of the past.

Henk-Jan Brinkman, Jan Willem Drukker and Chris Rümke assisted in developing a comparative method to analyse data on industrial disputes. Hans Erik van Helsdingen participated in the search for indicators of secularization.

I owe a lot to Sybrand Houtsma. He took part in the study as my student-assistant and contributed substantially to the construction of the time series in chapter III. His commitment exceeded the terms of his labour contract.

My friend and colleague Adam Szirmai deserves a special word of gratitude. He took the pain to listen to my ideas, to encourage me and to comment on the results of my thinking. His professional and emotional support meant a lot to me.

I am very grateful for the guidance Angus Maddison and Jan Berting gave me the last few years. They stimulated me to finish the study and patiently commented in detail on the several drafts of each chapter. Angus Maddison taught me to be precise in language. Thanks to him I learned to use time series in sociological analysis. Jan Berting challenged me to look for the assumptions hidden behind theory, to make them explicit and consistent.

Ellen van de Gevel and Tineke Tadema assisted me with word processing and printing. Ynte Jan Kuindersma helped with the final layout of the book. Klaas van Dijk kept smiling while photocopying the many pages I produced.

I owe much to the intellectual partnership and love of my wife Yvonne Dollee. With deep and warm feelings I thank her for accepting and understanding my many moments of psychic absence in our relationship. She made me a home that provided the balance to work and writing. I have to apologize to my son Allard for not always being there when he needed me. I hope that when he has grown up he will understand that writing a book is worthwhile.

CHAPTER I

INTRODUCTION

In the late sixteenth and early seventeenth century the Low Countries gained their independence from the Habsburg empire and became a republic. In the hundred years after its birth this "Republic of the Seven United Netherlands" attained enormous economic power based on its navy and merchant fleets and a large colonial empire extending from the Americas to Asia. In its "Golden Age" the country took a leading position in the arts.

In the seventeenth century, together with England and France, the Netherlands formed the centre of the Western world. Not only economically and politically, but also intellectually. Dutch men of science, like the jurist Hugo de Groot, the physicists Huygens and Boerhaave, the philosopher Spinoza and the politician and mathematician Johan de Witt, had a profound influence on the development of Western thought. The new republic was known for its atmosphere of tolerance: many authors from abroad published their works in the Netherlands because of the repressive climate in their countries of origin. The Dutch Republic became a refuge for those Europeans who where not able to speak, write and live in freedom in their home countries. Till today the labels of pluralism, tolerance and freedom are still associated with the Dutch nation.

Without denying its truth, the above picture of the Netherlands is a cliché and thus one sided. Soon after the period of struggle against the Spanish, major contradictions became manifest: differences regarding all kinds of interests and ideological disputes split the nation. However, they did not tear the country to pieces. In fact, the Netherlands never saw civil war nor revolution. One of the most striking features of Dutch society has ever been that despite its pluralism it has yet always remained its unity and has been free of disruptive outbursts of social conflict.

In Dutch history religious conflict, linked with social conflict, comes repeatedly to the fore. In the past many European countries saw often bloody conflicts between religious groups. The Netherlands, however, never experienced a "St. Bartholomew's night". From the seventeenth till the nineteenth century the Dutch Reformed church was dominant. However, dissents within Protestantism and other denominations were tolerated and informally granted religious freedom. The largest category of these religious deviants certainly have been the Roman-Catholics. Although treated as second rate citizens and disqualified from public office they were never prosecuted and many of them even reached positions of great wealth and influence.

In the second half of the nineteenth century Dutch society underwent major changes. The contradiction between rich and poor, labour and capital, was sharpened, or at least, became more apparent. New ideologies, socialism and liberalism, entered the political arena to compete with traditional political groups. In all segments of social and economic life processes of differentiation took place. Mass democracy rose and political parties were established.

These parameters of modernization were similar to those in other countries of the West. However, the institutional shape they eventually took, was different in many a respect. Instead of the overthrown of pre-industrial meaning systems, Dutch modernization as a complex of economic, social, cultural and political change meant the moulding of traditional pluralism into a new institutional pattern. Since the Second World War it is generally referred to as "pillarization" (in Dutch: "verzuiling").

If we disregard the nuances of all the definitions of pillarization drafted in the social sciences so far, there is one central element they have in common: the division of society along ideologically based vertical lines rather than along horizontal cleavages. In the 1950's when Dutch pillarization was considered to be at its peak, Dutch society was composed of ideological blocs denoted as "pillars" (in Dutch: "zuilen"). Some of these pillars comprised both lower, middle and upper classes. The socialist and the liberal blocs, however, respectively represented predominantly the lower and the upper strata. So Dutch pillarized society of that time is generally thought to contain a Protestant, a Roman-Catholic and a secular pillar, the latter usually being divided into a social-democratic and a liberal bloc. Each of these pillars formed a social world of its own embracing its own social organizations for the fulfillment of nearly every social function. The pillar's ideology shared by all its segments, geared the pillar to cohesion.

The above description of pillarization is preliminary and thus rather crude. We skipped many problems connected to the concepts of pillarization and pillar that will be dealt with in the chapters to come. For this introduction it is sufficient to regard pillarization as a particular form of institutionalised pluralism within a nation state. Because of the emphasis on vertical cleavage, the concept of pillarized society is analytically opposed to that of class society in which lines of social division are of a horizontal nature and stratification is based on clusters of similar economic positions.

As may have been understood from the preceding passages, Dutch pillarization is associated with late nineteenth century and twentieth century Dutch history. Since its origins in the sixteenth century, pluralism was ever the hallmark of Dutch society. Pillarization, however, came about only in the late nineteenth century. Around 1870 when the first political parties were established, pillarized structures arose. The institutionalization of pluralism into pillarization ended around 1920. At that time all pillars had matured and a set of rules with regard to the interaction between the pillars had been developed. In the 1950's pillarization is alleged to have reached a peak. In the 1960's a gradual erosion of pillarized structures set in. Today there is a difference of opinion on the issue to what extent Dutch society can still be regarded as pillarized.

At the moment there is not one generally accepted theory that explains Dutch pillarization. In an attempt to systematize the field, Daalder[1] has discerned four major approaches. The first one explains pillarization as the result of the emancipation movement of Roman-Catholics and the Protestant petty bourgeoisie. The second approach regards pillarization as a mechanism within a democratic political system by which political elites control their rank and files. The third category is rather descriptive. It is represented by Lijphart[2] who was interested in the issue how a divided society could maintain internal peace and at the same time could reach effective decision making. The last group of theories Daalder mentions, is comparative. It attempts to answer the question how in Western societies lines of political divisions developed during the nineteenth and twentieth century.

In chapter V I will review these hypotheses in detail. Here it will do to remark that they have an underlying theme. Emancipation, social control within democracy, decision making in a modern state and the development of nineteenth and twentieth century plural Western societies, all have in common that they refer to different aspects of the interdependent processes of economic, cultural, political and social change that started to develop in the era of industrialization. In a broad sense the total of these processes can be labelled as the "modernization" of society.

In 1984 Ellemers, applying a structural functionalist approach, put forward the hypothesis that the process of Dutch pillarization has been a specific response to the modernization of Dutch society[3]. In his view modernization produces problems that have to be solved by social differentiation. Pillarization is a specific type of social differentiation that occurs in small countries characterized by "segmented particularism" as a part of their national identities.

Ellemer's contribution is valuable because he links the explanation of pillarization to modernization theory. However, he does not go beyond the application of the structural functional paradigm to Dutch pillarization. Pluralism is taken for granted and described in structural functionalist terms instead of being treated as the subject of explanation. If pillarization is to be understood as a specific Dutch response to more general processes of modernization, the question is, why has the Dutch reaction been this characteristic.

A second major question to be dealt with in this study concerns the supposed process of de-pillarization as it is alleged to have developed after the 1960's. Contrary to the analysis of pillarization, the study of the de-pillarization of Dutch society may be considered as a rather underdeveloped domain. So far only empirical analyses of attitudinal changes are available[4]. Questions whether or not Dutch society of today is actually de-pillarized and to what extent, have not yet systematically been answered. Further, starting from the assumption that since the 1960's the Netherlands really de-pillarized, so far a satisfactory explanation has not been produced.

A related problem is that, similar to the late nineteenth century when pillarization came about, in the decades after World War II Dutch society again underwent major economic, cultural, political and social changes. Could it be that both pillarization and de-pillarization should be explained from progressive modernization? And if so, what are the differences between nineteenth century and post-war modernization that make understandable why pillarization came about in the former and eroded in the latter? These are some of the intriguing problems that follow from Ellemers' hypothesis and that will be dealt with in this study.

To summarize, two major issues are treated in this book:
1. How is pillarization as a specific type of Dutch institutionalised pluralism related to long term processes of modernization in the Netherlands?
2. To what extent did Dutch society de-pillarize in the last decades and, if so, how can de-pillarization be related to the progressive modernization of society?

These two questions can be put into more general terms. We are interested in the long term relation between modernization and institutionalised pluralism in Dutch society. In this respect "long term" refers to the period from the sixteenth century origins of the Dutch state till the 1980's.

The approach is multi-disciplinary. The analysis uses studies derived from the economic, the historical, the political and the social sciences. Each of these has produced studies on pillarization. Thus the problems studied below are not new, but are topics in a rich tradition. However, the uni-disciplinary approaches all have their inborn biases. The interest of economists in the problem is rather recent and is oriented to the problem of the costs of pillarization. The sociological view on pillarization generally lacks a sufficiently wide time perspective and usually concentrates on the 1950's, the alleged highlight of pillarization. Traditionally political scientists exclusively have studied the political dimensions of pillarization. Historians often fail to integrate their qualitative data into a theoretical framework. All four have their strong and weak sides. By putting these distinctive approaches together into one analytical frame I hope the sum total will gain. This implies that beside the presentation of new data existing material will be rearranged as to contribute to a broadening and systematization of knowledge on the subject.

Next to a multi-disciplinary approach this study attempts to illuminate the subject quantitatively as much as possible. By that I do not mean to apply sophisticated mathematical models that cover only a small time span. For the here intended long term analysis the construction of time series of the significant variables is more appropriate. In that way the development of the processes under study can be monitored over a relatively long period of time.

As this time period covers nearly four hundred years, it was not possible to construct time series that cover it all. Because of lack of relevant data before 1850, they could only be constructed for after that year. And even then, compromises had to be made because of data sources and definitions that were not compatible (The interested reader is referred to the appendices.). Where quantitative data were not available, the analysis is supported by qualitative material.

The advantage of this method is that a wide array of sources can enlighten an interesting aspect of Dutch history as it developed over a long period of time. However, the price to be paid is that statements on causality in a strict methodological and statistical sense cannot be made. The rich and colourful picture that emerged, made the price worth paying.

Another disadvantage of the method followed should be mentioned here. Dutch pillarization is often regarded by Dutch scholars as rather unique. Further, if one wants to relate Dutch pillarization to general modernization theory, a comparative approach is needed that makes clear in what respects the long term developments of Dutch society deviate from those in other countries. When I started to work on this study it was my intention to follow that way. However, soon it became obvious that the construction of time series in combination with a multi-disciplinary analysis for more countries and for several hundred years, exceeded the modest means at my disposal. I therefore had to restrict myself to Dutch society. Nevertheless, where needed references to developments in other countries are made.

In chapter II major theories of modernization are reviewed. It is meant both to develop an approach to modernization as well as to produce the key variables by which modernization can be measured over time.

Chapter III is on monitoring long term modernization of Dutch society. Indicators of the variables developed in chapter II are measured. At the end of the chapter conclusions are drawn with regard to accelerations, stagnation and decline of modernization in the Netherlands from 1600 till the 1980's.

Chapter IV concentrates on institutional change. It explores the origins of the Dutch civic culture and tries to explain how and why Dutch pluralism changed over time.

In the first part of chapter V I deal with the pillar concept and review theories of pillarization. On the basis of these an ideal typical model of Dutch pillarization is formulated. In the second part of chapter V pillarization is monitored.

Chapter VI monitors de-pillarization. Three central issues are treated. Firstly, the extent to which Dutch society de-pillarized since the 1960's. Secondly, the explanation of Dutch de-pillarization and, thirdly, the relation of both pillarization and de-pillarization to the progressive modernization of Dutch society. The latter is of a conclusive nature.

The construction of the time series and the sources used are accounted for in the appendices A, B and C. For reasons of convenience I present the full bibliographical annotations of every source directly in the appendix involved. The bibliography exclusively refers to works quoted in text and notes.

CHAPTER II

ON MODERNIZATION

Introduction

In the few studies that deal with de-pillarization of Dutch society since the late 1950's, it is conceived as the result of a process of modernization. Pillarization is regarded as fitting in a specific Dutch type of tradition that withered away as economic growth accelerated. There is an implicit or explicit assumption about causality[1]. Questions then arise as to the precise nature of the supposed causality as well as to the relation between the observed change of society as a whole and the change of the patterns of actual social behaviour within it.

Secondly, the assumption on the nature of the relationship between modernization and de-pillarization mostly implies the idea that societies develop along converging lines in more or less the same direction. In the course of this process dissimilar traits of different societies grow more alike. A tendency is alleged to exist from dissimilarity to similarity, from traditional to industrial or post-industrial society. Being identified with tradition, pillarization then disappears as society modernizes.

The relationship between modernization and de-pillarization is less self-evident as it is apparently believed to be. True enough, the traditionally pillarized political structure began to crumble in the early 1960's. On the other hand, these pillarized structures arose in the late nineteenth century when the country was also confronted with processes of enlargement of scale. In most theories on modernization the latter are identified with modernization. This seems to leave us with the contradiction that both pillarization and de-pillarization are explained by the same process.

To reach a solid explanation of the why and how of de-pillarization it is necessary to explore the relation between modernization, pillarization and de-pillarization more deeply. First of all, we should get a more precise idea about the nature of modernization processes in general. How do they develop, in what direction are they heading and what are the key variables involved? That task is attempted in this chapter. Secondly, we will have to understand how the long term process of modernization affected Dutch society in the second half of the nineteenth century and after World War II and how this is related to the prerequisites of both pillarization and de-pillarization. The latter problem will be dealt with in a chapter to follow.

The concepts of modernization and industrialization are often used interchangeably. To be clear, I shall use the term industrialization to denote exclusively the growth of the industrial sector of the economy. Modernization has a far wider connotation relating to social transformations in which social, political and economic changes are included.

The Idea of Modernization

The Social and Intellectual Origins of Modernization Theory

Modern social theories of modernization were inspired by different sources. The historical, social and intellectual backgrounds may enlighten some of the major problems inherent to modernization theory.

One source was the discrepancy between the ideas of the Enlightenment and social reality as it developed during the periods of the great social transformations in eighteenth and nineteenth century Western Europe. The social changes brought about by the French Revolution and the start of industrialization seemed to be both innovative as well as promoting chaos. Great classical thinkers like Comte, Durkheim, Marx, Saint-Simon and Spencer were all caught by this dilemma: they envisioned a future society in which humankind would be freed from the bonds of traditionality and magic, but, at the same time they feared increasing social instability and a growing number of people doomed to live in misery. Therefore, it is not surprising that their theories contain combinations of both conservative ideas and thoughts that can be traced back to the Enlightenment[2].

Evolutionism was important in the history of modernization theory. Although analogies of biological and social systems are now refuted in most modern social theory[3], in the nineteenth century the idea that social organisms develop by processes of differentiation was more common. Among the founding fathers of the social sciences it was strongly found in the works of Spencer and Durkheim. The latter understood the increasing division of labour as a process of social differentiation enhancing human freedom. By Sorokin and in post-war mainstream sociology as represented in the works of Talcott Parsons, functional differentiation is considered as the driving force of social development.

A third influence on modernization theory originates from the wave of de-colonization after World War II. Many new countries of the Third World were faced with the problem of gaining economic independence from the masters that once ruled them. It became clear that economic and social development were closely related. Because of the urge to catch up economically with the developed world, modernization of these countries became a policy problem of the highest priority[4]. This brought new dimensions to modernization theory: problems arising in this policy field led to new issues to be explained by theory. Modernization theorists tried harder to derive practical policy recipes from their theories.

Defining Modernization

The interest of the young social sciences in the great social transformations of the eighteenth and the nineteenth century expressed itself in studies and theories that did not make a distinction between the social and economic dimensions of modernization. For Durkheim, Marx, Weber and - to a lesser extent - Ricardo, the economic, social, cultural and political aspects of modernization were part of the same reality and were treated as such. Only in the twentieth century was a distinction made as to the economic and social dimensions of modernization.

In the nineteenth century the term modernization was not yet used to label the great transformations that took place in Western Europe at the time. After World War II the concept of modernization came in vogue[5] to describe the social evolution of societies.

> "It enabled one concisely to speak of those similarities of achievement observed in all modernized societies...as well as of those similarities of aspiration observed in all modernizing societies regardless of their location and traditions"[6].

The term development from then on tended to get an exclusively economic meaning. Thus, although the terms modernization and development are rather recent ones the field of interest for which they both stand has ever been at the heart of sociology and economics.

The rise in popularity of the concept of modernization certainly did not contribute to an increase of its clarity. As Fusé wrote:

"...confusion seems to stem largely from the omnibus nature of the term used in the past - i.e., it is a "catchall" concept - as well as from the different contexts in which it has been used in various branches of the social sciences"[7].

Thus, Durkheim thought of modernization as a change in the mode of social organization from mechanical to organic solidarity. Tönnies concentrated on the transition from Gemeinschaft to Gesellschaft. Maine saw modernization as leading to a shift from status to contract. Weber emphasized the rise of instrumental rationality. Although all these writers did not yet use the terms modernization and development their concepts and theories all relate to increasing structural differentiation as well as to institutional, economic and attitudinal change.

The distinction between modernization and economic development has been rather a matter of professional emphasis than of a rigid and exclusive restriction to content. To give a few examples, Rostow regards modernization as both the economic take-off as well as a set of social changes that create the conditions for the take-off[8]. Riesman's original theme is the change of psychological traits of people as a consequence of socio-economic and demographic changes[9]. McClelland observes a shift towards an achievement motivation as a prerequisite for economic development[10]. So we are not only confronted with a confusing mixture of economic, social and sometimes psychological variables, but also with supposed causalities of a different nature.

Although mere examples, the above cases are representative for the wide variety of approaches to the phenomenon of modernization. However, attempts have been made to reach an unequivocal definition. Among the most well known are those of Marion Levy and W. Arthur Lewis. From a sociological point of view Marion Levy defined modernization as *the growing ratio between inanimate and animate sources of power*[11] whereas Arthur Lewis stressed the economic dimension by defining modernization as *the growth of output per head of the population*[12]. These definitions put the question of causal factors aside and concentrate on measuring modernization. Their advantage is that they enable comparisons between societies as to the degree of modernity[13]. A major disadvantage is that the inner complexity of modernization processes is neglected. Both definitions restrain themselves to indicators of modernization and tell us nothing about the nature of modernization and the variables involved[14]. Besides, they have not succeeded in reaching general acceptance by the scientific community.

However lacking a consensus on the issue of a precise definition may be, a review of the literature yields a broad similarity of opinion on the nature of the phenomena of modernization and economic development. The latter is usually identified with economic growth, i.e. the rise of output per head of the population. The former is mostly understood as having a wider meaning referring to technologically induced economic growth in relation to structural, institutional, political and attitudinal changes. With regard to the sociological concept of modernization this relation between economic growth and socio-cultural variables is crucial. It is the watershed between economic development and modernization[15]. From now on we shall concentrate on modernization.

Taking all the nuances in the different concepts and theories of modernization for what they are - and indeed they are many - the hard core has always, implicitly or explicitly, something to do with the transformation of the world by technology that is indicated by economic growth. Even in such opposing paradigms as Marxism and Functionalism technologically induced economic growth is considered a key variable. Despite this consensus, sociological paradigms differ considerably as to the relation between economic growth and socio-cultural variables. This is not only the consequence of the fact that different variables are incorporated in the theories involved, but also because they contrast with regard to the causality patterns stated. So in Marxist theories a change of social institutions is usually explained by changes in the economic and technological infrastructure, while according to the Functionalist school the economy can only change after certain social prerequisites have been fulfilled.

We will take as the starting point for the formulation of our own views the idea that technological development and economic growth are both key elements in processes of modernization. In this respect it is important to notice that technological development can have different meanings varying from the invention of tools, like hammers and nails, to the achievement of knowledge to administer large organizations like multinational companies and modern states. The product of the first type of technological development is material. The second yields abstract knowledge. What these extremes have in common is the increase of the human potential to master the environment. Thus I will regard technological development as the increase in human ability to control the environment. Technology then is the set of tools, be it material or immaterial, available to reach such control. Knowledge that is not (yet) related to concrete problems usually is called science.

The modernization of societies implies that the application of knowledge is changed in such a way as to purposefully promote productivity. Historically this means a trend from the application of simple technology to the use of complex abstract knowledge for the production of goods and services. It is the trend from forge to conveyer belt, from witchcraft to neurosurgery, from turnpike-man to tax office. In modern society there is a firm consciousness that technology can raise productivity.

In the course of the process of modernization technology institutionalises in separate social domains. The development of medicine is a representative example. In ancient and mediaeval Western Europe medicine was strongly connected to religion. A medical profession did not exist. Priests and monks were engaged in the curing of the sick. In the late Middle Ages gradually a separation grew between religion and medicine. In our times this trend ultimately led to highly specialized professionalization, research institutes and specialized pharmaceutic industries. Contrary to the old days inventions in the field of medicine now barely come about by chance, but are strived for in a purposive attempt to improve health care.

To summarize, modernization is not so much dependent on the availability of science and technology as well as by the ways these types of knowledge are broadened and applied by society. The primary distinction between modern and non-modern societies lies in the strong connection between technology and productivity in the former. Although such a connection may also be present in non-modern societies, there it is much more dependent on chance[16]. In modern societies, however, technology is a separate institutional domain and is subject to the policies of private enterprises and the state with the ultimate aim to raise productivity[17]. It follows that in empirical reality the transition from traditional to modern society cannot be pinpointed to one specific moment in time as the institutionalization of technology is a gradual process.

The idea of modernization thus includes more than technologically induced economic growth alone. It refers to a strong tie between economic growth and changes of culture, structure, institutions and attitudes. Hence it has been identified with specialization of the division of labour, a shift from subsistence farming to the cultivation of crops for the market, the development of national markets and eventually of an international market, a decrease of the number of people employed in the primary sector of the economy in favour to the secondary and tertiary sector, a growth of government participation in the economy. As social dimensions of modernization factors are mentioned like the rise of the level of education and of professional skills, an increase of social and geographic mobility, urbanization and the spread of a cosmopolitan way of life, the diffusion of secular rational norms leading to the bureaucratization of society, a change in the nature and meaning of kinship ties and an increase of the quantity and density of the information spread over the population. As to the political dimension, modernization has been connected with a growth of national identity, an increase of political participation by the population and the development of intermediary bureaucracies aimed at the articulation of group interests.

This list of economic, social and political phenomena linked with modernity is not exhaustive and a review of the literature could easely yield a many more to add. All these processes can be seen as processes of economic and social change in their own rights. However, if we speak of modernization we regard them interdependently in their relation to processes of economic transformation of which technological development forms the hard core.

In our idea of modernization three elements therefore are central. Firstly, we have to do with the transformation of the economy induced by the purposive development of science and technology. Secondly, such a transformation is related to other processes of change in the economic, social, political and attitudinal domains. Thirdly, we only speak of modernization if all these processes are interdependent[18].

If we now add up all the aspects of modernization that we stressed before, we are able to define modernization as the process of interdependent changes in the social, cultural, political and attitudinal domains of a society rooted in the transformation of its economy induced by the purposive development of technology. Before proceeding it may help to prevent misunderstanding if we add some further remarks on the elements of purposiveness and technology that are incorporated in this definition.

The first question that immediately comes to our minds deals with the issue of whose purpose is involved. As we stated before, the purpose of technological development is the intention to reach higher levels of productivity. Such an intention may be reduced to individuals, organizations, nation states or supranational economic blocs.

This is not to say that the purposive development of technology is a voluntary process. The formulation of intentions to reach higher levels of productivity is a social process in which the actors involved are confronted with constraints that for a large part consist of the perceived intentions of others. Further, the formulation of intentions is influenced by the perceived probability that others may modify the course of relevant events. Think of the businessman investing in technological development because he feels his position in the market threatened by his competitors. Or, the large investments in military technology that nations make as a reaction to the perceived threat by foreign powers that counteract by increasing their investments, and so on. The result will be an arms race which can hardly believed to be wanted by the actors involved. Consequently, the purposive nature of technological development means that, although it can be reduced to particular intended actions, the possible outcome of such a process may be largely unintended and relatively autonomous.

The second question concerns the role of technology. Although most theories stress technologically induced economic growth as the hard core of the great transformations in eighteenth and nineteenth century Western Europe, they differ considerably on the explanation why technological development accelerated anyway[19]. I do not want to suggest that by my definition of modernization I have implicitly chosen for a particular school of thought whatsoever. In fact I don't think that any monocausal explanation of modernization can bear fruits, i.e. I reject the idea of technological determinism which states that technological development is the one and only driving force behind all social and cultural change[20]. It is more convincing that technological development has been the historical result of several interdependent factors. Thus Weber's idea that the acceleration of economic growth and technological development should be correlated to the combined effect of both a changed economic infrastructure and the spread of a value pattern based on legality and instrumental rationality seems quite attractive[21].

The relation between value patterns and technological development becomes the more complicated if one imagines that people may choose between alternative paths of technological development. Prevailing values may lead to the preference of one path to another. But even then, the purposive nature of such choices has to be regarded as relative autonomous and is often subject to market forces. So, from an ecological point of view wind energy has to be preferred to nuclear energy. Despite the Tsernobyl disaster and despite political party programs, many governments still build nuclear plants. They do so not only because of strong lobbies in favour of nuclear energy and the practical restrictions of wind energy, but also because they believe that abstinence of nuclear energy would eventually lead to a deterioration of their positions on international markets.

Thus one can often observe decisions on technological paths to follow that go against existing values. This can be explained by two interdependent factors. Firstly, today no society can be regarded as a closed system. Therefore, modernization is not only dependent on internal factors, but is also determined by the external environment that sets constraints on internal preferences. Secondly, societies have a strong inclination to maintain or even improve their relative power positions vis-à-vis others.

A further argument why societies cannot be regarded as closed systems beforehand, is the role of prevailing value patterns itself. Like in the case of technological development, modern societies are affected by cultural diffusion from abroad. Although we do not use the word *modern* in that sense, its connotation often relates to the patterns manifested by those groups or countries that are economically, socially or politically considered to be in a lead position. Modern stands for *being ahead* or *up-to-date*. It is this phenomenon that gives the concept of modernization its value laden character.

To get an idea of the way processes of modernization develop, circular causal reasoning is to be preferred to arguments of a monocausal nature. Once existent, technological development and economic growth lead to changes of social structure, culture, politics and attitudes. In turn these changes will affect technological development and economic growth, and so further. Therefore, modernization should be conceived as an endless chain of mutually reinforcing actions and reactions. Only the historically specific conditions of a society can explain how such a process of dynamic interdependencies is set in progress and how it develops.

Illustrations of the interdependencies between technological development, economic growth and institutional change have been presented by W. Arthur Lewis in his classic on the theory of economic growth. Lewis continuously stresses the complexity of the causal interdependencies involved and emphasizes that it is nearly impossible to isolate one factor as the exclusively independent variable. For Lewis the circular causal chain is set in motion by the historically unique circumstances of societies. It is the circular nature of these causal patterns that explain the economic cycli of growth and decline[22].

The historical circumstances that stimulate modernization may be internal or external to the society involved. As an example of the latter Lewis mentions the immigration of foreigners who bring with them new ideas, values and norms and may, because of their minority position, function as an economic avant-garde. Exploitation by foreign powers may be an external factor with either positive or negative effects. The latter seems relevant for the understanding of the "development of underdevelopment" in the Third World as Gunder Frank has called it[23].

Because of the colonial past and because of the fact that Third World countries have to modernize while in a disadvantageous position of competition vis-à-vis the highly developed Atlantic World, the nature of modernization in these countries is dissimilar from that of Western Europe in the last two centuries. The historical circumstances are divergent to a high degree. As we seek to explain long term social change in the Netherlands and as we do not pretend to formulate a general grand theory of modernization, we will not relate our analysis to the actual development of the Third World. This implies that the category of theories that attempts to enlighten the phenomenon of underdevelopment in the Third World will not be considered here. We now point at radical and Neo-marxist paradigms that have become known as *dependency theory*[24]. These have in common the idea that exploitation is the key variable that induces underdevelopment and failing modernization. Because of their exclusive concentration on relations of exploitation between countries, these models have been criticized for their neglect of the internal factors that could illuminate the difficulties of promoting modernization and a satisfactory rise of per capita income[25].

The Idea of Bipolarity: Tradition versus Modernity

From classical sociology modernization theory inherited the idea of bipolarity between tradition and modernity. The founding fathers of the social sciences tried to grasp the idea of modern society by contrasting it to tradition. As we have seen, these classical dichotomies were usually constructed by the use of one main variable that was considered to be the hallmark of distinction. Such models consist of the opposition of two ideal types and are rather descriptive than analytic[26].

The point of view inherent to bipolar models is that modernization has a logic of its own and develops accordingly. This is theoretically derived from the assumed logic of the process of industrialization: the technological and economic dimensions of industrialization are supposed to require social, cultural and political conditions to be fulfilled. The inevitability to meet these requirements leads to the development of modern society. It follows that modernization and industrialization are implicitly regarded as twin concepts and that processes of modernization are thought to have uniform characteristics[27].

The basic assumptions of the classical bipolar models have left their traces in many contemporary models of modernization. In these models modernization essentially is regarded as a process that develops between two poles: tradition and modernity. There is an assumed point of departure and one of arrival. Thus Marx, starting from the assumption about the productive nature of the human kind, thought society to develop logically into the direction of communist society. Although Marxist theory on the development of pre-capitalist and capitalist society is dynamic, the concept of communists society as the end of the process is rather static.

Compared with the earlier models, Parsons' structural functionalism is a major step ahead. Dissatisfaction with dichotomous classifications of social relations inspired Parsons to the formulation of *pattern variables*[28]. According to Parsons any action involves five dilemmas of choice and consequently can be described in terms of these aspects of choice: affectivity versus affective neutrality, self-orientation versus collectivity orientation, universalism versus particularism, achievement versus ascription and diffuseness versus specificity[29]. By the introduction of pattern variables descriptive classification was substituted by analytical classification therewith creating the possibility of the analysis of transitional societies, i.e. societies that are in between tradition and modernity[30]. However, the pattern variables themselves are bipolar types and therefore they do not overcome the problems inherent to bipolarity[31].

The school of thought that has become known as Post-Industrialism rejects the idea that industrial society is the end of all social evolution, but at the same time it creates a new evolutionary perspective which is called Post-Industrial society. This is so because the frame of analysis of the latter is still bipolar and as a result it faces the same dilemmas as any other bipolar model. Thus Bell thinks that industrial society is characterized by the exploitation of labour whereas the exploitation of knowledge is the essential trait of post-industrial society[32]. For Touraine the difference between industrial and post-industrial society lies in the focus on factory alienation in the former and the focus on the university as the source of theoretical knowledge in the latter. Therefore, the nature of exploitation and the patterns of social conflict are quite distinctive[33].

Bipolarity as a theoretical argument is also present in the convergence hypothesis. According to this line of thought, in dissimilar societies the logic of industrialization produces similar problems that ask for practical solutions. Some of the scholars who belong to this school believe that there is only one best solution to solve these problems[34]. As a result, the more technology develops, the more dissimilar societies will choose identical solutions to solve similar problems. Therefore, it is stated that in the course of modernization the significance of ideology decreases and societies will gradually show more similarity. The convergence school has mainly focussed on the patterns of development of capitalist and socialist societies. The bipolarity inherent to these models can be reduced to the poles of economic decentralism and centralism.

After World War II models have become popular that try to discern stages of modernization and economic development. Their roots go back to representatives of the nineteenth century German Historic School like List, Hildebrand, Bücher, Schmoller and Sombart. Contemporary stage models attempt to explain the transition from traditional to modern society by empirically connecting economic, social and political changes. The advantage of stage models to classical models is twofold: they are oriented towards empirical research and they try to relate variables of a different nature. Because they attempt to explain the transition from a first stage to an end stage, they too are basically bipolar. The most well known stage model is Rostow's which we shall treat below. As we shall see, most criticism that has been put forward to Rostow's ideas applies equally well to other stage models.

Theories of modernization based on the idea of bipolarity have to deal with five problems which they cannot resolve: the problem of homogeneity, the problem of chronology, the problem of finality, the problem of empirical boundaries and the problem of universality. Although these problems are related, we shal treat them separately.

The Problem of Homogeneity

Bipolar models reduce the complexity of reality to the distinction between traditional and modern society. Both are used as umbrella concepts: any society that by some criterion is not considered modern is labelled as traditional and vice versa. Thus both categories of the dichotomy are regarded as homogeneous while the criteria of classification are rather limited. Dissimilar civilizations like that of Neanderthal man, the Greek, the Romans, Indians, Pygmies, Saxons and so on, are shuffled together therewith neglecting the differences between these cultures. The same holds true with regard to the concept of modern society: major differences between highly developed societies are not taken into consideration.

One could argue that the reduction of the complexity of empirical reality is one of the aims of scientific description. However true this may be, the reduction of empirical reality has its limitations which are related to the usefulness of the theories that can be based on it. As we will illustrate below, we think that one of the aims of modernization theory should be the explanation of the variance of the courses and outcomes of modernization processes. Such an approach is seriously hindered by a classification that neglects variances because it reduces all phenomena of modernity to only one type.

The Problem of Chronology

The problem of chronology is related to that of homogeneity. With regard to Western Europe the end of traditional times has been located in the epoques of the French Revolution and the industrialization of England. From the logic of bipolar models it follows that the line between tradition and modernity has to be drawn somewhere. Because that line is drawn on the basis of a limited number of variables the problem arises that societies labelled as traditional may show signs of modernity and vice versa. A recent example has been Iran in the 1970's with its largely modernized urban centra and its traditional countryside. Thus a disequilibrium or strain between modern and traditional sectors within the same society is obscured by the use of dichotomous classifications.

Another aspect of the problem of chronology is that bipolar models assume an unilinear path of development. They cannot explain the historical fact that in the course of modernization some societies regress to tradition. Again Iran is a good example. From the point of view of modernization the rise of the Khomeiny regime has to be regarded as a drawback in the modernization of Iran. A less recent example is the fall of the Roman Empire which has been a set back in the social development of Western Europe.

The Problem of Finality

The bipolar conception of tradition and modernity implies that modern society is the supposed logical end of a process. A society is modern or it is not because bipolar classifications exclude the notions of continuity and degree. Whenever modernity is achieved, further social development stops. Thus modern society is a static one if we apply the concept as it is used by bipolar models.

Some authors have felt this problem. But, instead of leaving the bipolar scheme for what it is, they have broadened theory by inventing a new type of society which they have called post-industrial society. Thus the polar types of traditional and industrial society are substituted by those of industrial and post-industrial society[35]. Recently a new type of society became in vogue: the information society which is supposed to be dominated by the use of computer technology[36]. Although the names have changed, the problem of finality is not solved. Any bipolar conception fails to cope with the dynamics of continuous social and economic change within the supposed end of social evolution.

The Problem of Empirical Boundaries

In empirical research bipolar models yield problems with regard to the measurement of the boundaries between tradition and modernity. Before we put forward that stage theories of modernization are essentially based on a bipolar conception. Hoselitz summarized the empirical problems attached to stage theories. Firstly, it is not always clear whether stage theories are ideal typical abstractions or whether they pretend to be empirical generalizations. Secondly, the transition from one stage to another is difficult to explain. Thirdly, the scope of many a stage model is not clear[37]. The same criticism is valid for the more simple bipolar dichotomies. Hoselitz's objections can be illuminated by referring to Rostow's stage theory of modernization and economic growth[38].

Although Rostow discerns five stages of economic development and modernization he seeks to explain the transition from the first to the fifth stage, from tradition to modernity. His model is of a bipolar nature with a built-in theory on the transition. In the West this transition evolved in the nineteenth century. In non-western societies - with the exception of perhaps Japan - Rostows transition is a twentieth century phenomenon.

Rostow's merit is the connection of social, economic and political variables as well as the broadening of the bipolar conception to transitional societies. In this respect Rostow corresponds with Parsons[39]. Parsons' pattern variables too enable the connection of economic and social variables and the analysis of transitional types. Like Rostow, Parsons regards the nineteenth century as the age of transition. Later Parsons' scheme was specified and applied to contemporary developing countries[40].

Another point of similarity between Rostow and Parsons is that both are not clear and unambiguous as to the empirical references of their theories[41]. For Rostow the stages of economic growth and modernization are theoretical ideal types as well as empirical generalizations:

> "They are not merely a way of generating certain factual observations about the sequence of continuity. They have an analytical bone structure rooted in a dynamic theory of production"[42].

It is this mixture of theoretical and empirical pretensions that leads to serious problems.

According to Rostow the take-off is the turning point in the process of modernization. It is the line between traditional society characterized by Malthusian cycli of growth and decline and modern society with its self-sustained economic growth. The main variable that determines the watershed between these two worlds is the production and absorption of technological innovation:

> "...the take-off is a definitive watershed in a society's history: the innovational process has ceased to be sporadic and is a more or less institutionalised part of the society's life"[43].

Rostow discerns three major traits of the take-off stage: a rise of the productive investments from five per cent or less to ten per cent or more of national income; a rapid development of one or more strategic industrial sectors with high growth ratios; the creation of a political, social and institutional infrastructure that is able to utilize both the impulses to an expansion of the industrial sector and the potential external economies as well as to stimulate further growth.

Simon Kuznets, one of Rostow's major critics, has argued that on the basis of these three characteristics the take-off cannot be empirically separated from the preceding and the succeeding stages. Especially the distinction between the take-off and the second stage is not altogether clear: the development of a socio-economic infrastructure is not possible without one or more developed industrial sectors and without an investment ratio higher than five per cent of national income. The description of the take-off by the presence of the fast development of such a growth promoting infrastructure points to the fact that it could already have been in existence in the second stage. It is not clear what exactly is meant with socio-economic infrastructure[44].

A major proposition in Rostow's theory is the idea that a necessary prerequisite for the development of the take-off is the doubling of the investment ratio in a relatively short period of time. Empirical research has refuted this hypothesis of a short and vehement take-off[45]. Research to the industrialization of the USA[46], the UK[47], Germany[48], France[49], Japan[50] and Russia[51] rather points to a gradual rise of the investment ratio.

Rostow has countered this criticism by emphasizing the distinction between aggregate and sectoral analysis of economic growth and modernization. Modern economic growth is a sectoral process. It originates in the progressive diffusion of the production functions that modern technology can provide. The resulting changes can only be analysed by sector. The different sectors of the economy are closely related. Aggregate analysis, on the other hand, consists of the adding up of sectoral performances[52].

In Rostow's sectoral analysis or dynamic production theory the concept of leading sector is crucial. Economic growth is a concatenation of the rise and fall of leading sectors. Therefore, economic growth can only be understood by sectoral analysis.

Rostow's dynamic production theory specifies the question to the causes and the process of economic growth, but does not answer it. Rostow himself does recognizes this. He thinks that such questions cannot be answered as long as a general theory of human motivation is not available. Nevertheless Rostow developed some broad conceptions on the origins of economic growth. With regard to the second stage these origins lie in the development of the political requirements for the take-off. As to the fourth and the fifth stage he points at the self-sustaining nature of economic growth[53].

Although Rostow's model is empirically untenable, it is interesting that in later publications Rostow regards his model as a representation of reality from which it follows that economic growth causes universal problems for which in principle alternative solutions are available. His idea of a branchwise development deviates from bipolar thinking that implicitly postulates a fixed end of evolution, i.e. modern society. The change of Rostow's thinking is most clear as he describes three alternative developments within the fifth stage: a tendency to mass consumption, the growth of welfare society and the rise of the military state[54]. By this adjustment of the model its ideal typical and abstract character comes to the fore: modernization is conceived as a process that can develop into different directions.

The Problem of Universality

The idea of modernity as a universal phenomenon is strongly tied to bipolar thought. In this context *universal* means that if societies modernize, social change, technological development and economic growth inevitably produce societies that have uniform and similar characteristics. Although the points of departure may be different, the points of arrival are the same. The explanation for this trend towards universalism and similarity is based on the idea that technological development has a logic of its own. Thus in the course of technological development societies will change into the same direction.

What are the essentials of this universal conception of modernity? Despite differences in emphasis between different theories we are continuously confronted with the same key variables: traditional collectivism as it was manifested in the relatively homogeneous communities of the past, is substituted by individualism. There is a trend towards achievement motivation, to an increase of instrumental rationality which leads to increasing bureaucratization, a further division of labour and a progressive secularization. By the expansion of the mass media a mass culture is established that weakens the cultural barriers between categories of the population and nations. The progressive character of technological development produces an increase of capital intensive production techniques and, because of that, it leads to a shortening of the amount of lifetime people spend to labour. The level of education increases. The influence of governments in the economy expands and in politics the participation of people enlarges.

This summing up of the characteristics of modernization is not exhaustive, but it covers rather well the consensus on the issue as it existed in the 1950's and the 1960's. As we shall see below, since then the picture of modernization as an universal phenomenon has been amended to a substantial degree[55]. It is, however, remarkable how close this particular image of modernity comes to the actual characteristics of Western societies of the time. Therefore, it is not surprising that modernization theory has been accused of being ethnocentric and equalizing modernization to *Westernization*.

One could think of several reasons why modernization theory was so preoccupied with the idea of universal modernity. Firstly, the spectacular technological development with all its social, cultural, political and economic concomitants first took place in nineteenth century Western Europe. Scholars of those days were inspired by the tremendous changes of this part of the world in which they lived. Besides, the knowledge of non-western societies was rather limited which did not promote comparative thinking. Secondly, the influence of Darwin's evolutionism was strong: life forms and societies were regarded to develop continuously to a higher order. The Atlantic societies no doubt were historically unique. They were considered to be front runners that sooner or later had to be followed by those lagging behind. Thirdly, till the post-war period colonization implied scientific imperialism, i.e. so far as Western nations tried to modernize their colonies, these attempts were inspired by their own experiences. A fourth argument may lie in the missionary character of Christianity which is strongly connected to Western culture and which produced a strong inclination to regard the own cultural values as universal norms.

The idea of universal modernization has been severely affected by the process of de-colonization. De-colonization implied that many former colonies chose their own ways to development and modernization. Ways, that in many cases fundamentally deviated from those of the former colonial rulers. E.g. in Asia Singapore, Korea and Taiwan succeeded in promoting economic growth based on a mixture of traditional value patterns and a Western way of life. In this respect the rise of Japan's economic and technological power has been rather impressive, because this country's traditional value pattern was not substituted by imported Western individualism[56]:

> "Even Japan's massive industrialization after the Meiji era (1868-1912) does not support the case for an overwhelming "Westernization" impact, because in this very process of modernization...Japan has not become "less Japanese". It seems reasonable to assume that all the institutional changes and correlates ...have been additions to her traditional culture in the process of modernization, in order to attain a new dimension of cultural and institutional development without altering the fundamental fabric of Japanese society.... The ethnocentric assumption that the Western experience in modernization would serve as a universally applicable model for the rest of mankind is untenable"[57].

The Idea of Prerequisites and Consequences of Modernization

Science is intended to describe, explain and predict phenomena. In the nineteenth century social scientists and economists began to look for factors that could explain the great transformations of the times in which they lived. With different degrees of emphasis most of them also attempted to predict future developments. Their theories are often both explanations as well as visions of a society to come.

Because of their visionary characters many classical theories are, at least partly, value laden. On the one hand they often reflect a desire to the supposed harmony of the traditional past. On the other hand they echo the believe that out of the chaos of industrialization a future would rise that would bring hope and progress to the humankind. As a result of this value laden character, classical theories evaluate sometimes similar trends completely different. In this respect the expansion of the division of labour is a fine example.

For Marx the division of labour is the basic source of all social conflict as it involves the allocation and distribution of resources. Further, within the frame of capitalist society progressive specialization would promote alienation and the degradation of labour, a theme which is still central in Marxist thought[58]. Instead of being free, modern man would more and more turn into a slave of the economic world he himself produced. Only in communist society the natural state of freedom of the humankind would be restored.

Emile Durkheim took a quite opposite view. In his major work *De la division du travail social* he developed the dichotomy of mechanical and organic solidarity[59]. Characteristic of the former is a strong social consciousness or *conscience collective*, i.e. the total of *des croyances et des sentiments communs à la moyenne des membres d'une même société*. Durkheim believed that the only way of maintaining social cohesion in traditional society was by underlining continuously the collective consciousness by normative social control. As traditional societies are fairly homogeneous and because of the repressive nature of normative social control individual freedom was heavily restricted.

Modern society, however, is of a differentiated and heterogeneous nature that follows from a progressive proliferation of the division of labour. Social cohesion is reached by the mutual dependencies of social roles and thus can be traced back to the division of labour. This modern type of social cohesion Durkheim called organic solidarity. It can exist thanks to the division of labour and does not need repressive social control to survive. So according to Durkheim the expansion of the division of labour contributes to the weakening of the repressive nature of society and therefore increases human freedom and the potential for individual behaviour.

In the 1950's and 1960's the process of de-colonization urged scholars to build models that could be used for policy formulation. Both sociologists and economists began to present explanatory variables as prerequisites of modernization.

Again we refer to Durkheimian thought as an example. The repressive nature of mechanical solidarity explains why under such conditions criminal law is relatively important and dominant as compared with civil law. The differentiation of social roles under organic solidarity demands a broadening and specification of civil law as the division of labour can only function if the rights and duties of its participants are guaranteed. Thus according to Durkheim differentiation of the legal system is caused by the need to cope with the increased heterogeneity and complexity of social and economic relations.

In the 1960's this causal chain of Durkheimian thought was reformed by the structural functionalists into a set of prerequisites of modernization. For Parsons[60] modernization is a process of structural differentiation that leads to a growing independency of subsystems. The independence of politico-cultural and legal subsystems is regarded by Parsons as a prerequisite of the independence of economy and technology, of kinship and status systems. We can read similar ideas in Smelser's work[61]. In Lewis' classic on economic growth it is argued that a stable and proliferated legal system that guarantees rights and duties, is a major prerequisite of economic growth[62].

Marx' dialectic theory can be interpreted too as a model that stipulates the prerequisites and consequences of social and economic development. Each stage in the process of social and economic evolution can only progress if certain conditions are fulfilled. Each stage has inevitable consequences that in turn function as prerequisites of the one to follow.

In nineteenth century West-European thought on modernization the growth of individualism was regarded as a major issue. It was observed that traditional communities eroded as a result of new technologies and the inherent processes of enlargement of scale. For many individualism was the characteristic of a new society, an effect of industrialization that was not always evaluated as favourable. This fear of the break-down of the emotional component in human relations is behind Durkheim's anomia concept, Marx's ideas on alienation and Weber's description of the progressive rationalization and bureaucratization of society.

In the twentieth century individualism was regarded as a prerequisite of modernization. The economy will only grow if the individual is freed from his bonds to the collectivity and if he can seek his own gains in the market. The disappearance of the extended family is explained by the structural functionalists as a logical consequence of its disfunctionality in industrial society[63]. Parsons' pattern variables describe modernization as a tendency from collectivism to individualism. McClelland regards the achievement motivation of individuals as a prerequisite of economic growth[64].

In theories on modernization further conditions are stipulated such as the institutionalization of rational thought, the inclination to mobility, the institutionalization of property, labour, capital and exchange and the fostering of a spirit of entrepreneurship. Stage models are also good examples of the incorporation of prerequisites in theories on modernization as in such models every stage produces the prerequisites of the next.

The idea of the prerequisites and consequences of modernization is strongly related to the idea that modernization can be defined chronologically. Therefore, prerequisites and consequences are inherent to bipolar conceptions of modernity[65]. The idea that industrialization has a logic of its own, gave birth to the notion that, if industrialization had to take place, certain prerequisites had to be fulfilled and further, that industrialization and modernization had both universal and inevitable consequences[66]. Among these consequences are anomie, alienation, the degradation of labour, an ever increasing economic surplus[67], a decreasing mortality rate, urbanization and so on.

Of course, this summing up is far from being exhaustive and neither does it pretend. A review of the literature on modernization, however, makes one thing very clear: although there is a broad consensus that the hard core of modernization is technologically induced economic growth, there exists a deeply rooted dissensus on the causes, prerequisites and consequences of modernization.

The idea that modernization has universal causes, prerequisites and consequences leaves us with two problems. Firstly, the before mentioned ethnocentric bias of many a theory of modernization originates from the fact that the historical circumstances of West-European societies have been implicitly considered as universal conditions of modernization. This ethnocentric bias not only refers to the modernization process itself, but also to the supposed prerequisites and consequences. As empirical research progresses this bias becomes apparent. Again Japan is the outstanding example.

If theory hypothesizes that individualization is a prerequisite of modernization and economic growth, it is clearly refuted by the Japanese case. Contrary to the West, Japan seems to have generated its social and economic development from its traditionally strong collectivist culture. The essential traits of modern Japanese society are modifications of social and cultural patterns that can be traced back to its feudal past[68]. This certainly is a point in favour of Berger's statement that, instead of being a universal feature of modernization, individualism is rather a Western phenomenon related to the tradition of Christianity[69].

A same kind of argument can be put forward with regard to the emphasis of the legal system in general and civil law in particular as necessary conditions to modernization. Although a legal system developed in Japan, it functions quite differently from the legal systems in the Atlantic world. The Japanese will only go to court to solve their problems as a last resort. In the Eastern collectivist tradition going to court is regarded as a sign of social inability and is connected with feelings of shame. As a result the legal profession in Japan is less developed and has considerably less prestige than that of the West.

Secondly, the theoretical distinction between prerequisites and consequences on the one hand and the hard core of modernization on the other appears to lead to difficulties in empirical research. We elaborated on this issue when we treated the criticisms to Rostow's stage model. Similar problems arise if one tries to draw a line between prerequisites and consequences. Thus on the one hand individualism is regarded as a prerequisite, on the other hand it is stated that individualism leads to negative consequences like anomie and alienation. One wonders whether anomie and alienation are not rather components of individualization than dependent variables or whether the three concepts overlap. Whatever, it is evident that in empirical research such questions are tricky. Another example refers to social and geographical mobility which is sometimes regarded as a prerequisite of modernization and economic growth, sometimes as a consequence or even as belonging to the hard core characteristics of these processes. In the latter case career orientation and the orientation to upward mobility are considered as fundamental traits of modern man[70].

The above examples indicate how difficult it is to isolate prerequisites, hard core and consequences. These difficulties originate from the unilinear conception of modernization. If we would start from a circular causal conception of modernization without an ethnocentric bias we would not run into such troubles because we could better cope with all kinds of variants of modernity that can be observed in reality.

The acceptance of a circular causal conception of modernity, however, does not solve all the problems automatically. A major problem that has to be coped with is, that as long as certain phenomena - e.g. individualism - are regarded as essential dimensions of modernity, it seems possible on the basis of some dimensions to conclude to modernity, while concluding to tradition on the basis of others. Or, what conclusions do we draw if the values measured on the dimensions of the modernity scale differ from or even contradict each other. What to think, for example, of the simultaneous appearance of a high rate of social mobility, a dominant collectivist culture and a weak developed division of labour. Although the probability that we will observe such a mix of contradictions in empirical reality is non existent, from a theoretical point of view it is a strategic issue that has to be dealt with.

The solution we propose is the introduction of the concept of compatibility. The term compatibility is frequently used in the world of computer technology where it describes the phenomenon that software does not fit to hardware. Let us suppose that we know what variables are involved in modernization processes. We will no longer think in terms of prerequisites, hard core and consequences because that would not be in line with the circular causal conception we adopted just before. We could imagine that by the measured values of some or all of the variables involved, some of these variables are by theoretical inference incompatible towards each other: the values of some of these interrelated variables are not corresponding with our theoretical expectations while others do. In other words, their values are not compatible.

The concept of incompatibility may be enlightened by a few examples. For the sake of clarity we made these examples highly extreme and therefore empirically unrealistic. There only purpose, however, is to demonstrate the meaning of our concept. Thus, a highly developed division of labour is incompatible with an educational system that is not specialized. Or, a low orientation to mobility is incompatible with nearly all variables that induce economic growth. Or, more specific and historical, a religious ideology that regards the exploration of a dead human body as a violation of God's creation is not compatible with the development of medical knowledge.

By the introduction of the concept of compatibility, modernization is conceived as the whole of changes within a set of interdependent variables. These variables may, but need not change equally into the same direction. The incompatibilities between the variables involved may result in all kinds of strain within the total complex of modernization and thus lead to acceleration, stagnation or even regression of the modernization process. It is the major task of empirical research to detect such loci of strain and to investigate what level the variables involved should take to remove the incompatibilities observed.

Of course, the idea of strain is not a new one. Actually, our conception of incompatibilities within the modernization process has been inspired by the structural functionalist theory of strain within the social system. This does not imply, however, that we developed another structural functionalist model of modernization. We substituted the idea of prerequisites, hard core and consequences to that of incompatibilities and as we see modernization as a progressive process of circular causation, we are able to regard modernization as an endless chain of incompatibilities which are responsible for conflict, change, growth, stagnation and decline. Contrary to structural functional thought, in our conception dynamics rather than statics are the key features. However, ours is essentially an open system approach. Technological development, social and cultural change are not regarded as phenomena that exclusively originate from internal system dynamics. They may equally well be caused by changes in the whole of relations of which the society under study forms a part, e.g. changes in the international power structure or in the economic world system.

The Idea of Convergence

In this section we will come back to convergence theory. We have two good reasons for doing so. Firstly, in its most pure form convergence theory represents a line of thought that treats modernization as an universal phenomenon. A further exploration of convergence theory and its critics may contribute to a specification of our own ideas on modernization especially with regard to the notions of circular causation and compatibility that we enfolded before. Secondly, till now we did not give a conclusive statement as to what variables are part of the process of modernization. The results of empirical research done to convergence theory enable us to select those variables.

The convergence hypothesis states that from dissimilar positions societies develop into the same direction. In the course of that process they will become more alike. Thus there is a trend from dissimilarity to similarity. Some specifications of convergence theory even claim that in the end all societies will be similar as to their main characteristics.

The most well known variant of the convergence hypothesis relates to economic development and the trend towards similarity of economic systems: capitalism and socialism are believed to move towards each other in the course of history. Here the latent variable is the degree of centralization of economic decision making: market versus plan. The implications of convergence theory, however, are much broader and are anchored in the traditions of the social and economic sciences. It would go to far to treat the roots of convergence theory in this chapter extensively. The interested reader is referred to Clark Kerr who analyses the historical origins of this line of thought[71].

Convergence theory originates from the assumption that industrialization has a logic of its own. Once industrialization has started, certain conditions have to be fulfilled otherwise it can not proceed. These conditions are independent of the type of society in which industrialization evolves. Kerr and associates have identified four such basic conditions: the creation of an industrial labour force that must be recruited, motivated, trained and maintained in a state of productivity. The establishment of large scale production units and urban centers, a substantial participation of government in the economic process and a minimal consensus about what should be done and what not. The development of leadership able to execute policies that make at least an adequate use of the available potential of production. The development of a labour force that accepts the economic structure[72].

Before, we discussed at length the problems involved in the idea of prerequisites of modernization. Convergence theory is somewhat different from other theories that use the idea of prerequisites, because here the conditions are never completely fulfilled. They are succeeded by other conditions belonging to another stage of industrialization. Thus each of Kerr's conditions remains a source of ongoing concern.

Without any doubt Jan Tinbergen is a representative of convergence theory in its most pure manifestation[73]. Tinbergen's ideas have become known as the theory of the optimum regime. His basic assumption is that economic systems of whatever nature have fundamentally the same tasks and aims: the maximation of wealth and the realization of a future optimum growth rate. The two most important means to reach these aims are market and plan. These are not thought by Tinbergen as contradictory but rather as complementary. The market is a suitable tool for decision making in situations of decreasing returns to scale and no substantial external effects. The state plan has to be preferred in situations of increasing returns to scale leading to monopolies and if substantial costs or benefits have to be taken into account. In their search for optimal results economies rather attempt to find an optimal mix of market and state plan instead of exclusively applying market or state plan principles. According to Tinbergen, as economic development progresses, economic systems would rather converge than diverge as to the mix of market and plan.

Tinbergen's ideas have been of great significance to the development of the thought that the effects of ideologies were of minor importance for the outcomes of modernization processes. Modernization meant *the end of the ideological age*[74]. Although Tinbergen's theory is generally regarded as the beginning of modern convergence theory, his work is in many respects not revolutionary as it is built on earlier ideas.

In 1953, six years before Tinbergen published his famous article on the optimum regime, Robert A. Dahl and Charles E. Lindblom issued a book titled *Politics, Economics and Welfare*[75]. These authors resist the ideological thinking on socio-economic management. Instead they argue that in any economic system the aims can only be achieved by *rational social calculation*.

> "In economic organization and reform, the "great issues" are no longer the great issues, if they ever were. It has become increasingly difficult for thoughtful men to find meaningful alternatives posed in the rational choices between socialism and capitalism, planning and the free market, regulation and laissez faire, for they find their actual choices neither so simple nor so grant[76]....Plan or no plan is no choice at all; the pertinent questions turn on particular techniques: who shall plan, for what purposes, in what conditions and by what devices ? Free market or regulation ? Again, this issue is badly posed. Both institutions are indispensable"[77].

According to Dahl and Lindblom the price system, hierarchy, bargaining and majority vote can be analysed both as potential complements of and as potential substitutes for one another. An optimal mix of those four depends on the conditions at hand, the goals of the socio-economic system and the issues that need allocation and/or distribution. According to Dahl and Lindblom the question of how the future can be reached is not ideological but technical. The aims, however, can only be reduced to universal values.

The ideas of Tinbergen were followed throughout the whole Western World. In the USA writers like Bell, Galbraith, Lipset, Parsons, Riesman, Schlesinger Jr. and Shill represent the end of ideology school. In Europe their colleagues were Berlin, Dahrendorf, Duverger and De Jouvenel. Tinbergen's disciple, Hans van den Doel, amended his ideas substantially. According to Van den Doel, Tinbergen had neglected the fact that societies differ as to their goals. Societies thus will converge to a substantial degree but not totally. The dissimilarity of goals prevents total convergence[78].

Ellman even criticized Van den Doel's theory of partial convergence

> "noting that any nation, in the absence of adequate knowledge and in the midst of internal political conflict, would have difficulty maximizing any fixed set of goals. He noted also the varying availability of resources and the differing nature of social constraints in the environments that surround different nations, even if they have the same or similar goals. The concept of an optimum, or of alternative optima, is an elusive one"[79].

Post-war convergence theory has been strongly affected by the East-West conflict. It became popular during the cold war and one could say that in that era convergence theorists were the optimists proclaiming that the perceived contradictions between capitalism and socialism were temporarily and relatively irrelevant as both systems had to cope with the same problems. Later on, many adherents of convergence theory moderated their argument. In 1977 Lipset stated that it was only the intention to weaken the *the most passionate attachments to both right wing and left wing revolutionary ideologies*[80].

Another criticism of convergence theory regards its reduction of the complexity of empirical reality to a bipolar abstraction. Kerr and associates[81] have rejected the idea that there were only two options of development, capitalism and socialism. As Kerr wrote later, he and his co-authors were convinced that "there were several intermediate and changing solutions between the two extremes; instead of pure socialism, there were several mutating forms of pluralistic industrialism in various relations to one another. The future lay with a diversity of economic arrangements rather than with any single uniformity or with opposing dual uniformities"[82].

From the Marxist point of view, convergence theory has been heavily criticised too. The Soviet elite regarded it as a source of corruption for students, intellectuals and for the leaders and potential leaders of developing countries. Convergence theory was treated as an expression of anti-communism with the aim to undermine the Soviet system, to sow discord between the Soviet Union and its allies, to prevent the emancipation of the Third World and to repress the class consciousness of the masses in the developed West[83]. With the exception of a few[84], Western Soviet oriented Marxists have echoed the official Soviet view. This Marxist critique concentrates too much on the aspect of communism fading away. The inevitable development of Western capitalism to a more centralized economic system is the complementary part of the message of convergence theory that has been largely neglected by its Marxist opponents.

Non-marxist critics have attacked convergence theory for its neglect of the importance of ideologies. According to this criticism national leaders often believe their own ideologies to be true, they often take them as guides to their actions and they do try to preserve them. Thus Brzezinski argued that the actions of the Soviet elite are heavily affected by ideology[85]. The same argument can be put forward to US leadership although differences of degree are observed between Democratic and Republican presidencies.

The approach to the role of ideology in the process of modernization has been differentiated by Gerschenkron to an important degree[86]. Gerschenkron argues that the influence of ideology on policy is especially strong in the initial phase of industrialization, but decreases as the latter progresses. Then the function of ideology turns over: instead of being a guideline to the development of policy, ideology serves to justify and legitimate policies that are executed to solve practical problems which have little to do with any ideological argument whatsoever.

This kind of criticism of convergence theory has been further elaborated by Wilbert Moore. The cornerstone of convergence theory is the assumption that policy makers are rational human beings who will choose for rational solutions if confronted with practical problems. The aggregated effect of such rational behaviour will be convergence. The practical problems, however, are strongly related to the society's goals, values and belief system. I.e. what is a problem and what is not, is largely a matter of perception. It follows that convergence will only be reached if the goals, values and belief systems of different societies are more or less similar. As goals, values and belief systems are ideologically determined, convergence will not occur as long as there remain ideological differences between societies[87].

Other critics of convergence theory have concentrated on its bipolar nature. They state that there is no such thing as one best solution, but instead to any problem more liable solutions exist. From research to the economic performances of countries that differ as to the degree of economic centralization, Pryor concludes that the differences in economic performance do not seem very large. Thus there are neither empirical nor theoretical arguments that unambiguously point to the desirability of either a low or a high degree of economic centralization[88].

After all we will introduce an important critique to convergence theory that is related to the one on the neglect of ideology. It has been put forward by, among others, Joseph Schumpeter[89] and Bertram D. Wolfe[90]. The nucleus of this critique is that convergence theory treats economic and social development as if it evolves in a historical vacuum. However, a society's social structures, its institutions and patterns of values and attitudes are the product of its historical past. Such historical patterns are interwoven with present behaviour. Although it is admitted that any rational action can only evolve within certain preconditions, it is the society's historical heritage that affects the choice between possible alternatives. As Wolfe has put it:

> "Every land moves towards its future in terms of its own past, its own institutions and traditions"[91].

The Explanation of the Variance of Modernity

Statements like Wolfe's are rather vague and suggest that every society's path to modernity is frozen by the genes of its historical past. Thus the pragmatic determinism of convergence theory is exchanged for historical determinism. Therewith the establishment of more general theories of modernization is theoretically prevented. Historical determinism neglects

> "the possible relations between different degrees and mechanisms of continuity and discontinuity and patterns of change in "pre-modern" settings of...societies on the one hand and patterns of their development on the other"[92].

The explanation of the outcomes of modernization processes is a problem of both uniformity and variance. Uniformity, because it can not be denied that modernizing nations tend to show similarity in some aspects. Variance, because significant differences tend to continue despite such similarities.

We now shall concentrate on this mixture of similarity and dissimilarity. To gain a better idea of this phenomenon we will use the theoretical conception of Eisenstadt[93] who has tried to cope with both these sides of the coin of modernization without plunging into a new determinism.

Eisenstadt's point of departure is the idea that every process of modernization produces problems that have to be solved by the society involved. So he agrees with convergence theorists that modernization has some logic of its own. However, the fact that modernization in different societies produces similar problems does not necessarily imply that the responses to these problems have to be similar too.

We will emphasize this thesis by elaborating an example. One of the problems industrialization produces is the requirement to discipline the labour force in such a way that it can be fitted into a standardized and highly structured production process. In Western Europe and the USA industrial discipline was reached by, what Etzioni has called[94], utilitarian or renumerative power. To put it simply, by the thread of being fired and the promise of being promoted. This has led to a predominantly calculative relationship between employers and employees in which conflicts are solved rather by negotiations than by the use of force or moral appeal[95].

In Stalinist Russia, on the other hand, coercive power applied by the state was used as a means to discipline the labour force. Its effect has been an orientation of the workers different from that in the West and to the inability to establish organizations that are truly representing the employees.

A third way of disciplining the labour force has been and still is practised in Japan. The key to understand Japanese industrial discipline is the moral involvement of the Japanese worker to Japanese society in general and to his employing organization in particular. Much more than renumerative and coercive power the effect of normative power has been the establishment of a set of mutually moral obligations between employers and employees that can be reduced to the culture and structure of the Japanese feudal household. As a result of this predominantly moral relation between employer and employee Japanese industrial conflicts are of a rather ritual nature and unions are not organized on the national level but on the company level as the company is the main locus of moral identification.

Eisenstadt argues that the nature of the response to social problems depends on the social evaluation of the social reality of which the problem at hand is a part. This evaluation process is determined by social codes - also labelled by Eisenstadt as ethics - that are derived from the symbolic dimensions of a society's culture. These symbolic dimensions have to do with fundamental issues like *the evaluation of human existence (ritual, politics, economics) and the definitions of boundaries of political and cultural collectivities.*

According to Eisenstadt, the social function of social codes is the reduction of uncertainty that exits in any situation of social interaction. Only by the reduction of this uncertainty social stability is possible, a theme which is frequently encountered in structural functionalist thinking. In Eisenstadt's conception codes therefore provide the groundrules of social organization. They make up the "hidden" or "deep" structures of a social system.

If we take a closer look at the nature and the meaning of these social codes, we see that this conception of Eisenstadt's is a better fit to social and economic reality than convergence theory. Next to a similarity in problems, the latter implicitly assumes similar goals for different societies and for different power groups within these societies. In elaborating the idea of social codes, Eisenstadt gives full attention to the phenomenon of conflicting interests within societies. Thus social codes refer to the specification of goals, the attributes of similarity and criteria of membership, the criteria of the regulation of power over resources in different social situations and institutional spheres and the legitimization of criteria in terms of distributive justice. Further, social codes determine the intensity of conflicts, the perception of the acuteness of conflicts, the range of flexibility of the response and the regressive or expansive nature of policies aimed to cope with such conflicts.

Although social codes affect the respons to modernization problems to a substantial degree, Eisenstadt's conception is not totally deterministic in this respect. He argues that despite the presence of social codes, in any situation of social interaction uncertainty and degrees of freedom remain to exist. Thus man is not a puppet on a string, although social codes guide his actions substantially. Further, similar social codes in different societies may produce different institutions and therefore dissimilar responses. The knowledge of the content of social codes enables the researcher only to formulate probabilistic hypotheses.

The attractiveness of Eisenstadt's theory lies in the relation between structural problems evoked by modernization on the one hand and the historical and cultural influences on the responses to these problems on the other. In different societies problems of modernization may be identical, the responses to these problems may be not. In this way the variance in the outcomes of modernization processes is explained despite the common nature of these processes.

The Nature and Key Elements of Modernization

It is now time to put things together and to enfold our ideas on modernization in a more systematic way. First of all, we defined modernization as the process of interdependent changes in the social, cultural, political and attitudinal domains of a society rooted in the transformation of its economy by the purposive development of technology. We emphasized the interdependencies of the changes involved contrary to mutually independent types of social change. I.e. modernization has to do with a complex of intertwined variables in which either congruent or discongruent changes may appear as a result of a progressive technological development.

The variables involved in the process of modernization relate to each other in a circular causal way. The purposive nature of technological development is the nucleus of the whole process. In this context "purposive" means the deliberate development and application of technology with the aim to raise productivity. "Purposive" does not imply that elements of chance and human creativity are completely deleted from the process of invention, but rather that existing technological knowledge is rationally and calculatively applied in the economy and that the development of technology is institutionally anchored in specialized organizations, educational systems and professional groups. The degree of institutionalization of technology, the tempo of its development and the degree and rate in which it is absorbed by the economy are the gauges of modernization.

Modernization is a progressive process. There is no such thing as a drawing line between traditional, modern and future society. Such abstractions tell us nothing about reality because they deny its heterogeneity and therewith deprive it from its historical and social meaning. According to the criteria enfolded above, we prefer to regard modernization as a concept of relativity and to speak of scales or degrees of modernization. Thus Antiquity should be regarded as less modern than nineteenth century Western Europe because in the former knowledge was largely sought for its own sake while in the latter the relation between technology and economy was much stronger.

From the history of thought on modernization we learnt that it is misleading to speak of universal prerequisites and consequences of an universal modernization process. First of all, because prerequisites, hard core and consequences of modernization are both theoretically and empirically difficult to distinguish. Secondly, the postulated universal nature of modernization is empirically untenable and seems to represent a kind of ethnocentrism.

We introduced the concept of compatibility. The process of modernization should be conceived as a complex of variables that under specific historical conditions may be mutually more or less compatible. The degree of compatibility between the variables involved determines the degree of tension in the complex as a whole. The idea of compatibility can help to explain acceleration, stagnation and decline in modernization processes.

A major task in the study of modernization processes is the analysis of incompatibilities and their consequences for social and economic development. Thus the formerly negatively regarded consequences of modernization such as anomie, should be studied as particular types of incompatibilities between parts of the modernization process. Therefore, instead of being considered as a negative consequence of modernization, anomie should be understood as a manifestation of the incompatibility between on the one hand economic growth and technological development and, on the other hand, particular types of social organization that integrate the individual in society. Such an incompatibility can be lessened by either a stagnation of economic growth and technological development or the growth of new and more adequate institutions of social integration.

It may be clear that we regard modernization as an endless and continuous chain of incompatibilities: in the search for equilibrium - or: the attempt to erase existing incompatibilities - new incompatibilities arise. Both the search for equilibrium and the manifestation of continuous incompatibilities are indissolubly linked to modernization. Modernizing societies then are open systems that are subjected to both internal dynamics as well as the exigencies forthcoming from the relations to the external environment. The former and the latter are closely related. In this respect one could think of incompatibilities produced by the increased intertwining of external economies, by changes in the stock volumes of raw materials, by the growth of international value patterns and so on.

We do not deny that processes of technological and economic development have a logic of their own. However, we seriously doubt if this logic goes as far as Tinbergen suggested in his theory of the optimum regime. Although technological and economic development can only proceed if the incompatibilities they provoke are levelled some way or other, this does not logically imply that the way in which such incompatibilities are solved is given by their generators. The nature of the response to incompatibilities depends on a society's social codes and is therefore the product of its historical past. Because of the differences between societies as to these codes, the responses may be dissimilar too. This is why there is so much variety in the outcomes of modernization processes.

Within the complex of variables that together make up the modernization process, economic growth and technological development both have effects on and are dependent of the social, cultural, political and attitudinal domains of a society. Despite the fact that economic growth may stagnate as a result of various factors - even under conditions of a developing technology - the heart of modernization is the institutionalised and purposive development of technology aimed to raise productivity. Because of that, modernization is indicated by a rise of investments in research and development as a percentage of national income, a progressive specialization within the educational system, an increasing accessibility to higher levels of education for larger proportions of the population and a progressive division of labour.

These developments can be observed in all modernizing societies irrespective of the nature of their cultures and historical pasts. Of course, big differences exist between societies as they are not all at the same level of modernization at one specific moment.

In a recent study on the merits of convergence theory Kerr gives an empirical and comparative analysis of those characteristics of societies that tend to converge as societies modernize[96]. Here we are interested in such common developments because they give us a closer look on the process of modernization and the incompatibilities involved. Below we will use Kerr's analysis with gratitude.

Kerr concludes that as modernization progresses the content of knowledge gets a more universal character and is no longer dependent on national boundaries. The development of mass media, the spread of education and of scientific institutions promote a constantly speeding up of the diffusion of knowledge.

A second domain in which convergence can be noted is, what Kerr calls the mobilization of the resources of production. In all societies subject to Kerr's analysis the investment in research and development as a percentage of national income has increased substantially. So did the endeavours to explore and exploit the national stock of natural resources. There has been a mobilization of labour and capital. The latter is connected to a decrease of employment in agriculture, an increase of the quality of education, an increase of the number of professionals and higher educated and a greater accessibility of the educational system in general. Many of these changes can be labelled as investments in human capital.

A third domain of convergence is the organization of production. Modernization is characterized by the creation of manufacturing, construction and transportation industries and service industries to aid the industrial sector of the economy. Production moves to urbanized areas. There is a substantial enlargement of scale of enterprises. The number of self employed decreases. Within the labour force the attribution of status by inheritance is substituted by status based on achievement. Intergenerational mobility increases. So does geographic and social mobility.

A fourth trend of convergence relates to the patterns of work. From the developments in the domain of production it follows that ever larger parts of the population are working in larger organizations with predominantly formal patterns of social relations. The employing unit becomes more bureaucratic. The use of worktime is subject to strict control. In all countries a web of rules is gradually established that specifies the number of working and leisure days, the hiring, firing and promotion of personnel, the treatment of grievances and the amount of work that is supposed to be done. To summarize, there is a strong tendency to standardize and to bureaucratize labour.

The fifth domain Kerr discerns is that of the patterns of living. The number of years an individual spends throughout his life on education, labour and pension tends to converge in the different countries. Relatively more people live in urban centers. The households become smaller: the nuclear family is the dominant pattern. The number of children per household decreases, net reproduction rate approaches or falls below the replacement rate. Mobility by travelling increases and so does the use and consumption of mass media information. From the comparisons between the countries involved it turned out that life styles are converging too: a world wide cosmopolitan life style based on standardized mass consumption is established.

In the pattern of distribution of economic rewards two significant trends are observed. In the first place, rewards are increasingly based on merit and seniority instead of on inherited status or political acceptability. Secondly, the state fulfills a growing redistributive function reflected by the increase of welfare expenditures.

The above six domains of modernization can be summarized into six basic parameters connected with economy and technology and with the social, cultural, political and attitudinal domains of societies. Firstly, the expansion of the division of labour. Secondly, enlargement of scale, both economically (size of the firm) and socially (urbanization). Thirdly, an increase of social and geographic mobility. Fourthly, bureaucratization and formalization of social and economic relations which includes a shift from ascription to achievement. In the fifth place, a growth of the public sector of the economy and at last a standardization and greater accessibility of information.

The compatibility of these six parameters towards each other and towards the logical needs of technology is of great influence on the progress of technological development and economic growth. As technology advances an increasing amount of people with specific skills and knowledge are needed to apply that technology purposefully to raise productivity. Compatibility between the division of labour and technology is a necessary, but not sufficient, condition for modernization to proceed. To reach an optimal allocation of talents to positions in the economy the educational system should be compatible with the division of labour. Regarding the former, not only a great accessibility is required but also a sufficient degree of differentiation and specialization.

As the rate of technological development accelerates, the nature of the division of labour will have to change accordingly. The resulting change in the demand for labour can only be met if the labour force is characterized by dynamic attitudes aimed at mobility and a change of the content of work.

A high level of technological development demands an allocation of manpower based on knowledge, skills and abilities: on achieved instead of on inherited status. For the development and application of advanced technology large investments are required that can only be realized by large organizations. Such organizations can only be controlled by formal and standardized procedures. The control structure of large scale bureaucracies is only able to function properly as long as its members have attitudes that promote the commitment to such ways of social control. The embedding of bureaucratic attitudes is effected partly by the organizations themselves and partly by a change of the patterns of primary socialization. Thus bureaucratization is not restricted to the domain of organizations alone, but pervades the whole of society.

The role and function of modern government should also be understood from the level of technological development. On the one hand because the development of advanced technology requires an infrastructure that is heavily dependent on national or even supranational policies; on the other hand because the social and ecological side effects of technological development and of a highly developed economy can only be countered by national governments.

The above trends and relationships are produced by the logic of technological development. If a society wants to proceed on the path of modernization it will have to ensure compatibility. This does not mean that the logic of technology prescribes one fixed and logical configuration of elements. Within the attempt to reach compatibility alternative solutions are available. The choice to be made depends on the social codes of the moment that are the products of the historical past. These codes are not static in itself. They too are part of the complex of relations we call modernization, i.e. they adapt and modify as a result of historical experience.

Social codes should be distinguished as to the degree in which they are resistent to change. The emancipating effects of technology that are the unintended consequences of the diffusion of education and information, the increase of mobility and of legal-rational decision making have especially affected social codes defining the boundaries of political and cultural collectivities: however present they may be, the exclusive nature of subcultural groups and minorities tends to disappear as a result of technological development. Furthermore, the social distance between the different categories of the population is lessened.

Next to the conclusion that in six domains substantial convergence has occurred, Kerr concludes that in the domain of values and belief systems a high degree of heterogeneity between the countries studied continues to exist. Thus, the cultural elements that guide the selection of goals and the ways technology is applied do not converge; this certainly contributes to the variance in the outcomes of the processes involved[97].

At the end of this chapter we wonder how technological development is related to the social consciousness of the individual. Looking back at history, the spread of an attitude of rational calculation has been most dramatic. Perhaps even more dramatic than Max Weber ever could have supposed. Life in modern technological society requires a mentality aimed at planning, often long term planning[98]. Rational calculative attitudes, the importance of educational achievement and the dominance of the ever present bureaucratic society have led to what Peter Berger has called a *quantitative atomising life style* that has spread to all areas of social life. The increased geographical and social mobility makes the engagement in long lasting all embracing social ties more difficult than ever before. Besides, the rise of real income and the spread of bureaucratic social security makes people less dependent on such ties.

CHAPTER III

MONITORING DUTCH MODERNIZATION

Modern Economic Growth and Technological Development

For a long time there has been little challenge of the thesis that the Netherlands modernized relatively late compared to other European countries. Economic growth linked to industrialization and associated technological development was considered not to have started before the last quarter of the nineteenth century[1].

For Simon Kuznets modern economic growth is primarily related to the rise of Gross Domestic Product per capita (GDP) as it took place in the currently developed countries over the last two centuries. According to Kuznets, modern economic growth is associated with significant demographic changes: a new and unique combination of low mortality with low fertility allowing for a much greater long-term rate of natural increase than that over the preceding centuries of high birth and death rates. As Kuznets remarks:

> "It had to be new because the opportunities for reducing the death rates to the low levels attained were new and unparalleled in the past"[2].

It follows that, before the start of modern economic growth, population remained stable within certain margins. This was the case from the agrarian revolution of 8000 BC onwards till the eighteenth century.

For the Netherlands Maddison[3] presents a figure of per capita GDP of 440 (1970 US dollars) in 1700 and corresponding figures for France and the UK of respectively 275 and 288. Maddison, who distinguishes between "the growth potential of the "lead" country, which operates nearest to the technological frontier, and that of "follower" countries, which have a lower level of productivity", concludes that since 1700 there have been only three lead countries in the history of capitalist development: the Netherlands, the UK and the USA. The Netherlands took the lead position until the 1780s[4].

According to Maddison's analysis, the Netherlands was the first country to embark on modern economic growth. As early as the sixteenth century it was far ahead of its European neighbours. This conclusion is confirmed by other facts as well. At the end of the sixteenth century the Dutch provinces were modern compared to the rest of Europe. In property relations vassalage was practically absent. Farmers owned or rented the land. Agriculture and cattle breeding were advanced and market oriented. A substantial part of agricultural production was sold to nearby customers: the population of the numerous cities. In the seventeenth century, Holland was one of the most urbanized areas of Europe. While in other parts of Europe about 70 per cent of the population was engaged in agriculture, in Holland about 54 per cent lived in the cities[5]. Both economically and politically the centre of gravity lay in the merchant class. Absolute rule by princes had given way to some degree of democratic order.

Until the late eighteenth century, the Dutch were able to profit largely from entrepôt trade, thanks to favourable geographic conditions and the pursuit of mercantilist policies. The Republic's eighteenth century economic downfall was largely due to a change in external conditions. The fact that other countries followed a mercantilist policy, affected Dutch trade substantially[6]. Further, the Republic lost its

35

trade monopoly due to a number of conflicts with England and France[7]. Finally, the Napoleonic dominance at the end of the eighteenth century and the beginning of the nineteenth century and especially the Continental System meant the final blow for Dutch trade as the main source of national welfare.

The era of glory of the Dutch economy in the seventeenth century is commonly called the *Golden Age*. Although there is much discussion about the exact dating of the end of Dutch prosperity, it is generally agreed that it must have been in the last third of the seventeenth century. As to the development of the Dutch economy, the debate concentrates on eighteenth and nineteenth century Dutch growth paths. Since for this period data are scarce, most scholars have to rely on estimates.

Table 3.1. Population Increase of Eleven European Countries, 1500-1820 (Percentages and Average Annual Compound Growth Rates)

	1500-1600 (% 1500)	1600-1700 (% 1600)	1500-1700 (% 1500)	1500-1700 (growth)	1700-1820 (growth)
Austria	26.8	16.7	47.8	.20	.3
Belgium	-6.7	42.9	33.3	.14	.5
Denmark	8.3	7.7	16.7	.08	.4
France	11.0	16.0	28.8	.13	.3
Germany	25.0	0.0	25.0	.11	.4
Italy	26.7	0.0	26.7	.12	.3
Netherlands	57.9	26.7	100.0	.35	.2
Norway	33.3	25.0	66.7	.26	.6
Sweden	38.2	65.8	129.1	.42	.6
Switzerland	25.0	20.0	50.0	.20	.4
UK	54.5	36.4	110.8	.37	.7
Mean	27.3	23.4	57.7	.22	.4

Source : Accounting of the percentages derived from Angus Maddison, 1982, table B1, p. 180; the average annual compound growth rates were derived from A. Maddison, 1982, table 3.4., p. 49.

In a brilliant article on the subject, the American economic historian Jan de Vries made an inventory of the different positions in the debate[8]. These positions are summarized in figure 3.1. In an attempt to illuminate this figure, we will follow De Vries' line of argument. As we want to detect the changes in the tempo of the modernization process and the corresponding patterns of social change, we are particularly interested in the trends or growth paths of the Dutch economy.

Derksen estimated national income at 10-year intervals back to 1860. He has, however, never published his study and consequently, as De Vries remarks, it is not possible to evaluate his estimates and lend them much credence[9].

Teyl estimated Dutch national income by two series extending back at five-years intervals of 1850[10]. The first series is based on Dutch government tax revenues and the second on the estimation of a Cobb-Douglas production function for the Dutch economy. Although different from decade to decade, they end up in close agreement in 1850. Teyl described the Dutch economy as growing at 0.7 per cent per year over the period 1850-1910. Based on his criticism to Teyl's method and assumptions, De Vries suggests the possibility of a more rapid growth.

Figure 3.1. Alternative Proposed Growth Paths of Dutch National Income, 1688-1900

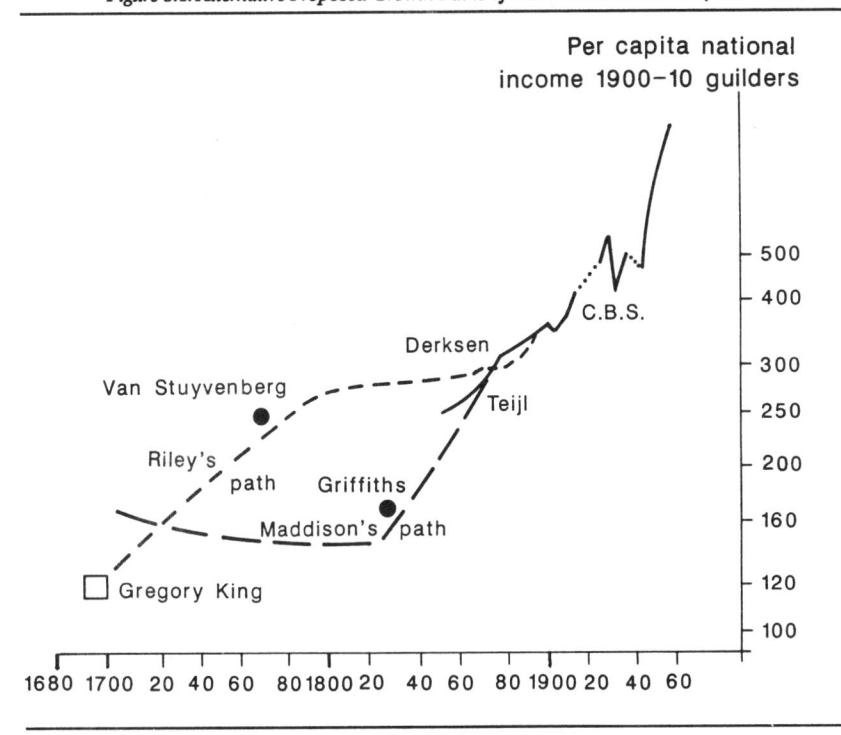

[figure showing per capita national income 1900-10 guilders, with paths labeled: Gregory King, Riley's path, Van Stuyvenberg, Maddison's path, Griffiths, Derksen, Teijl, C.B.S., spanning 1680 to 1960, with y-axis values 100, 120, 160, 200, 250, 300, 400, 500]

Source : Jan de Vries, 1984, p. 161.

Van Stuijvenberg represents the view shared by many historians that the Dutch economy broke out of a long-term stagnation only after 1850. He guesses that real 1770 net national income was about equal to Teyl's estimate for 1850[11].

Richard Griffiths asserts that the growth path of 1850-1910 should be extended back to 1830, since there is no basis for supposing the mid-century to be an economic turning point. Although he offers no national income estimates, he criticizes Teyl for underestimating growth in the period 1850-1910[12].

The above estimates are all backward projections and their result is that, as De Vries remarks, *the view backward from 1900 is still shrouded in mist*[13]. Consequently De Vries' next step is to examine forward projections from the Golden Age onwards. A central figure in this respect is Gregory King who estimated Dutch and French national incomes for 1688 and 1695 in conjunction with his efforts concerning England[14]. Scrutinizing King's method, De Vries concludes that King's estimates for 1688-1695 cannot be justified[15].

Maddison estimates Dutch seventeenth century production levels 50 per cent above those established by King for England in 1688. Maddison bases this proposition on 5 arguments[16]:

a. upwards of two-thirds of the Dutch labour force found employment outside of agriculture, where incomes are usually higher;
b. the high-income foreign trade and service sectors were equal in size to those of England and on a per capita basis much larger;
c. agriculture was highly specialized and presumably more productive;
d. the industrial structure was diversified and benefited from a low-cost energy source in the form of peat;
e. the investment rate was higher in the Republic than in England, and this must have provided some measure of technical progress.

After establishing the Dutch income level in 1700, Maddison proceeds by accepting Johan de Vries' assertion that in the eighteenth century Dutch national income remained about the same, but per capita GDP, which excludes income from foreign investments, fell by about 10 per cent between 1700 and 1760, remaining constant from then on until 1820. For the period after 1820 Maddison makes no direct estimates of Dutch growth.

As Jan de Vries rightfully remarks, if one links Maddison's growth path for the period 1700-1820 to the available twentieth century data, the result is an amazingly high growth rate in the nineteenth century.

The decline in Dutch economic growth in the last quarter of the eighteenth century is generally explained by two factors: the lethargy and self indulgence[17] of the Dutch economic elites, and the decline in Holland's entrepôt function due to external factors. Riley[18] found that the Amsterdam capitalist market showed a rapid growth of Dutch foreign investment in the second half of the eighteenth century. Further, reviewing studies on Dutch shipping, Riley argues that the decline in the entrepôt function was compensated in the rise of direct Dutch shipping services that bypassed the Republic's ports. Riley argues that the first half of the nineteenth century was characterized by economic stagnation. Consequently the apparent doubling of real income between 1688 and 1860 must have been chiefly an achievement of the eighteenth century.

So far we have summarized Jan de Vries' analysis of the different positions in the debate on Dutch economic growth. The conclusion must be that apart from a consensus on the advanced position of the Republic in the sixteenth century and the acceleration of growth in the second half of the nineteenth century, there is still much confusion as to the years in between. In this respect Maddison's and Riley's growth paths are sharply opposed.

We now come to the hypotheses Jan de Vries himself put forward. He starts with the observation that both the estimates of King, Maddison and Riley are based on a comparison of the Dutch economy to that of the U.K.[19]. Reviewing new evidence of the UK's growth path[20], De Vries concludes that:

"the English economy before 1760 was more industrial and generated a higher national income than has hitherto been accepted".

Then De Vries puts forward his own hypotheses[21]:
1. "the Dutch economy declined absolutely during the period 1675-1813, with most if not all of the decline occurring before 1750;
2. from 1813 to 1850 the economy recovered from severe depression in the Napoleonic period, but could not have grown rapidly;

3. per capita income in the third quarter of the seventeenth century stood at double the level identified by Gregory King - a revision that is based in roughly equal parts on a revision of England's income in 1700 and an argument that the Republic's income could have exceeded that of England by two thirds".

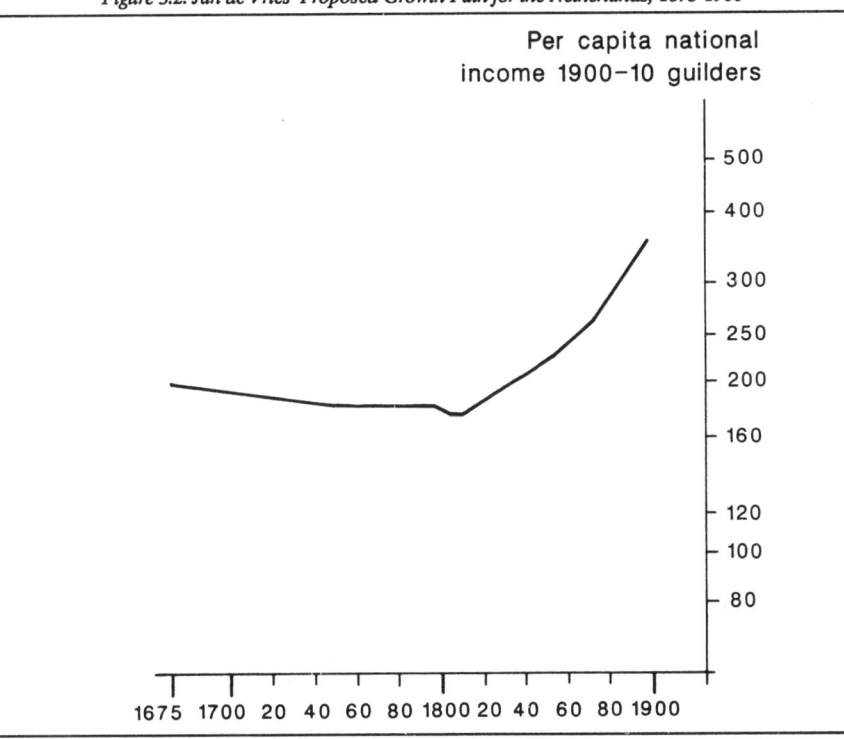

Figure 3.2. Jan de Vries' Proposed Growth Path for the Netherlands, 1675-1900

Per capita national income 1900-10 guilders

500
400
300
250
200
160
120
100
80

1675 1700 20 40 60 80 1800 20 40 60 80 1900

Source : Jan de Vries, 1984, p. 168.

De Vries advances three kinds of arguments for his hypotheses: the development of sectors of industry, of population and of income and wages.

Basing himself on a wide range of historical case studies, De Vries asserts that commodity production declined in the course of the eighteenth century. So did agricultural output between 1660 and 1740 in the west of the Netherlands, but this found some compensation in the eastern provinces. In the second half of the eighteenth century, agricultural output expanded. Commercial, shipping and banking services formed important economic activities in the Republic. Its centre, Amsterdam, did not decline to 1780 although, as De Vries remarks,

"there are many indicators that eighteenth century Amsterdam attracted to itself the remnants of a previously more distributed commercial life"[22].

On a more aggregate level, De Vries concentrates on transportation and population. Intercity passenger travel by barge formed the backbone of the Republic's transportation system. De Vries found earlier[23] that the demand for intercity passenger travel by barge was largely dependent on urban per capita income. From 1675 to 1745, intercity passenger travel by barge declined.

> "Thereafter the demand for passenger travel revived slightly, suggesting that in the second half of the eighteenth century Dutch cities did not experience further income decline"[24].

Modern economic growth is strongly related to population. From 1520 to 1650 Dutch cities grew considerably. From 1675 to 1810 population was minimal. The nineteenth century was, however, characterized by substantial growth. As the 1800 level of urbanization was exceeded only after 1870, in the first half of the nineteenth century population growth must have been absorbed by the rural sector[25].

Concerning the trends in real wages De Vries rests his case on several studies and draws the following conclusions[26]:

1. "Real wages rose erratically from the mid-sixteenth century to the 1680s, the rise being interrupted most notably by the many years of high prices during the wars of the 1650s and 1660s. The wage increase dominated the pattern to the 1640s; thereafter price changes generated the changes.
2. From the real wage peaks extending from the 1680s through the 1730s, a new trend of declining real wages set in that reached its nadir shortly after 1800. Price changes dominated this trend, but from the 1770s onwards various stratagems to increase earnings reduced the erosion of purchasing power for at least some workers.
3. The recovery of real wages after the Napoleonic era was only partial and was determined by price changes. Only after the 1860s does a combination of rising wages and falling prices push up the real wage to unprecedented levels. This breakthrough was concentrated in the 1870s and 1880s".

The second part of these conclusions on the development of real wages seems to be contradictory to De Vries' main hypothesis that between 1680 and 1750 Dutch national income declined. De Vries explains this contradiction from income redistribution effects. As capital investment withdrew from labour-intensive industries, commodity production decreased. Since high costs pressed on farmers, forcing the abandonment of land, the adoption of less labour-intensive production techniques and bringing about the collapse of rental values, agriculture was depressed too.

> "When a fall in prices is not associated with output growth and cost-reducing investment, it takes the form of an exogenous shock to the economy. In the absence of adjustments in nominal wages, such exogenous price changes have the effect of redistributing income, in this case from capital and rents to the wage bill. The high real wage bill is compensated for in depressed land values and low returns to capital and, eventually, unemployment"[27].

Reviewing the debate on modern economic growth in the Netherlands, it is difficult to decide in favour of one of the positions mentioned. They all have their pros and cons. Thus Riley seems to overestimate eighteenth century growth and to underestimate the nineteenth century trend. The rapid spread of steam power in the Dutch economy from 1853 to 1890 (see table 3.2.) is barely reconcilable with Riley's growth path for that period. If Riley should be wrong on the nineteenth century, we consequently have to reject his eighteenth century estimates too.

Turning to Maddison, his thesis that the seventeenth century Dutch economy was very advanced, is not contradicted by other scholars, although the discussion on the exact level of this advantage continues. But if one agrees with Jan de Vries' conclusion that Maddison's 1688 per capita income is still too low and one further accepts Maddison's eighteenth century trend, then the problem of the high nineteenth century growth rate resulting from Maddison's sixteenth and seventeenth century estimates, disappears and one is left with a trend that more or less approaches Jan de Vries' growth path.

In a recent paper Maddison[28] revised his estimates of both British and Dutch growth. Having adjusted his estimates of the UK growth levels, Maddison now points Dutch 1700 GDP 30 per cent higher than that of the UK, while in his earlier work[29] he estimated the Dutch GDP 50 per cent higher than the UK's. However, Maddison did not change his assumptions on the Dutch growth path: (1) from 1700 to 1780 a 10 per cent decline of GDP, (2) further decline and restoration between 1780 and 1820 resulting in a GDP of 1820 which equals that of 1780, and (3) a relatively high growth rate from 1820 to 1870. As a result of Maddison's adjustment of the Dutch GDP of 1700, his previous estimated growth rate of 1.5 per cent a year for the period 1820-1870 is now flattened to 1.1 percent.

Both De Vries and Maddison start from the same assumptions on the Dutch growth path. They differ, however, as to the estimate of Dutch GDP in 1700. Since De Vries' estimate is higher than Maddison's, the outcome of the former projections is a growth rate of 0.6 per cent a year from 1820 to 1870 which is about half of Maddison's growth rate for the same period.

Twentieth Century Technological Development and Economic Growth

In his book on the Dutch economy in the twentieth century Joh. de Vries considers five structural characteristics from which its developments can be viewed[30]. The first is the growing significance of industry and services. The second is the expanding connection to the world economy, especially after World War II. Thirdly, the growth of industry and services affected all regions be it not to the same extent. The fourth structural characteristic De Vries mentions, is the decrease of the number of small firms (1 to 10 employees). In the fifth place, there has been a significant expansion of the role of government in the economy especially after the 1960's.

The economic structure of the Netherlands in the twentieth century is characterized by the continuation of processes that developed in the nineteenth century. Economic growth is the underlying force[31]. In this respect the twentieth century can be pictured as an enlargement of dimensions. World War I, the crisis of the 1930's and World War II have to be regarded as periods of retardation in a progressive process. Especially after the 1950's the Dutch economy saw spectacular growth, enlargement of scale and dynamization[32]. This suggests that after World War II Dutch modernization accelerated for the third time since the sixteenth century.

An increase of labour productivity means a more efficient employment of labour as a factor of production. This can be due to technological innovation, a more efficient use of capital goods available, an increase in the level of education of the labour force, stability in industrial relations and changes in the demand size of the economy. All these factors may in themselves or interdependently contribute to the explanation of a rise in productivity.

According to Maddison[33] the growth of productivity accelerated to a substantial degree after World War II. For all the industrialized countries, with the exception of the U.S., in the period 1950-1973 the growth of productivity on average was 2.5 times higher than in the years from 1850 to 1950. As important factors contributing to this rise in average productivity Maddison mentions a high and stable level of demand, an acceleration in the growth of the stock of capital goods, improvements in the allocation of the means of production and the levelling of international trade barriers. Regarding technological innovation Maddison concludes that Western Europe and Japan caught up with the U.S. rather than moving the technological frontier ahead.

Compared with other West European countries the increase of Dutch labour productivity after World War II is not deviant[34]. The average compound growth rates of Dutch productivity amounted to respectively 1.5 for the years between 1913 and 1938, 3.4 for the period 1950-1960 and 5.3 for the years from 1960 to 1973. Similar to other industrialized countries, after 1973 a decrease in the Dutch productivity growth rates set in[35].

The factors Maddison mentions as contributing to the rise of productivity in the industrial world after 1945, may also apply to the Dutch case. In the Netherlands, however, the factor of technological innovation attracts special attention. At the end

Figure 3.3. Dutch Per Capita National Income (Net Market Prices, in Guilders of 1900, x 1000), 1900-1986

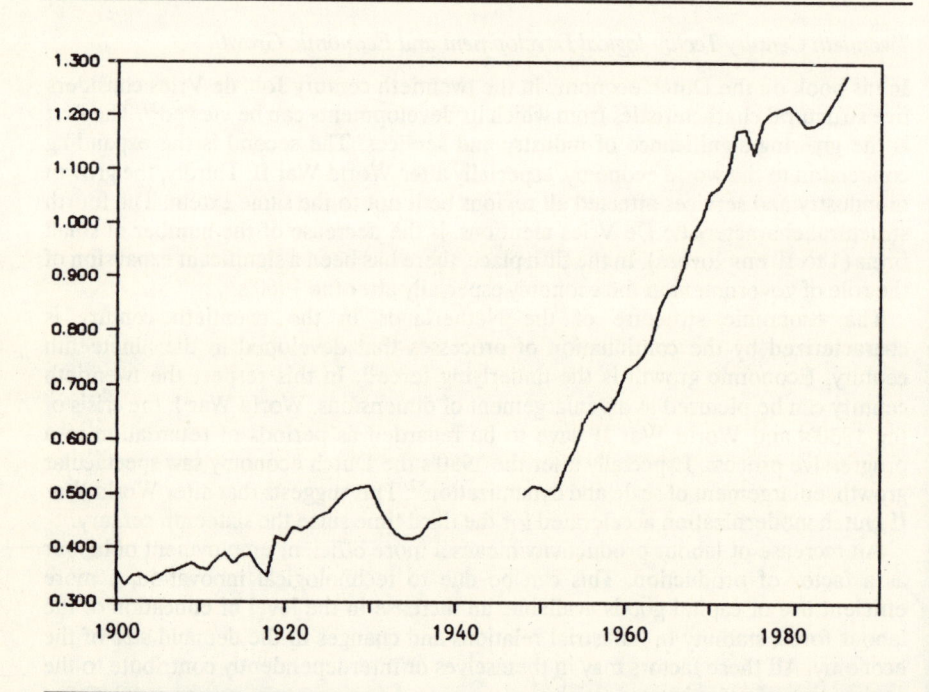

Source : Appendix A, table A.1.I

of the nineteenth century the Dutch economy technologically lagged behind that of other industrial nations. In this respect the whole twentieth century can be seen as a gradual process of catching up, accelerating in the sixties[36].

More than in the twentieth century, in the nineteenth century the Netherlands largely had to rely on imported technology. Only after 1890 significant inland production started in the fields of engine-building, shipbuilding, means of transportation, food processing and electric bulbs[37]. After 1920 a fast introduction of electricity in households took place[38]. After World War II the pace of technological development was higher than ever before. The 1950's were years of preparation. According to some the dominant hallmark of the 1960's was the *technological explosion*[39]. Between 1959 and 1964, a period of merely five years, Dutch expenditures in real terms for research and development rose by 30 per cent[40]. Both capital equipment and the types of consumer goods produced changed profoundly.

The acceleration of labour productivity in the 1960's is reflected in the speeding up of economic growth. The rise of per capita income in the 1950's and 1960's can indeed be called spectacular if one compares it to the pre-war period (see figure 3.3.). In the 1960's growth was higher than in the 1950's (see appendix A, table A.1.I.). After 1970 it slowed down and even declined in 1974. In the second half of the 1970's the Dutch economy grew again, but at a much lower rate than in the two decades before.

Economic Growth and Technological Development Summarized

Concerning the historical trend of Dutch technological development and economic growth I come to the following conclusions:
1. The seventeenth century was very prosperous. With regard to both economic growth and technological development the Dutch Republic must be considered a "lead country". At the end of this *Golden Age*, decline set in.
2. The eighteenth century was characterized by continuing decline, reaching its lowest point during, or somewhat after, the Napoleonic era.
3. Nineteenth century growth started already in the first half of that century, according to Maddison with 1.1 per cent a year since 1820. De Vries' estimated growth rate of 0.6 per cent is more in line with the conventional view on the period 1820-1870. In any case, the first half of the nineteenth century cannot be understood as an era of stagnating economic growth. After 1870 Dutch growth accelerated. So did technological development after 1890.
4. The development of the Dutch economy in the twentieth century is a continuation of the nineteenth century trend. After 1950 an acceleration of economic growth and technological development started which was particularly high in the 1960's and lasted till 1973. After that year Dutch economic growth devaluated substantially.

Having reviewed the main trends of economic growth and technological development, I will now describe the social concomitants of economic growth that I considered in chapter II as integral parts of modernization processes: the division of labour, enlargement of scale, social and geographic mobility, the formalization of social relations, the growth of the public sector of the economy and the standardization of information.

The Division of Labour, Inequality and Social Mobility

Our knowledge of the Republican stratification system is limited. We may, however, suppose that the social ranking order did not change very much till 1795, although the social distance between categories increased over time[41]. The upper class consisted of regents and rich merchants, lawyers, doctors, rich free-holder farmers and owners of large enterprises. Shopkeepers, artisans, free-holder farmers and ships officers formed the middle class. Within the lower class an upper-lower class comprising industrial workers, sailors and soldiers may be discerned from a lower-lower class of unskilled workers and the poor.

From the beginning of the eighteenth century onwards, the ranking order of Dutch society showed phenomena of contraction, which is pretty well in line with both Maddison's and De Vries' views on that period. After 1770 the Republic was confronted with massive unemployment and increasing pauperism. As Dutch foreign investments were proportionally very large as a percentage of national income[42] and Dutch agriculture was less struck by economic downfall than trade and industry[43], the upper classes deriving their incomes from these sources, were relatively successful in maintaining their life styles. The middle classes were most affected by the crisis. Their impoverishment had started as early as the second half of the eighteenth century and in the first half of the nineteenth century they had vanished almost completely. At that time Dutch stratification was polarized between the rich and the poor.

For nineteenth century Rotterdam, Van Dijk[44] constructed six social strata by using data for tax levies on heads of households. He discerned a lower class (I) of the unskilled. An intermediate class (II) between the lower and middle class that consisted of skilled workers, craftsmen, small shopkeepers and teachers. A middle class (III) of self-employed and civil servants. Class IV consisting of professionals, merchants and the clergy, and an upper class (V) of rich merchants, bankers and high ranking civil servants. By using samples from the census, Van Dijk was able to describe the structure of inequality from 1830 to 1880. It is clear that middle class positions were relatively few: in 1830, 1840 and 1849 classes I and II combined comprised respectively 91.8, 91.7 and 91.8 per cent of the samples drawn. In 1859 and 1879 these percentages decreased to 87.1 and 87.3 respectively. Van Dijk regards the latter figures as proof of increased welfare. Comparing his findings with those of Armstrong for York and those of Daumard for Paris[45], Van Dijk concludes that in those cities the middle classes were considerably larger[46].

The combination of a stagnating economy and sharper social stratification meant decreased upward social mobility[47]. As Van Tijn remarked, the picture of Dutch society in the first decades of the nineteenth century is that of an estate society characterized by a large degree of inequality between the lower and the upper status groups[48]. In 1817 half of all inhabitants of Leiden depended on poor relief. In the years from 1840 to 1850 about 30 percent of the Dutch population had to rely on some kind of welfare[49]. Around 1850 60 to 70 per cent of the Dutch population was potentially eligible for poor relief[50]. Thus around 1840 the Netherlands was, on the one hand, a country of the unemployed, farmhands and manual workers, living under miserable conditions, and, on the other hand, of a small upper class of traders, rentiers and bourgeois aristocrats. Large scale industries, labour organizations and social welfare acts were lacking[51].

Only after the mid nineteenth century these social and economic conditions changed. The attempts to restore the economy, however, had already started after the defeat of Napoleon in 1813 when the Netherlands became a kingdom. The first Dutch king, William I of Orange, had a profound influence on economic reform. His economic aims were the promotion of industry and the creation of a modern infrastructure. The latter was more successful than the former, because he met much resistance from the established money circles and the traditional trade-oriented merchant elite. In 1840, at the end of his reign, the Bank of the Netherlands and the Dutch Trading Company had been established. In the Dutch colony Indonesia a system of forced agricultural levies had been introduced. The merchant fleet had been rebuilt, the inland canal system expanded. However, these initiatives came mainly from the government. A modern entrepreneurial spirit was seldom observed[52].

Had a reform of institutions set in in the first half of the nineteenth century, the years 1848-1875 were characterized by the start of modest technological innovation in the economy. Major improvements in the transportation system were achieved. In 1856 Amsterdam and Rotterdam were directly connected with Germany by rail. However, the Dutch economy could only fully gain from the introduction of railways when in 1880 the railroad system was completed and the whole country was connected with Germany. In the second half of the 1860s technological innovation started modestly in many branches of industry. Above all it was related to the production of consumer goods. Exceptions were the paper and cotton industries. Especially the latter underwent a rapid mechanization: from 1861 to 1871 the amount of raw materials processed, doubled.

Van Tijn represents the view that the years 1850-1870 must be characterized as a period of modest technological innovation. The rise of output was especially determined by an increase of inland demand. In general, the production techniques remained traditional. Thus Van Tijn concludes that it was *more, but mainly according to the old ways*[53].

Table 3.2. Average Annual Compound Growth Rates of Steampower Application (numbers of steam engines and horse power), 1853-1895

	steam engines	horse power
1853-1858	13.8	n.a.
1858-1864	8.7	n.a.
1864-1872	5.0	n.a.
1853-1872	8.5	6.2
1872-1880	6.9	6.7
1880-1890	3.1	4.5
1890-1895	2.0	3.4

Source : Computations based on data derived from J.A. de Jonge, 1968, p. 495.

After 1870 economic growth accelerated. Employment shifted away from agriculture to industry. New industries arose and the traditional sectors of production were modernized by the introduction and application of new technologies. The average annual compound growth rate of steam-power application (horse power) rose from 6.2 in the years 1853-1872 to 6.7 in the period 1872-1880 and then declined to 3.4 in the years 1890-1895 (table 3.2.). After 1870, population increased rapidly and many moved from the countryside to the cities. After 1890 geographic mobility even accelerated[54].

These transitions in the Dutch economy evoked changes in the division of labour. From 1849 to 1909 employment in agriculture dropped from 44 per cent to 28 per cent in favour of a rise of employment in the industrial and the service sector of the economy. Unfortunately, reliable data for 1860-1880 are not available. However, the main trend is one of increased specialization and growth. Especially the growth of employment in the sectors of civil service, education, banking, insurance and the professions, was remarkable[55].

In his work on the industrialization of the Netherlands De Jonge observes the rise of the middle class in the nineteenth century. From 1849 to 1889 the average annual compound growth rate of clerks as a percentage of the labour force was 2.4 and from 1889 to 1909 2.0 (De Jonge defines clerks as administrative and controlling personnel)[56].

Table 3.3. Numbers of Clerks as a Percentage of the Labour Force (1849-1909)

	Absolute Numbers	Percentage of Labour Force
1849	40,000	3.5
1889	155,000	9.0
1909	298,000	13.5

Source : De Jonge, 1968, p. 297.

In the mid nineteenth century the revival of the Dutch middle class did not yet offer possibilities for upward social mobility to the lower strata of society. Firstly, the number of administrative positions was still scarce. Secondly, the educational level of working class children was too low to enable successful competition with children from the higher strata who got a superior education in expensive private institutes. Thirdly, office jobs yielded social esteem. For that reason the higher strata used to reserve such positions for their own offspring. The rapid extension of the number of office jobs after 1850 and the improvements of lower class education in the second half of the nineteenth century meant the growth of clerks with a working class background. It implied the increase of upward social mobility[57].

Lack of reliable data makes the description of nineteenth century social mobility difficult. We have only available a few historical case-studies. The only one that covers a large part of the nineteenth century is the before mentioned study of Van Dijk on Rotterdam[58]. He accounted the average probabilities of upward and downward intragenerational social mobility for three decades and concludes: (1) the average probability of upward social mobility increased over time; (2) there is some tendency to *objective proletarization* between 1830 and 1850: a decreasing probability of upward social mobility for class I (lower) and an increasing probability of downward social mobility for class II (intermediary between lower and middle class); (3) increasing mobility of class III (middle) over time: in all periods the probability of upward social mobility increases and so does downward mobility from 1850 to 1860; (4) only from 1870 to 1880 a rise of the probability of upward mobility for class I can be observed[59].

Van Dijk's conclusions only refer to Rotterdam and one should be careful with respect to generalization. Another problem is that Van Dijk's classification system is based on taxes and thus related to income. So, when Van Dijk concludes that the highest probabilities of upward social mobility are observed between 1870 and 1880[60], this may be partly due to the push up of real wages which was, according to De Vries, concentrated in precisely the 1870s and the 1880s[61].

Thus, Dutch economic development in the second half of the nineteenth century brought social differentiation instead of a convergence. However, class struggle did not increase. In fact, as we shall see in the next chapter, political and social organization on religious criteria prevented this from happening.

In his essay on the rise of modern class consciousness in the Netherlands, Van Tijn postulates that the economic and technological developments that shocked Dutch society in the second half of the nineteenth century implied the proletarization of the working class. The growth and modernization of the economy led to intergenerational upward mobility of the skilled segment of the working class. Because of the growth of educational facilities and the growth of administrative positions, this segment provided the germs of a middle class. At the same time, expanding markets, increasing competition and technological innovation led to enlargement of scale and in many lines of production made craftsmanship obsolete. Teenagers were now able to handle machines and thus replaced more experienced workers. Where such processes took place it meant a decrease of the wage level. The result was a split of the working class. Further, as a side-effect, the patriarchal relationships between bosses and skilled hands disappeared[62].

Van Tijn's description of these developments in the working class are rather general. Although the changes were indeed significant, many sectors of the economy were not affected. Although mechanization was introduced in agricultural production, it still had a small scale nature. Only after the Second World War agricultural innovation accelerated and caused a significant loss of jobs in that sector. In the textile industries of Twente, patriarchal relationships remained powerful despite mechanization. Further, many production processes still had to rely on craftsmanship for the simple reason that adequate machinery had not yet been invented at the time.

Despite these nuances, I can agree with the general pattern to which Van Tijn points: an enlargement of scale, standardization and rationalization of production processes that de-humanized labour to a substantial extent. The negative effects of this modernization process became clear from a survey of the conditions of the working class ordered by Parliament in 1886[63]. The results shocked the nation, especially with regard to the employment of women and children.

Twentieth Century Changes in the Division of Labour, Inequality and Social Mobility

In the twentieth century the Dutch division of labour saw two major changes. Firstly, a shift from agricultural to industrial employment which had already started in the nineteenth century. During the whole of the twentieth century, agricultural employment as a percentage of total employment decreased steadily. This process accelerated after 1947. When we look at the absolute number of people employed in agriculture we can see that despite the relative decrease, the absolute number increased till 1947. After that year a dramatic downfall in agricultural employment took place.

Secondly, a steady growth of the service sector which accelerated significantly in the period 1971-1981.

Figure 3.4. Employment by Sector of Industry (agriculture (a), industry (b), services (c)) as a Percentage of Total Employment, 1899-1981

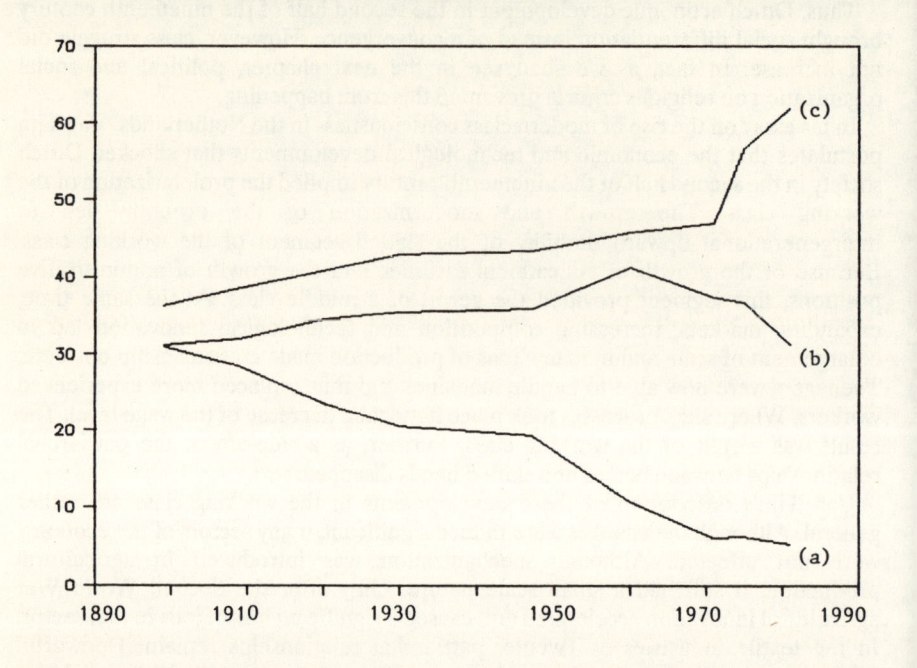

Source : Appendix A, table A.2.II.

In figure 3.4. the relative changes in the pattern of employment by sector of industry in the Netherlands are presented. In table 3.4. the average annual compound growth rates based on the absolute figures are given. Table 3.5. presents the average annual compound growth rates based on percentages.

Table 3.4. Average Annual Compound Growth Rates of Employment by Sector of Industry, 1899-1981

	Agriculture	Industry	Services
1899 - 1909	0.8	1.9	2.2
1909 - 1920	0.0	2.4	2.2
1920 - 1930	0.2	1.8	2.1
1930 - 1947	0.8	1.2	1.2
1947 - 1960	-4.0	1.7	1.0
1960 - 1971	-3.9	0.3	1.4
1971 - 1981	-0.5	-1.0	3.5

Source : Appendix A, table A.2.I.

Table 3.5. Average Annual Compound Growth Rates of Employment by Sector of Industry as a Percentage of Total Employment, 1899-1981

	Agriculture	Industry	Services
1899 - 1909	-0.8	0.3	0.6
1909 - 1920	-1.7	0.7	0.5
1920 - 1930	-1.3	0.3	0.5
1930 - 1947	-0.4	0.1	0.1
1947 - 1960	-4.5	1.1	0.4
1960 - 1971	-5.1	-0.9	0.2
1971 - 1981	-1.3	-1.7	4.7

Source : Appendix A, table A.2.II.

What happened in the Netherlands after 1945 was not so much the set in of industrialization as well as the coming of an expansive process of modernization of the Dutch economy. Especially in agriculture and industry new technologies were introduced. With regard to modernization by the application of new technologies, in the pre-war period the Netherlands lagged behind other European countries. From 1900 onwards until 1940, technological innovation can be observed. However, the process of innovation accelerated tremendously after the war and reached a peak in the 1960's. The equipment of production units as well as the nature of production itself had changed profoundly, especially in the domain of chemicals, metallurgics, transportation and electronics[64]. In agriculture the introduction of new technologies made many a farmhand redundant. Many moved from the countryside to more urbanized centres where they found new jobs in the expanding industrial and construction activities. For the efficient application of these new technologies the extension of cultivable area per unit of production was a prerequisite. This combination of rationalization and enlargement of scale explains the downfall in agricultural employment after 1947. One may wonder whether the new patterns of employment that developed in the course of the twentieth century had their effects on social mobility and social inequality. It is plausible that proliferation of the division of labour will promote social mobility. Empirically such a causal link is difficult to state. The same holds with regard to social inequality. The least we can do, however, is to monitor twentieth century trends of social mobility and inequality.

In the social sciences prestige, wealth and power are commonly applied dimensions of social inequality. With regard to the study of long term trends of social mobility in the Netherlands only estimations of changes of the prestige dimension are available[65].

In 1954 Van Tulder[66] measured differences in occupational prestige between 1919 and 1954. Already in 1953 a scale for the prestige of occupations in the Netherlands was constructed by Van Heek and Vercruysse[67]. For this occupational prestige scale Van Heek and Vercruysse asked 500 respondents to rank 57 occupations. The resulting rankings were treated as interval scores[68]. From these scores an occupational prestige stratification of six layers could be constructed.

The first stratum consisted of the academic professions, directors of large firms, secondary school-teachers and high ranking civil servants. Directors of small firms, non-academic technicians, large farmers, white-collar civil servants and high ranked employees formed the second stratum. The third stratum consisted of the upper middle class, middle ranked civil servants and employees. The lower middle class, clerks, lower educated workers, owners of small agricultural enterprises, and low ranked civil servants were grouped as the fourth stratum. The fifth stratum comprised skilled workers and low ranking white-collar employees. The sixth stratum consisted of mainly unskilled workers[69].

Van Tulder used this occupational prestige scale to measure social mobility between 1919 and 1954. He made a distinction between group mobility, i.e. the changes in the distribution of the labour force as to the six strata, personal mobility, i.e. the mobility of a person, and intergenerational mobility, i.e. the differences between fathers and sons with respect to their positions on the occupational prestige scale.

With regard to group mobility Van Tulder concludes that from 1919 to 1954 a shift from the lower to the higher strata occurred[70]. In the domain of personal mobility there were only minimal changes[71].

With regard to intergenerational mobility, Van Tulder compared the occupations of sons in 1954 with both the occupations of fathers in 1919 as well as with the last occupations of fathers. As to the occupations of fathers in 1919 intergenerational mobility was high: 60 percent. In 1954 37 per cent of the sons had experienced upward social mobility, 23 percent turned out to be downward mobile and 40 percent took the same position on the occupational prestige scale as their fathers[72]. If one compares the occupations of sons in 1954 to the last occupation of fathers intergenerational mobility is somewhat lower: 58.9 percent of which 32.7 percent was upward mobile and 26.3 percent downward mobile[73].

Ganzeboom and De Graaf replicated Van Tulder's study with 1977 data. They compared son's occupations with the last occupation of fathers. They concluded that between 1954 and 1977 the number of mobile sons had increased: 66.1 percent of which 43.6 percent was upward mobile and 22.6 was downward mobile[74].

Ganzeboom and De Graaf make a distinction between structural social mobility and circular social mobility. The former includes social mobility that results from changes in the occupational structure caused by external factors like technological development. Circular social mobility excludes these externally caused changes and thus refers to the question whether the father's occupational status determines that of the son, given structural mobility[75]. Van Tulder's research and the comparison of his results with those of Ganzeboom and De Graaf are restricted to structural social mobility. By the application of log-linear analysis Ganzeboom and De Graaf could also measure the change of circular mobility between 1954 and 1977. They conclude that it had increased and that Dutch society had become more open[76].

In another study Sixma and Ultee compared circulation heterogamy between 1959 and 1977. Heterogamy is the degree to which marriage partners differ with respect to the level of education. As with intergenerational social mobility, a distinction can be made between heterogamy caused by structural changes in the educational system (structural heterogamy) and heterogamy corrected for such changes (circular heterogamy). Sixma and Ultee conclude that between 1959 and 1971 circular heterogamy increased. Between 1971 and 1977 it hardly changed. They conclude that with regard to the relation between marriage and level of education, Dutch society has become more open in the 1960s, but not in the 1970s[77].

The study of Sixma and Ultee gives some idea about the openness of Dutch society in the 1960s and 1970s, but it does not enable us to trace down developments in the twentieth century as a whole. For that aim the works of Van Tulder and Ganzeboom and De Graaf are more useful. They point to the fact that structural intergenerational mobility increased from 1919 till 1977, but that between 1954 and 1977 it was higher than before. Further, after World War II upward mobility was relatively higher than downward.

Social mobility is related to social inequality. Social mobility patterns refer to the chances to be unequal. Although data on income mobility are lacking, much research has been done to the skewness of the Dutch income distribution, especially after World War II[78]. According to calculations of Pen and Tinbergen between 1938 and 1976 income inequality in the Netherlands declined by some 50 per cent[79].

Although other researchers used different measures or different concepts of income distribution they confirmed the views of Pen and Tinbergen. Reviewing recent research, Szirmai concludes that the fall in the income share of the top ten per cent of households over the whole period 1938-1976 was dramatic. All other income categories have increased their shares at the expense of this top decile. Further, from the second half of the 1960s onwards and especially in the first half of the 1970s, the process of redistribution underwent a marked acceleration[80]. Szirmai explains the above trend by increased government intervention in the primary distribution of incomes. Notably the introduction of minimum wage legislation in 1969 turned out to have an unintended but strong redistributive effect[81].

Enlargement of Scale

The impact of enlargement of scale on the nature of society was already understood by the founding fathers of the social sciences. Thus Durkheim thought that a cause of the proliferation of the division of labour was the increase of society's "material" and "moral density". By material density Durkheim meant population density and the degree to which the population is dispersed over the area of society. In its turn, material density brings about more intensive interaction. New ways of communication and transportation decrease the distance between the different sectors of society. This latter effect of material density Durkheim calls "moral density" because it is associated with a society's culture.

Given a plural society, increasing concentration of the population will not only increase the number of interactions per individual, but also the probability that different cultural segments come in close contact to one another. Such a break-through of cultural isolation could be imagined to lead to four alternative social responses: (1) growing intolerance and an associated struggle between competitive values and ideas resulting in the victory of a dominant culture and the repression of subcultural elements; (2) the gradual merging of existing cultural patterns into one new culture; (3) increasing cynicism, alienation and discontent, and (4) increasing tolerance and cohabitation. The latter increase of cultural heterogeneity promotes individual freedom. This is an important reason why in every society the cities traditionally have been the breeding grounds of new ideas as well as a refuge for those who could not live elsewhere because of deviant ideas and behaviour.

In the preceding chapter I pointed out that enlargement of scale is a major concomitant of modern economic growth. Part of this enlargement of scale are the demographic changes associated with economic growth that lead to population increase which usually implies increasing concentration of the population. By this mechanism modern economic growth is directly linked to the cultural system of society. However, measuring this process is rather difficult, especially as it developed in the past, since quantitative data of the cultural climates of cities in past times are mostly not available. For that reason we can only measure changes in the degree of population itself.

A second indicator of enlargement of scale is the average number of persons employed per enterprise. The choice of this variable is based on the same ideas as that of urbanization. I assume that within an enterprise the degree of possible cultural heterogeneity is dependent of the number of workers. Further, in economic organizations criteria of recruitment, of the evaluation of behaviour and behaviour itself change as the number of personnel increases. Dependent on the kind of organization and its environment involved, such a change may develop either in the direction of more centralized bureaucratic control or in the direction of a higher degree of de-centralized power. Whatsoever, there are substantial differences as to the organization of production between the blacksmith and his mate versus an iron mill employing five hundred workers.

In a study on the evolution of the Dutch urban system Van der Knaap[82] selected 502 Dutch municipalities which in 1970 had more than 5000 inhabitants. From 1840 onwards, for every decade Van der Knaap counted the number of inhabitants of each municipality selected and classified the total population of all municipalities under investigation in classes ranging from municipalities with less than 2,500 inhabitants to cities with 250,000 to one million inhabitants.

In order to measure changes in the degree of concentration of the Dutch population, I used Van der Knaap's data to compute the coefficient of variation for every decade (see appendix A.1.). A change of the coefficient of variation over time indicates a change of the dispersion of the population over the municipalities selected. A rise of the coefficient of variation points to a higher growth rate of the larger municipalities relative to the smaller ones, whereas a declining coefficient indicates the reverse. The result of the computations are presented in figure 3.5.

From figure 3.5. it follows that from 1850 till about 1900 the larger municipalities grew faster than the smaller ones. Especially between 1880 and 1890 the growth of large municipalities accelerated. After 1930 a reverse trend set in. From then on, smaller municipalities grew relatively faster than the big cities[83].

This process becomes the more clear if one looks at the distribution of the total population to type of municipality and the balance of inland migration. From table 3.6. it follows that in the Netherlands urbanization reached its peak in 1960 as 55.5 per cent of the population lived in large urban municipalities. After 1960, the degree of urbanization decreased. Considering the years from 1930 to 1980, the percentages of the population living in big cities do not show major changes. In that sense one may conclude that in the Netherlands the urbanization process had already been completed before World War II.

Focusing on the years after World War II we see a decrease in the percentage of the population living in rural municipalities and an increase in the percentage of inhabitants of urbanized rural municipalities. Although these changes can already be discerned at the beginning of the century, they accelerate in the years between 1960 and 1971.

The difference between rural and urbanized rural municipalities not only lies in the urban character of the latter, but also in its occupational structure, which is more diversified and less oriented to the agricultural sector than that of the rural municipality. The decrease of the percentage of the population living in rural municipalities must be connected to the before mentioned changes in the structure

Figure 3.5. Dispersion of the Inhabitants of 502 Dutch Municipalities, 1840-1970 (Coefficients of Variation)

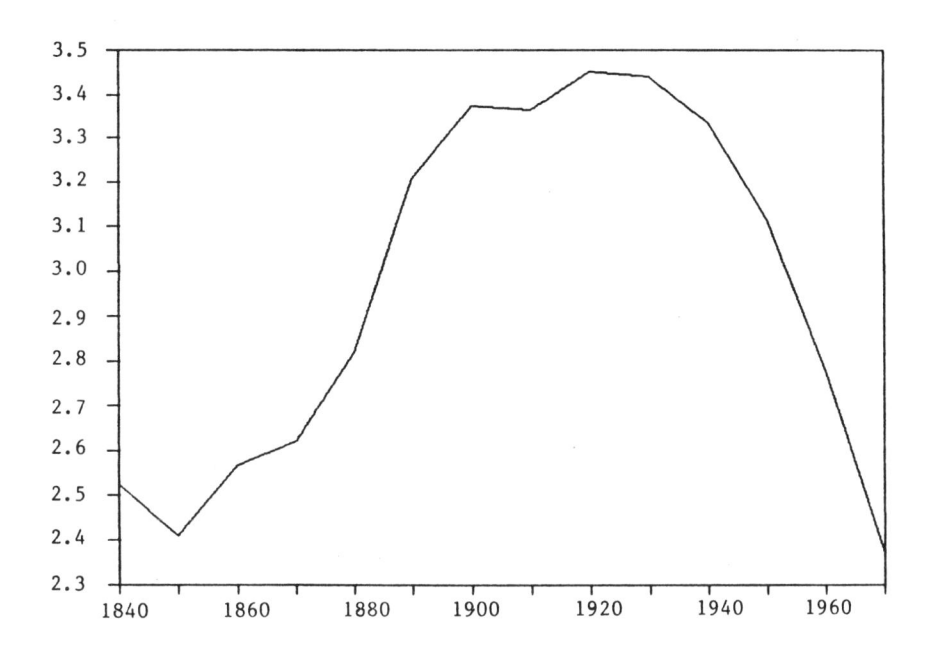

Source : See Appendix A.3., table A.3.I.

of the Dutch economy, notably the decrease of employment in agriculture which implied a change of the occupational structure of many a rural municipality. Not only increases in the number of inhabitants but also changes in the occupational structure lead to the classification of former rural municipalities as urbanized rural municipalities since one of the criteria for classifying a municipality as urbanized rural is, that over 80 per cent of the economically active population is employed in non-agrarian occupations. Thus these shifts in the classification of rural municipalities are not merely a matter of definition, but point to a change in the socio-economic outlook of the Netherlands.

Table 3.6. Dutch Population by Degree of Urbanization as a Percentage of Total Dutch Population, 1899-1980

	Rural Municipalities	Urbanized Rural Municipalities	Urban Municipalities
1899	51.5	3.3	45.2
1930	40.9	6.8	52.3
1947	29.4	16.2	54.4
1956	24.7	20.6	54.7
1960	21.8	22.7	55.5
1971	11.1	34.0	54.9
1980	11.6	36.3	52.1

Source : For 1947, 1956, 1960 and 1971 I used CBS, *Typologie van de Nederlandse gemeenten naar urbanisatiegraad op 28 februari 1971*, The Hague, Staatsuitgeverij, 1983, p. 13, staat 3. For 1899, 1930 and 1980 I used P. Kooij, 1985, tabel 2, p. 97.

Note : *Rural municipalities* have less than 300 inhabitants to the square km, whereas the largest population cluster has less than 5000 inhabitants. In terms of the occupational structure they have a predominantly agrarian character.
Urbanized rural municipalities have a predominantly non-agrarian occupational structure. Over 80 per cent of the economically active male population earns a living in non-agrarian occupations.
The general characteristics of *urban municipalities* are: a densely populated, contiguous built-up area with over 2,000 inhabitants to the square km. For the municipal territory as a whole, including the city itself, over 500 persons to the square km. At least 70 per cent of the total population residing within the central built-up area.

Table 3.7. Average Annual Compound Growth Rates of Dutch Population by Degree of Urbanization as a Percentage of Total Population, 1899-1980

	Rural Municipalities	Urbanized Rural Municipalities	Urban Municipalities
1899-1930	-0.7	2.3	0.5
1930-1947	-1.9	5.1	0.2
1947-1956	-1.9	2.7	0.1
1956-1960	-3.1	2.4	0.4
1960-1971	-6.1	3.7	-0.1
1971-1980	0.5	0.7	-0.6

Source : Table 3.6.

Table 3.8. Balance of Dutch Inland Migration by Type of Municipality (per Thousand Inhabitants), 1957-1980

	1957	1964	1969	1973	1980
Rural Municipalities	-5.6	-0.9	6.3	16.9	4.5
Urbanized Rural Municipalities	7.0	6.9	10.3	13.4	3.2
Dormitory Towns	17.1	11.1	17.3	10.9	9.6
Small Towns (less than 30,000 inhabitants)	11.5	6.1	8.4	13.7	2.5
Medium Sized Towns	2.2	0.0	3.2	-11.1	-2.7
Big Cities (more than 100,000 inhabitants)	-3.3	-7.7	-17.0	-25.8	-9.6

Source : P. Kooij, 1985, table 6, p. 108.

Besides the change of character of rural municipalities, a process of suburbanization can be observed in the 1960's. Already at the end of the 1950's, the big cities began to loose inhabitants. In the second half of the 1960's and the first half of the 1970's, the negative migration balance of the big cities increased in favour of the small provincial towns, the urbanized rural municipalities and the dormitory towns (see table 3.8.). Many, especially young families preferred to live in dormitory areas or in the countryside because of the lower costs of housing and the fact that the conditions of living were considered by many as outstanding compared to those in the big city. Because of a rise in real wages, car ownership increased so that living at a certain distance from the place of work was no longer a privilege of the well-to-do[84].

The picture emerging from these changes in the spatial environment of the Dutch population is that of a change of character of community life. In small rural municipalities, the autochthonous population was confronted with changing patterns of employment and the import of new life styles brought by newcomers with a big city background. In many cases cultural and political conflicts between traditional and new life styles resulted which, in extreme cases, led to a dual social structure[85]. For the traditional rural community the coming of new men had a profound innovative impulse on social life and increased social and cultural heterogeneity to a substantial degree.

For the big city, on the other hand, suburbanization implied a separation of place of living and place of work. In the sixties, the centers of the big cities became more dominated by office buildings and business activities. Population was expelled to the outskirts where large dormitory areas arose. The characteristic features of these areas are the high degree of mobility and the relative absence of community life. The stereotype is that of people being next-door neighbours without really knowing each other. The keyword for this kind of living conditions is anonymity. The dormitory neighbourhood is an artificial one, lacking a spontaneous social life. State-subsidized welfare agencies were established in order to provide for these functions.

The destruction of the city neighbourhood was not only effectuated by the taking over of business but, to a substantial degree, also by the arrival of foreign, predominantly Mediterranean, workers, students, drug addicts and the very poor. They took the places of those who moved to the better houses on the outskirts of the towns. This meant the impoverishment of housing areas in the inner city and, at the same time, an increase of their cultural heterogeneity. At the end of the 1960's, some big cities in the Netherlands saw riots of which the parties involved were the Dutch still living in the city and the Mediterranean immigrants.

Regarding enlargement of scale of enterprises reliable data are scarce. Because of that we could only construct nineteenth century time series data for the industrial sector from 1859 onwards and for all enterprises from 1930 onwards. With regard to the industrial sector we had to restrict ourselves to the percentage of the industrial labour force employed in medium and large scale industrial enterprises. As we had to use different sources, the definitions of medium and large scale enterprises are not exactly the same for all years investigated. For a full account of the sources used and the differences between them the reader is referred to appendix A.4. However, in general the different years are comparable: enterprises with 10 to 50 persons employed are considered of medium size. Those with more than 50 persons employed are regarded as large size enterprises. The data are presented in the tables 3.9. and 3.10.

Taking medium and large size enterprise together, enlargement of scale manifested itself most clearly from the 1880's till the first decade of the twentieth century. The same conclusion can be drawn with regard to large scale enterprises taken as a separate category

Table 3.9. Number of Employed Persons in Medium and Large Size Industrial Enterprises as a Percentage of the Total Number of Employed Persons in Industry, 1859-1978

	Percentage employed persons in medium and large size industrial enterprises	Percentage employed persons in large size industrial enterprises
1859	20.0	n.a.
1889	23.5	15.2
1909	44.5	29.2
1930	64.5	47.2
1950	74.1	54.3
1963	84.4	63.3
1978	88.5	66.1

Notes : See for the definition of medium and large size enterprises appendix A.4.
Sources: See appendix A.4.

Table 3.10. Average Annual Compound Growth Rates of the Number of Persons Employed in Medium and Large Size Industrial Enterprises as a Percentage of the Total Number of Persons Employed in Industry, 1859-1978

	Growth of employed persons in medium and large size industrial enterprises	Growth of employed persons in large size industrial enterprises
1859-1889	0.54	n.a.
1889-1909	3.19	3.26
1909-1930	1.76	2.29
1930-1950	0.70	0.70
1950-1963	1.00	1.19
1963-1978	0.32	0.29

Source : Table 3.9.

Table 3.11. Average Number of Employed Workers per Enterprise Establishment by Sector of Industry, 1930, 1950, 1963 and 1978

	1930	1950	1963	1978
Industry	6.9	10.5	15.5	17.7
Commerce	2.5	2.9	4.0	4.5
Transportation	6.3	7.1	10.9	11.6
Hotel and Catering	2.2	2.9	3.5	3.2
Other services	3.7	3.5	4.5	3.9
Total	4.4	6.1	8.4	8.4

Source : Accounting based on CBS, 1899-1984. Vijfentachtig jaren statistiek in tijdreeksen, The Hague, Staatsuitgeverij, 1984, p. 88.

Table 3.12. *Average Annual Compound Growth Rates of the Average Number of Employed Workers per Enterprise Establishment by Sector of Industry, 1930-1950, 1950-1963, 1963-1978 and 1930-1978*

	1930-1950	1950-1963	1963-1978	1930-1978
Industry	2.1	3.0	0.9	2.0
Commerce	0.8	2.3	0.8	1.2
Transportation	0.6	3.3	0.4	1.3
Hotel and Catering	1.5	1.3	-0.6	0.8
Other Services	-0.3	2.0	-1.1	0.1
Total	1.6	2.4	0.0	1.3

Source : table 3.11.

Our main argument is that processes of rationalization and enlargement of scale tend to formalize social relations. The increase in the size of personnel per unit of production is in this respect a special case. The larger the number of people that work together in an organization, the more a tendency can be observed to the formalization of social contacts. This is not to say that in large economic organizations informal social relations are absent. However, informal organization uses to function as a countervailing power against the formal structure and as a means to fulfill basic psychic and social needs which cannot be satisfied by the formal organization. So informal organization is a residue to formal organization. In small scale organizations, where only a few people work together, the distinction between formal and informal organization tends to be diffuse. Enlargement of scale of economic organization, as measured by the size of personnel, does not mean the withering away of informal relations, but it does effectuate an increase in formal social relations.

Table 3.13. *Average Annual Compound Growth Rates of Enterprises with Five Employees or Less and Enterprises with More than Five Employees as a Percentage of All Enterprises, 1930-1978*

Number of Type Workers	1930-1950	1950-1963	1963-1978
0 -5	-0.28	-1.10	-0.17
>5	2.25	4.66	0.47

Source : Appendix A.4., table A.4.I.

The before mentioned enlargement of scale of Dutch agriculture meant the extension of cultivable area per unit of production and not the increase in the size of personnel of agricultural enterprises. On the contrary. In the Northern and Eastern provinces of the Netherlands the application of new agricultural technologies led to the disappearance of the phenomenon of the farmhand. In the South it meant that the number of relatives contributing to the production of the farms decreased. Thus the main importance of the rationalization of Dutch agriculture lies not so much in the formalization of social relations in the countryside because of an increase in the number of agricultural workers, but in the flow of people from countryside villages to more urbanized areas that are more characterized by formality in social relations than the small settlements dominated by farmers.

Figure 3.6. Enterprises with Five Employees or Less (a) and Enterprises with More than Five Employees (b) as a Percentage of All Enterprises, 1930-1978

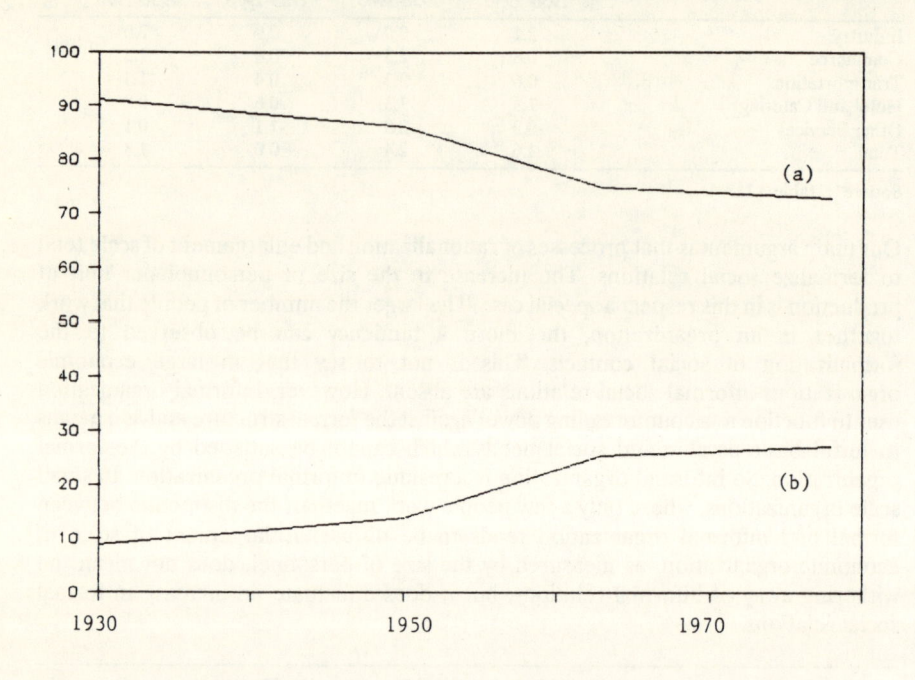

Source : Appendix A.4., table A.4.I.

Differently from the agricultural sector, in the other parts of the private sector the decrease of the total number of enterprises can be explained by the growth by size of personnel of production units. As can be seen from the tables 3.11. and 3.12., this development can already be observed before the Second World War.

Instead of comparing time periods with regard to the average number of enterprises' personnel (tables 3.11. and 3.12.), a more precise method is to look at the changes in the number of enterprises classified to the size of their personnel (table 3.13. and figure 3.6.). Because of lack of data this is only possible for 1930 and for the years after World War II (See appendix A.4., table A.4.I.).

Both approaches indicate a substantial enlargement of scale in the period 1950-1963. Between 1950 and 1963 the percentage of enterprises with more than five employees increased significantly, while the number of small firms with less than five employees decreased.

Bureaucratization and the Formalization of Social Relations

In the former chapter on modernization theory I concluded that there is a broad consensus among scholars that bureaucratization and the formalization of all kinds of social relations are important dimensions of modernization. It is the adagium of such classical writers as Durkheim and Weber that inspired many social scientists since then.

For Durkheim the change from mechanical to organic solidarity was associated with increasing social differentiation, especially as to the division of labour. Modern social cohesion was primarily based on the division of labour under the condition that contracts stipulating mutual obligations were guaranteed by law. Thus, the increase of organic solidarity could be identified by the importance of civil law in society.

Weber's preoccupation with modernization involved a change in types of authority. Authority in modern society is primarily legitimized by legal rationality, i.e. by the objectivation of social rules. Participants in social relations derive authority from principles of rationality and formalized rules. In this context the formalization of rules can generally fulfill its social functions if such rules are written down in some form or another. Legal rationality as a dimension of modernity implies the increase of the number of laws and written regulations at all levels of society.

Having in mind the ideas of Weber and Durkheim, at first sight the most simple method to measure the degree of formalization of social relations is to account the number of civil laws and regulations at a given time. Because we actually want to measure the degree to which social interaction is guided by formal rules, two objections should be made to that procedure. Firstly, among the number of laws measured, there may be many to be considered as dead letters that had no impact on social life whatsoever. Secondly, laws usually differ considerably in scope. Take for instance such extremes as, on the one hand, laws on taxation that concern all income earners and property owners, and, on the other hand, legal prescriptions concerning mining activities. Such differences of scope could easily distort conclusions on the impact of formalization on social life.

To measure the degree of formalization of society I took two proxies. The first one is the number of personnel employed in the ministries of the Dutch national government. The choice of this proxy is based on the assumption that there is a relationship between the number of laws and the number of persons employed in ministries, because laws need to be prepared, controlled, adjusted and executed by people at the governmental level. The increase of the total number of Dutch civil servants was much larger than the number of ministry personnel alone. In his dissertation on civil servants and bureaucracy in the Netherlands, Van Braam writes that between 1899 and 1950 almost half of the increase of national government personnel was caused by the growth of utilities like post, mining and public works. Only ten per cent of the total growth of the number of civil servants can be accounted for by the expansion of the ministries in The Hague[86].

Apart from the formalization of social relations, there are other causes for the growth of government personnel which generally can be associated with changing needs of the economy, a broadening of government tasks and incidents like the need for a planned distribution system during the Great War[87]. Because activities in these fields were mostly concentrated in the hands of officials outside the ministries, the number of ministry personnel is a fairly reasonable proxy for measuring formalization, although I must admit that other causal factors may interfere to some extent. For that reason I will use a second proxy, i.e. the number of civil cases brought to courts of justice. This variable indicates the extent to which people try to solve their conflicts by appealing to formal legal rules.

Figure 3.7. presents ministry personnel as a percentage of both the population and the labour force for the period 1849-1975. The related average annual compound growth rates are listed in table 3.14.

Concerning the growth of Dutch ministry personnel I draw three conclusions:
1. Before World War II one can distinguish two periods of strong growth: 1849-1877 and 1900-1929.
2. In the post war period only after 1960 ministry personnel regained its growth.
3. Even the growth rates after 1960 are generally lower than those before 1929. This seems very remarkable indeed, as the creation of the welfare state is usually identified with an expansion of the public sector. However, this is not to say that in the Netherlands the creation of the welfare state did not produce the associated employment: it did not spectacularly in terms of governmental ministry personnel, because the Dutch government mainly attempted to reach the goals of welfare society by subsidizing private organizations. This phenomenon has been strongly related to the pillarized nature of Dutch society to which we will come later in this study.

Figure 3.7. Dutch Ministry Personnel as a Percentage of the Labour Force (a) and of Total Population (b), 1849-1975

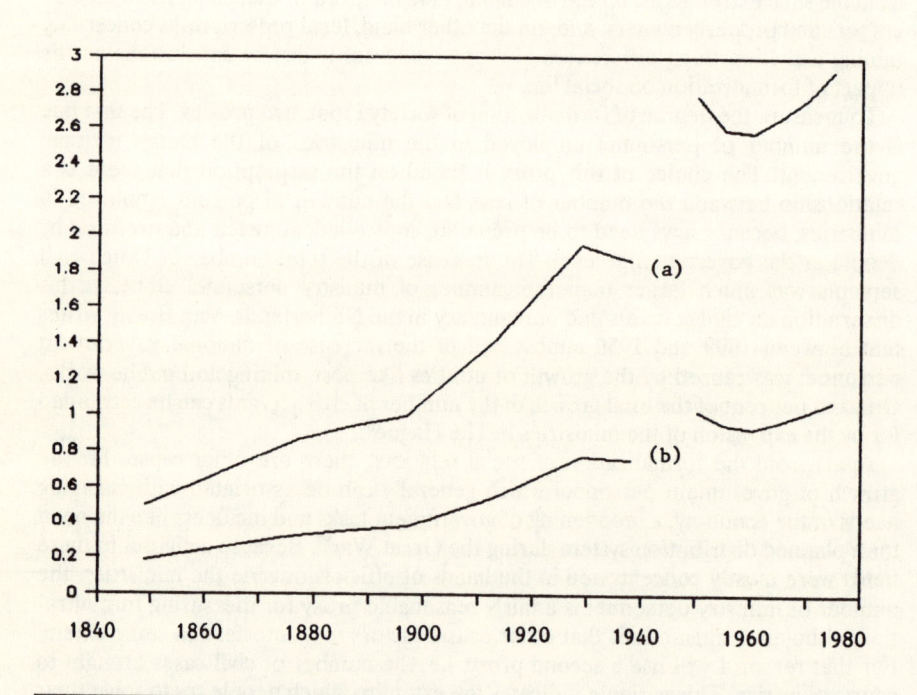

Source : See Appendix A.5., table A.5.II.

Table 3.14. *Average Annual Compound Growth Rates of Dutch Ministry Personnel as a Percentage of the Labour Force and of Total Population, 1849-1975*

	Labour Force	Total Population
1849-1877	2.01	2.03
1877-1900	0.98	0.95
1900-1929	2.06	2.22
1929-1938	-0.52	-0.25
1950-1960	-0.78	-0.97
1960-1975	0.86	0.98

Source : See Appendix A.5., table A.5.II.

The number of civil cases brought to courts of justice, the second proxy of formalization, is much more difficult to interpret than the first one. This is so, because the nature of the Dutch judicial system has been rather dynamic than static and adjusted itself to changing social and economic conditions. Before the Napoleonic codification in the beginning of the nineteenth century, Dutch law was a mixture of traditional Roman and German legal principles. In the highly de-centralized Republic that lasted till the end of the eighteenth century, the judicial system was dispersed to a large extent. The administration of justice differed from province to province and even from town to town. No wonder that citizens often felt it arbitrary[88]. During the French occupation in the first decade of the nineteenth century, codification started and rationalization of the judicial system continued after the Netherlands had become a kingdom in 1813.

The number of civil cases brought to courts is affected by the dynamics of the judicial system, especially by the administration of justice in society. At the end of the nineteenth century the Dutch system changed to the extent that conflict regulation was partly delegated to administrative bodies outside the judicial system. This phenomenon has become known as *withdrawal of the legislators*[89]. Further, at that time a tendency set in which implied that judges became less tied to the letter of the law. The introduction of a new concept of tort in 1919 strengthened this development.

It may well be that these dynamics of the judicial system affected the inclination of people to appeal to the courts. A greater flexibility of judges vis-à-vis the letter of the law may have led to a changed perception of people as to their chances to be successful in court. Whatever, such phenomena probably interfere with the measurement of formalization. For that reason one should be careful with the interpretation of the data presented here.

In figure 3.8. the number of civil cases brought to all courts of justice are expressed per thousand of the population for the period 1852-1985. Since 1967 data on civil cases brought to all courts were collected differently than before (see appendix A, note to table A.6.I.). Therefore I could not construct one continuing time series from 1852 onwards. However, a continuing time series from 1852 to 1985 could be constructed if civil cases brought to districts courts were excluded. In figure 3.8. I present the latter too as a check on the trend of the former.

Figure 3.8. The Number of Civil Cases Brought to All Courts of Justice (a) and to All Courts of Justice Minus Districts Courts (b) Per Thousand of the Population, 1852-1985

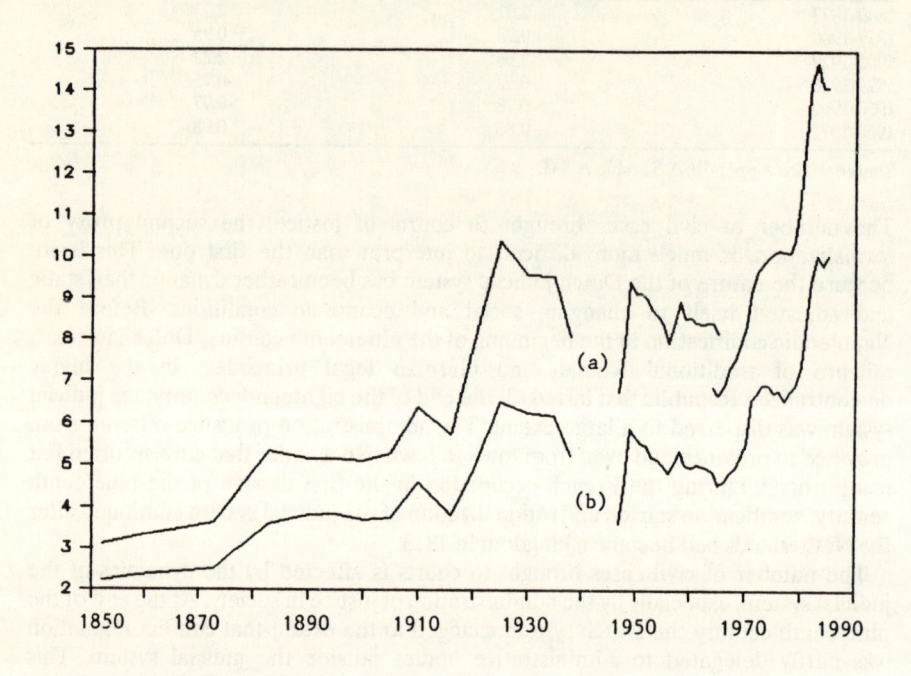

Source : Appendix A.6., table A.6.II.

Table 3.15. Average Annual Compound Growth Rates of Civil Cases Brought to All Courts of Justice and to All Courts of Justice Minus Districts Courts Per Thousand of the Population, 1852-1985

	All Courts	All Courts Excl. District Courts
1852-1873	0.71	0.59
1873-1883	3.88	3.61
1883-1900	-0.35	0.06
1900-1925	2.86	2.26
1925-1938	-1.92	-1.66
1950-1965	-0.72	-1.46
1965-1980	3.69	3.31
1980-1985	4.14	6.15

Source : Appendix A.6., table A.6.II.

Reviewing figure 3.8. three periods of significant growth can be discerned. Period one, the decade from 1873 to 1883, shows an increase of the number of civil cases brought to the courts. In period two, 1883-1900, growth stabilized and even decreased slightly. A new increase took place from 1900 to 1925, whereas the period from 1925 to 1938 again shows decrease. After World War II growth started in 1965 (see table 3.15.).

Because of the above mentioned dynamics of the Dutch judicial system, the period 1900-1925 is the most difficult to interpret. It is reasonable to suspect that in this period the observed rise of the number of civil cases is caused by changes to an unknown extent within the judicial system itself, notably the introduction of the concept of tort in 1919. Therefore it is not possible to draw firm conclusions from these data as to the degree of increasing formalization from 1900-1925. One could prudently remark that changes in the legal system itself partly provoked an increase of the number of appeals to that system.

In my opinion, the interpretation of the period 1873-1883 is much easier, because the judicial system was more stable and thus the growth of the number of civil cases can be largely accounted for as an indication of formalization.

It seems plausible to me to associate the growth after 1965 with the rise of the welfare state which brought an increase of regulations in all social domains and therewith a growth of the demand to legal interpretation and conflict solving.

If one compares the increase of ministry personnel to the growth of civil cases, one can observe that in the second half of the nineteenth century the former preceded the latter. After 1900 both indicators of formalization show similar trends. Taking the dynamics of the judicial system into consideration, one can identify two major periods of formalization: the 1870's and the years after 1965.

Public Sector Growth

According to the model presented in the preceding chapter the increasing involvement of the government in the economy is one of the main dimensions of modernization processes. We could think of the nature and quantity of economic legislation, the number of state-owned enterprises and so on. Thus "involvement of the government in the economy" is a broad concept indeed. As I wanted to measure government participation in some quantitative way and over a long period of time, I was forced to narrow down the concept to the volume of public expenditures. I consider the latter as a proxy indicating the degree of total government involvement in the economy.

Table 3.16. presents public expenditures as a percentage of Gross Domestic Product (GDP). Figure 3.9. is based on these percentages. In table 3.17. the average annual compound growth rates of public expenditures as a percentage of GDP are given.

In figure 3.9. and table 3.17. one can see that the trend of public expenditure as a percentage of GDP accelerated from 1880 to 1890, from 1910 to 1921, from 1930 to 1938 and from 1960 onwards.

We used public sector growth as an indicator of modernization. From that point of view, 1910-1921 and 1930-1938 take an exceptional position since those were years of national crises (World War I, the Great Depression). Special circumstances urged the Dutch government to increase its expenditures. Although the Netherlands remained neutral during the Great War it had to maintain the strength of its armed forces far above peacetime level. As a result of the war going on abroad, the country was confronted with increasing scarcity which forced the government to respond by market intervention. Although during the 1930's the official policy aim of the Dutch government was cutting the budget, our figures make clear that it did not succeed. In this crisis, market intervention was evoked.

Table 3.16. Public Expenditures (excl. Social Insurance) as a Percentage of Gross Domestic Product (current market prices), 1850-1982

Year	Public expenditures as a percentage of GDP
1850	6.37
1860	5.19
1870	5.13
1880	5.78
1890	8.77
1900	6.65
1910	6.66
1921	18.17
1930	11.81
1938	19.19
1950	25.16
1960	23.36
1970	28.89
1975	34.76
1982	40.48

Source : Appendix A.7., table A.7.I.

Table 3.17. Average Annual Compound Growth Rates of Public Expenditures as a Percentage of GDP, 1850-1982

Years	Public expenditures as a percentage of GDP
1850-1860	-2.05
1860-1870	-0.11
1870-1880	1.19
1880-1890	4.18
1890-1900	-2.78
1900-1910	0.02
1910-1921	9.12
1921-1930	-4.79
1930-1938	6.07
1950-1960	-0.74
1960-1970	2.13
1970-1975	3.70
1975-1982	2.18

Sources: Table 3.16. and appendix A.7., table A.7.I.

Figure 3.9. Public Expenditures as a Percentage of GDP (current market prices), 1850-1982

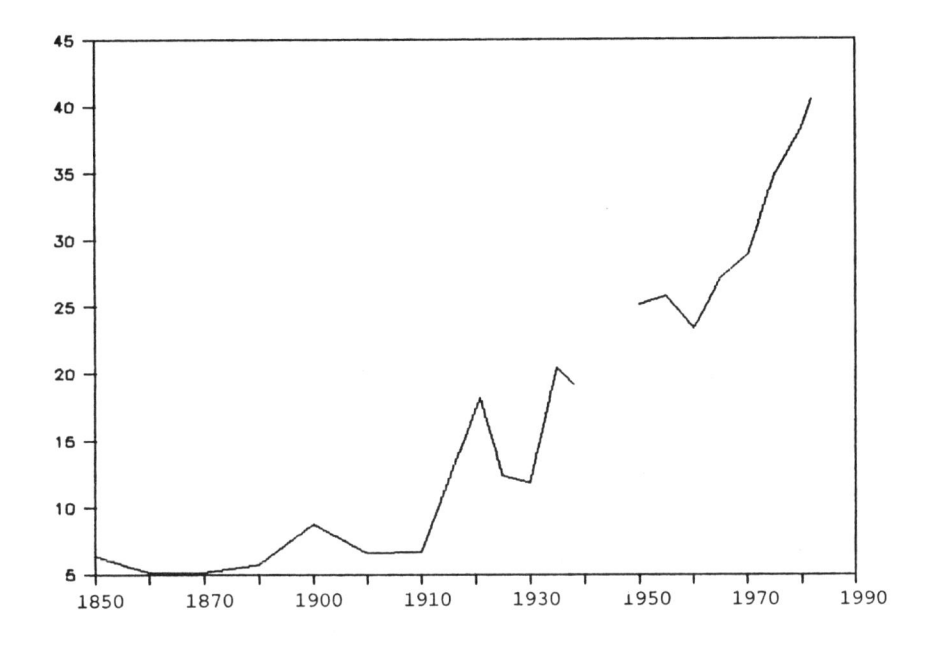

Source : table 3.16.

These are facts on which one could decide to exclude the 1910's and 1930's from the analysis because in those years the causes of public budget growth were only partly related to modernization. However, the forced budget growth in the two periods of crisis had as side-effect that, against one's will, one learned to live with relative high levels of public expenditure. This may have prepared the way to the growth of the public sector in the period after World War II. Therefore, my conclusion is threefold:
1. public sector growth in 1870-1890 and after 1960 should be regarded as a genuine indication of modernization;
2. the rise of public expenditures in the periods 1910-1921 and 1930-1938 are heavily affected by exceptional circumstances and thus are biased indicators of modernization;
3. the forced public budget growth during World War I and the Great Depression may have smoothed the way for the expansion after 1960.

Standardization of Information

In the Republic education was highly decentralised. Every Dutch province had its own regulations as to hours of teaching, holidays, the content of curricula, the admission of teachers and their salaries. These were often thwarted by rules issued by the towns and cities[90]. Primary education was strongly connected to the church and focussed mainly on matters of religion and morality.

Figure 3.10. Numbers of Participants in Formal Education per Thousand of the Population Aged Between 3 and 25 Years of Age, 1850-1960

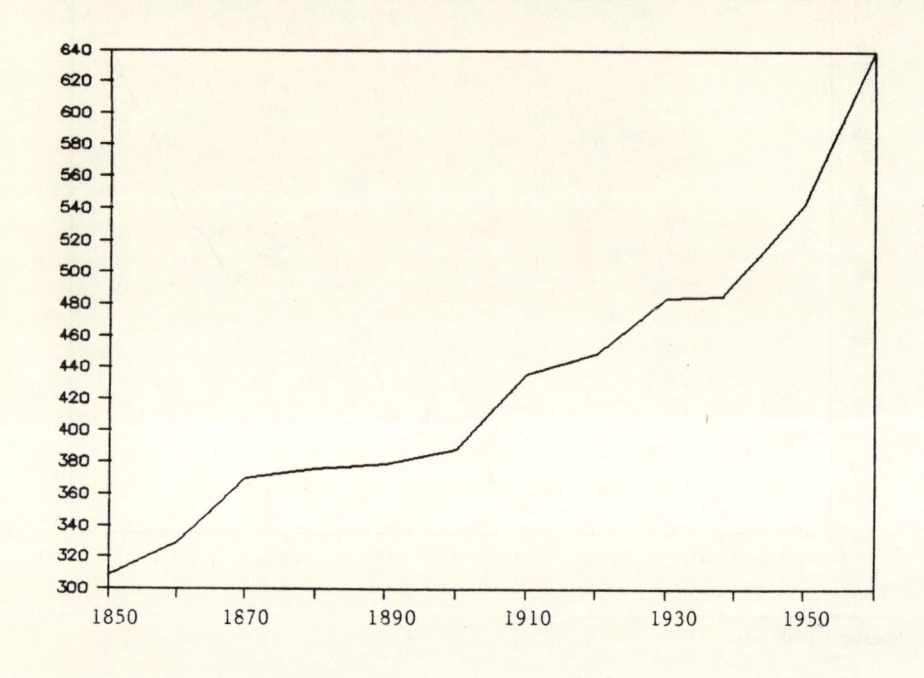

Source : Appendix A.8., table A.8.I.

Offspring of well-to-do families were send to private schools or got private tuition. After finishing primary education these children went to grammar-schools which prepared for academic training. In the eighteenth century French schools were established as a preparation for high functions in business and trade.

Only at the end of the eighteenth century an interest arose to the improvement of national education. This was primarily provoked by increasing pauperism and inspired by a concern of the perceived decline of morals and the consequent threat to the social order. In 1798 education was officially made a care of national government. In the first half of the nineteenth century several measures were taken as attempts to increase educational participation. These concentrated mainly on the poor as the sanctions involved had to do with the stoppage of poor relief. The effects of these measures are not exactly known. However, in the last forty years of the nineteenth century school absenteeism decreased considerably[91]. According to Idenburg it dropped from 24 per cent to 9 per cent of all children between 6 and 12 years of age[92]. Nevertheless it lasted till 1900 before compulsory education was introduced.

The second half of the nineteenth century saw two important developments in education. Firstly, government control of the content of curricula and professional qualities of teachers was broadened. The former led to severe tensions between the

Figure 3.11. Participation in University Education (a) and Higher Vocational Education (b) per Thousand of the Population Aged Between 18 and 24 Years of Age, 1930-1984

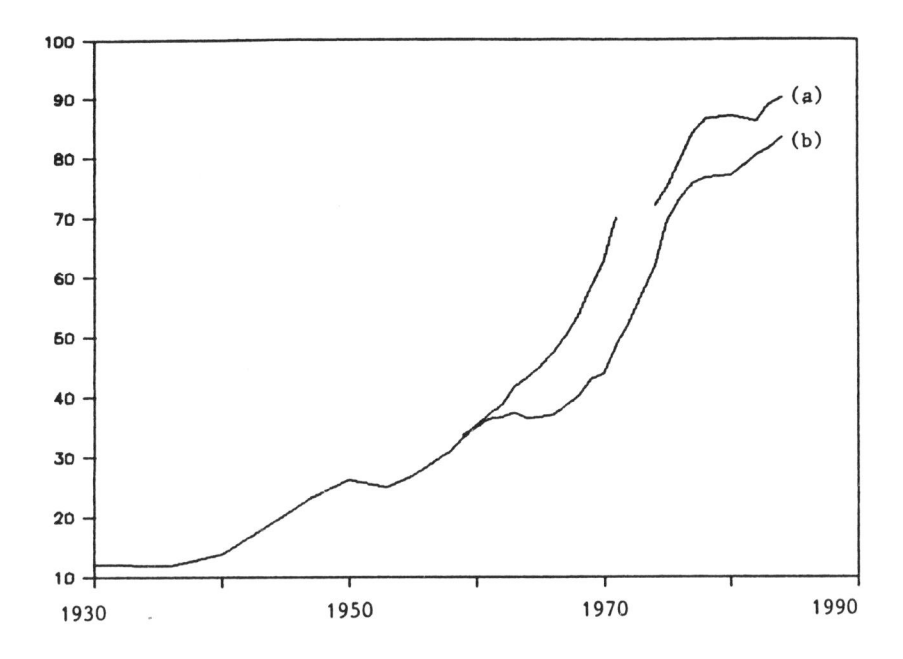

Source : Appendix A.8., table A.8.III.

different ideological blocs in Dutch society. This so-called school issue will be treated in the next chapter.

Secondly, as a response to the needs of industrialization which required functional skills, the educational system was proliferated. In 1857 advanced elementary education (MULO) was introduced. In 1863 it was followed by the secondary modern school (HBS), the polytechnic school and agricultural instruction. According to the liberal government of that time, vocational training had to be reserved to private initiative. Only in 1919 a law on vocational education was enacted.

Table 3.18. Average Annual Compound Growth Rates of Participation in University Education and Higher Vocational Education per Thousand of the Population Aged Between 18 and 24 Years of Age, 1930-1984

	Higher Vocational	University
1930-1940	n.a.	1.37
1947-1959	n.a.	3.00
1959-1968	2.00	5.33
1960-1971	2.99	6.24
1969-1974	7.35	4.23
1974-1984	2.98	2.25

Source : Appendix A.8., table A.8.II.

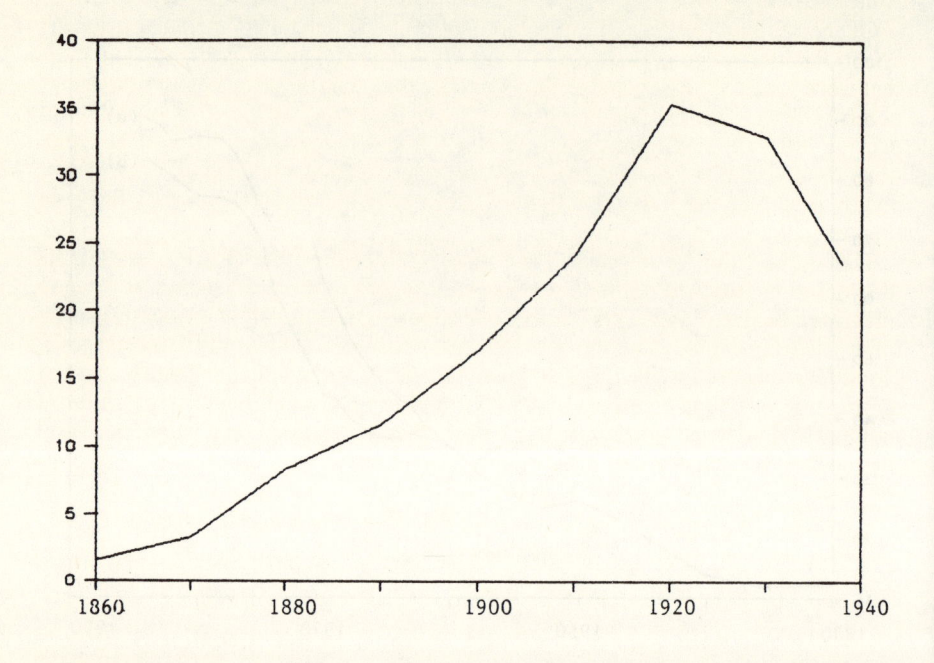

Figure 3.12. Per Capita Number of Newspapers Sent by Inland Mail, 1860-1938

Source : table 3.19.

The broadening of government control and the proliferation of the educational system in the direction of secondary schools implied both a standardization of curricula and an increase of educational participation. The latter is presented in figure 3.10.

In the twentieth century the most remarkable development in education certainly was the spectacular increase in higher education's participation rate (figure 3.11.). From 1960 to 1971 participation in university education rose from 33 to 70 per thousand of the relevant age category. In the first half of the 1970's a significant increase of the participation in higher vocational education took place (see also table 3.18.).

A second source of the progressive standardization of information in the nineteenth century was the rise of mass media, i.e. the big newspapers. Several factors contributed to this: (1) the improvement of the quality of education which led to a decrease of analphabetism; (2) population growth; (3) urbanization; (4) technological innovations in the printing industries and (5) the development of information transmission by telegraph and telephone. These factors are interrelated and it is difficult to distinct between causes and effects[93].

Since the Republic the Netherlands had a rich tradition of pamphlets. These were issued on a small scale and were aimed at spreading a political or social message. Therefore, the pamphlet should rather be regarded as a form of propaganda than as a medium of communicating "value-free" information. So far as pamphlets contributed to the distribution of information, the degree of standardization of their contents was very low.

Of special importance to the development of the Dutch newspapers has been the abolition of the tax on newspapers in 1869. This opened the way to bigger newspapers that were issued on a national scale. Hemels writes that the abolition of newspaper tax also led to a change of themes treated. Journalists now had more opportunities to provide differentiated information on Dutch and foreign topics. After 1869 topics were treated like family and school, work, wages, poverty and poor relief, public health, justice, the arts and civil rights. Before 1869 these issues occurred hardly in the newspapers[94].

The above is not to say that in the nineteenth century the Dutch press lost its ideological character. Preliminary to our conclusions in the next chapter, it became integrated in the pillarized system that developed in the last half of the century: each pillar had its own press that presented the news plus the ideologically based exegesis. However, compared to the past the developments of the nineteenth century implied an increase of the standardization of information because all papers treated more national and international items than before, they were issued on a larger scale and the number of readers grew.

Table 3.19. Number of Newspapers Sent by Inland Mail. Absolute Figures (x 1000) (1) and per Capita (2), 1860-1938

Year	Absolute Numbers (x 1000) (1)	Per Capita (2)
1860	5,107	1.54
1870	11,987	3.33
1880	33,681	8.32
1890	52,834	11.64
1900	88,062	17.13
1910	142,741	24.20
1920	241,201	35.36
1930	259,767	32.95
1938	204,245	23.52

Source : See appendix A.9.

Table 3.20. Average Annual Compound Growth Rates of the Per Capita Number of Newspapers Sent by Inland Mail, 1860-1938

Years	Growth Rate
1860-1870	7.73
1870-1880	9.16
1880-1890	3.36
1890-1900	3.86
1900-1910	3.46
1910-1920	3.79
1920-1930	-0.71
1930-1938	-4.21

Source : table 3.19.

The latter is illustrated by table 3.19., table 3.20. and figure 3.12. The figures should be interpreted with care. Before the second half of the nineteenth century newspapers generally were sent to subscribers by mail. Due to the fact that after 1850 mailing to individual subscribers became more expensive than sending whole packages of papers by rail to local agencies or shops, the number of newspapers mailed directly to subscribers decreased relatively to those sold by agents or shops. In 1882 one of the largest national newspapers, the Nieuws van de Dag, started selling locally. The locally sold newspapers are not included in our figures. This implies that after 1880 the total number of inland newspapers sold is much larger than our figures do suspect. The decline in the 1920's of the number of newspaper mailed does only indicate a major shift in favour of the number of papers directly sold.

Monitoring Dutch Modernization: Conclusions

This chapter monitored the structural dimensions of Dutch modernization since the origins of the Dutch state in the sixteenth century. I did not include in the analysis the cultural developments. These are treated in the next chapter.

In monitoring long term modernization of Dutch society I was confronted with the problem of insufficient empirical data. The construction of time series was only possible for the period after about 1850. For the years before I was forced to restrict myself to estimates and analyses of a more qualitative nature.

With regard to the key variable of modernization, economic growth, I paid ample attention to the works of Jan de Vries and Angus Maddison. Their research casts serious doubts on the main stream hypothesis that the first half of the nineteenth century is one of economic stagnation and that Dutch GDP started to rise only after 1860. Instead, they are convincing in their statements that nineteenth century Dutch economic growth started as early as the 1820's and accelerated after 1870. After 1950 a new era of economic growth and technological development set in. Acceleration took place in the 1960's. After 1973 economic growth slowed down.

My general conclusion is that from the sixteenth to the twentieth century Dutch society saw three waves of modernization. The first was concentrated in the sixteenth and seventeenth century. The second wave lasted from 1870 till the 1920's. The period 1820-1870 is transitional between these two waves. The third wave started in the 1960's and lasted till 1973.

As to the first wave, in the sixteenth and seventeenth century the Netherlands was one of the most modern countries of the world. Maddison even speaks of the Netherlands as a "lead country". This does not only refer to the level of Dutch per capita income, but also to the modernity of its institutions, its high degree of urbanization and population growth, the development of economic infrastructure, the state of Dutch industry and the service sector, and the relative openness and proliferation of the Dutch stratification system.

In the last quarter of the seventeenth century stagnation, and later decline, set in. This period lasted till the end of the eighteenth century. It was not only distinctive by a drop in economic performance, but also by phenomena of contraction in the other domains of modernization. From 1700 till 1820 population growth decreased relative to the period 1500-1700 while in all other Western European countries it increased. The growth of cities stagnated. Opportunities of upward social mobility decreased substantially. The developments in the stratification system as a whole showed a polarization to the extremes: the erosion of the middle classes led to the division of

Dutch society in two status groups: the poor and the very rich. Growing poverty reached a peak in the first half of the nineteenth century. These conditions of living were legitimized by education which was aimed at the reconciliation of the individual to the state of the social order.

Within the complex of the structural components of modernization no sufficiently endogenous explanation of Dutch decline can be found. Some have pointed to a change of mentality of the Dutch population. Others have emphasized changed external economic conditions by which the Dutch economy was, as it were, outsmarted. If one accepts the latter, Dutch decline has to be conceived as a phenomenon of maladaptation or perhaps the inability to cope with changed external conditions. Such inability points in the direction of incompatibilities between the economic, the social and the cultural system caused by changes in the countries external environment. In this chapter we neglected this problem which we will treat in the chapter to follow.

Whatever the causes, the first quarter of the nineteenth century was a new turning point in Dutch modernization. In the 1820's per capita income started to rise. GDP growth accelerated from about 1870 to 1920. Reviewing the other indicators of modernization, all showed accelerated growth in the second half of the nineteenth century. Urbanization increased from 1850 to 1900 with an acceleration between 1880 and 1890. The enlargement of scale of industrial enterprises largely took place from 1880 to 1910. Public expenditures grew substantially from 1870 till 1890. We could see an increase in the number of civil cases brought to courts, an indicator of the formalization of social relations, since 1873. The standardization of information also increased after 1860 by educational reform and a progressive spread of newspaper reading. The mid of the century saw a revival of the middle class. Lower class' social mobility increased after 1870.

Within the complex of modernization variables the growth of government personnel takes a somewhat exceptional position. The first growth period has to be dated as early as 1849-1877. However, it may well be that government personnel growth started earlier because at the beginning of the nineteenth century the French took several measures which implied centralization of Dutch administration. After 1813 the Dutch king William I executed a policy of reform of economic institutions. Both may have resulted in a growth of government personnel. Because of lack of data, however, we can not be sure.

Putting the nineteenth century trends of economic growth and social development together, my first conclusion is that the period from about 1820 to 1860 is one of transition. Economic stagnation came to an end. Economic growth started. However, modernization as a whole was characterized by incompatibilities. Among the structural factors, the educational system draws special attention. Before 1870 it was not yet ready to produce manpower adequately trained to satisfy the needs of the economy. For example, in the first half of the nineteenth century engineers from abroad had to be used for the construction of Dutch railroads. In the 1860's educational reform gradually lifted this situation. But, as we shall see in the chapter to follow, these reforms caused severe social and political tensions.

After 1870 all indicators of modernization show accelerated growth. Because of this parallelism, I consider the second wave of Dutch modernization to start around 1870. Thus the distinction between a transitional period from 1820 to 1870 and the second wave of modernization from 1870 to 1920 is based on differences in tempi of growth of the respective indicators on the one hand and differences in incompatibilities within the modernization process as a whole on the other hand. Further, Dutch nineteenth century modernization started with economic growth in the transition period. This growth later on promoted a further structural modernization of Dutch society.

In the twentieth century the nineteenth century trend continued. The two world wars and the crisis of the 1930's can be considered as periods of temporary stagnation. Especially after 1960 economic growth accelerated. Although to different degrees, all other dimensions of the modernization process developed into the same direction. So in contrast to the second wave of Dutch modernization in the nineteenth century, the third is not characterized by timelags or incompatibilities within the complex of the structural dimensions of modernization.

CHAPTER IV

LONG TERM MODERNIZATION AND INSTITUTIONAL CHANGE IN DUTCH SOCIETY

Introduction

Many authors who have studied Dutch twentieth century pluralism, have pointed to its historical roots. In the early seventies, Christopher Bagley wrote of Dutch society:

> "Historically, individualism has been a leading feature of Dutch life. This individualism has been of groups rather than of individuals: the individual has afforded deference, often extreme deference, to the elders of his group. But the group to which he belonged was often independent of, and perhaps in conflict with, other groups"[1].

And in the mid sixties a leading Dutch political scientist, Hans Daalder, stated:

> "Even present-day social cleavage, between classes and masses and between latitudinarian Protestants, orthodox Protestants, and Roman-Catholics have their roots in sixteenth and seventeenth century divisions"[2].

According to Daalder, the highly dispersed power structure of the Republic and the absence of a powerful national bureaucracy gave rise to a culture of elite bargaining by which different interests could be accommodated.

> "All decision-makers, therefore, learned to live in a climate of reciprocal opposition, which fostered the habit of seeking accommodation through slow negotiations and mutual concessions"[3].

So Daalder argues that under conditions of pluralism gradually a cultural pattern arose which permitted the accommodation of conflicting interests. In the late nineteenth and in the twentieth century this pluralist tradition gave rise to the *pillarization* of Dutch society.

Although I generally agree with the above line of interpretation, it should be remembered that Dutch twentieth century pillarization was primarily built on denominational cleavage. Religious pluralism became only a social and political issue in the late nineteenth century. Before that time, Dutch pluralism had a different content. This fact raises questions as to how Dutch pluralism developed and how it affected the social codes of Dutch society.

Ellemers[4] is one of the few scholars who has explained Dutch pillarization as the outcome of a more general process of modernization. The core elements of Ellemers' thesis on the relation between modernization and Dutch pillarization are threefold. Firstly, *segmented pluralism*, i.e. "the society is characterized by an extraordinary degree of social cleavage", is part of the national identity. Secondly, modernization produces on the one hand social differentiation and, on the other, a need for national integration because of the necessity to promote economies of scale and to achieve an internationally competitive economic position. Thirdly, in a society characterized by segmented pluralism, pillarization is the way in which these two prerequisites can be fulfilled.

Besides Ellemers' rather abstract and functional reasoning, his explanation leaves us with some tricky questions. First of all, we may wonder why it is that in the Netherlands segmented pluralism is part of the national tradition. Ellemers takes this phenomenon more or less for granted when he remarks:

"It seems that these nations (Belgium, the Netherlands and Switzerland) were difficult to control and absorb within a larger framework because of their cultural, religious, linguistic or ethnic diversity, which in some way formed part of their national identity"[5].

If one studies Dutch nation building, the concept of national identity itself is not without problems. According to historical studies, it was not until the late eighteenth century that appeals to some national identity began to appear in pamphlets[6]. Before that time the Dutch themselves did not feel a national identity, at least not in the contemporary sense.

Secondly, if there are good arguments to doubt the existence of a Dutch national identity in the sixteenth, seventeenth and eighteenth century, why is it that the different provinces which formed the Republic, came together in the first place? And further, why did segmented pluralism not lead to a disintegration of the Dutch state?

Finally, the nature of nineteenth and twentieth century Dutch segmented pluralism was different from that in the centuries before. Questions then arise as to whether these two types of pluralism are related, in what way and what factors caused the transition of one type to the other.

These are the main questions I will treat in this chapter. I will analyse them by focussing on the long term process of modernization in Dutch society and the associated institutional changes.

Segmented Pluralism in the Dutch Republic

The Origins of the Dutch Republic

The Dutch state developed in the sixteenth century in the course of the revolt against the Spanish. There were three major components: resistance against the centralizing tendencies of the Spanish monarchy; tax-resistance as an expression of struggle for greater economic freedom; and opposition to Catholicism.

Dutch nation building involved strong resistance to centralized authority. Since the fall of the Roman Empire, the Netherlands, including present day Belgium and Luxembourg, were part of the German Empire. Formally Dutch noblemen were vassals of the German emperor. In fact, however, successive German feudal lords showed little interest in their northern countries. They were much more oriented towards their southern interests. Contacts between the German emperor and his Dutch vassals were scarce. Besides, by contrast with England and France, the Medieval German empire was only loosely integrated and lacked a central power structure embodied in a national dynasty. Within this historical framework the Dutch nobility was able to gradually achieve a high degree of autonomy.

In 1515 this situation changed. The German emperor of that time, Charles V, inherited the Netherlands as part of his crown lands. Thus the Netherlands became strongly tied to the House of Habsburg. When Philip II succeeded his father as Lord of the Netherlands and King of Spain, the low countries became part of the Spanish Empire.

One could characterize the history of the Netherlands before the end of the sixteenth century as one of loose ties with princes far away. In this respect the Netherlands differed from other European countries like England, France, Denmark, Sweden, Spain and even Russia. These countries had already achieved unity under a national dynasty in the early middle ages. From a historical point of view, one could consider the Netherlands as backward in the process of nation building.

Economically and socially, however, the sixteenth century Netherlands were very modern. Thanks to their high degree of autonomy, the low countries could fully gain from favourable geographic conditions. The relatively independent position of the cities vis-à-vis feudal lords permitted the early development of a bourgeoisie strongly involved in entrepôt trade. So what these Dutch regions had in common was, besides their language, the modernity of their economies, their reliance on foreign trade and a degree of freedom and autonomy which was relatively high.

In the second half of the sixteenth century the Spanish Habsburg monarchs tried to unify their empire. Not only in their own interests, but also to respond to the requirements of an expanding economy in need of increasing integration and better systems of communications[7]. The revolt originated from the attempt of the Dutch nobility to protect their autonomy against Spanish centralization. It was an example of conservative change aimed at the protection of medieval privileges. Although not successful at the beginning, the threat of excessive Spanish taxes broadened its support. Notably, the opposition to the "tiende penning", a turnover tax of ten per cent on every transaction involving movable goods, proposed by the Spaniards in 1569, linked the interests of the nobility with those of the merchant class. The "tiende penning" was a fixed tax rate. It therefore violated the privilege of the traditional political institutions of the Netherlands to negotiate tax incidence with the king. At the same time it was a serious menace to the trade interests of the merchants.

The third element of the revolt consisted of opposition to Catholicism. In Dutch history schoolbooks, this aspect of the rebellion is given special emphasis. In the minds of the Dutch people the Reformation was the main ideological justification of the Revolt. The concepts of Calvinism and patriotism became almost synonymous in Dutch national thinking. Because of its leading role in the revolt against Catholic dominance, the House of Orange has since then always been strongly associated with the Calvinist interest.

The Power Structure

In the beginning of the seventeenth century when the Netherlands gained independence through the peace of Munster (1648), the new republic turned out to be a divided political entity. During the revolt the seven allied provinces had not succeeded in achieving national unity. As Smit remarked:

> "The bourgeoisie of Holland had carried through exactly the degree of reform it needed to promote economic expansion and yet feel free from overcentralization"[8].

In fact the new state was only a federation of seven politically autonomous provinces: Holland, Zealand, Utrecht, Gelderland, Overijssel, Groningen and Friesland. The territory of the Republic also comprised Brabant and Limburg. They had no sovereignty like the other provinces, but were directly governed by the Staten-Generaal in which they did not have representatives.

How was the government of this federation organized? Three levels of governmental control can be discerned: the city, the provinces and the national level. Cities were ruled by "vroedschappen", city councils dating from the middle ages. Membership of the vroedschap was reserved for the "most wise and richests" burghers. Office allocation was the prerogative of the vroedschap. Usually these offices were held by the members of the vroedschap itself or members of their families. In the sixteenth century these vroedschappen were relatively open. Outsiders who had acquired wealth and prestige could enter the vroedschap. When Spanish control ended, the cities had increased autonomy. As a result, local city officials gained prestige and influence. Gradually, an oligarchic class of "regenten" (regents) developed that remained powerful till the end of the Republic in 1795[9].

On the level of the province, the "Staten" formed the highest authority. They consisted of representatives of the cities, the clergy and the nobility. However, although all seven provinces had their Staten as highest political authority, the composition and working of these institutions was dissimilar (see table 4.1). Besides the delegates, everyone of the Staten had a Stadtholder and a "raadspensionaris", a paid legal advisor and influential first executive. The Stadtholder originally represented the king in the Staten but after the revolt he became "first servant of the Staten" and commander in chief. Further, the Stadtholder had the prerogative to grant free pardons and to nominate city magistrates from a select list. With the exception of the years from 1650 to 1672 and from 1702 to 1747, when there were no Stadtholders, these positions were held by members of the House of Orange.

Finally, the Staten-Generaal, officially the highest authority in the Republic, consisted of delegates of the seven provinces. Chairmanship changed every week. In cases of truce, peace, war and taxation, unanimity was required for decision making. Thus formally the provinces had equal influence in general policy and no important decision could be taken if one of the provinces was against.

Table 4.1. Distribution of Power in the Seventeenth Century Dutch Republic. Dutch Provinces by Percentage Share in the Republic's Budget and by Percentage Share of City Votes in the Provincial Staten

Province	Percentage Share in Republic's budget*	Percentage Share of City Votes**
Holland	58.0	95
Zealand	9.0	86
Utrecht	6.0	33
Groningen	5.5	50
Friesland	11.5	25
Gelderland	5.5	50
Overijsel	3.5	50

Note : * Percentages do not add to hundred due to rounding.
　　　　**Remaining percentage share of votes contributed to nobility, landed gentry and clergy.
Source : Data derived from H. Wansink, 1980.

Although the above outline of the Republic's political structure is rather general, it nevertheless makes clear that it was an assembly of city-republics:

"In so far as any political consciousness was present, it was oriented to the interests of the province and, in the first place, to the interests of the city. The province, and especially the cities were ... the political entities that were most familiar and near to every one"[10].

Since the city oligarchies of regents dominated the Staten of Holland and Zealand and as these two provinces contributed the larger part of the Republic's budget (see table 4.1), they in fact formed the ruling class of the Dutch Republic in the seventeenth and eighteenth century. Thus, although started by the nobility, during the Revolt, the ruling power shifted from the nobility to the bourgeoisie. In the seventeenth century the nobility no longer dominated the political power structure, although influential positions in army, navy and diplomacy were reserved to noblemen[11].

The Nature of Segmented Pluralism in the Republic

The new Republic was internally divided into cities and provinces which had dissimilar or even conflicting interests. The nature of these often rested on the kind of economic activity which dominated the city or region involved. In this respect, however, the important distinction is that between Holland and the other provinces. Holland represented above all the interests of the merchant class. As Amsterdam contributed one third to the budget of the Staten of Holland, the interests of Amsterdam's regents were articulated pretty well. Because Zealand had also a strong merchant class orientation, Holland and Zealand can be regarded as allies within the Dutch federation.

Holland's policy until 1795 can be considered as isolationist. It looked upon the other provinces as allies under the condition that they accepted its leadership. Since every province had one vote in the Staten-Generaal, Holland could never dominate the other provinces completely. But as it contributed 58 per cent of the federal budget it had a large influence. The other provinces had to restrict themselves to at most the obstruction of Holland's policies. They lacked the strength to formulate a policy of their own. Under this balance of power the Orange Stadtholders often tried to use conflicts between provinces to strengthen their own positions[12].

It should be noted that the interests of the Stadtholder and the merchant class of Holland and Zealand were opposed. The Orange princes had strong dynastic ambitions and promoted the expansion of the Republican territory. The merchant class was primarily interested in conditions of trade which it thought would be damaged if an Orangist policy were pursued[13]. This latent conflict gave rise to two political parties, the Staatsgezinden and the Orangists, the former promoting the republican ideal whereas the latter supported the ambitions of the House of Orange.

As Roorda[14] remarked, in this context the word "party" should be understood in its seventeenth and eighteenth century meaning and not in its current one. It is not comparable with the twentieth century idea of a political party characterized by a high degree of organization. The seventeenth century party was only a political group that attempted to mobilize the largest part of the population behind a political ideal. Such parties manifested themselves only in times which were generally felt as national crises. Their significance in every-day seventeenth and eighteenth century politics was limited.

Everyday politics in the cities of the Republic was controlled more by factions than by parties. Roorda defines factions as groupings of local (city) rulers and their clients which aimed at the promotion of group interests[15]. These were primarily the acquisition of offices from which power and the further accumulation of wealth could be derived[16]. Faction membership usually was based on family ties. Competition between factions within the cities should thus primarily be understood as a struggle for power between families. It follows that the city's regent oligarchy was in fact made up of family groupings or coalitions of family groupings facing the opposition of other factions.

Factions of city regents were essentially an upper class phenomenon. In the Republic the middle and lower classes were not politically involved. Only in national crises did the Staatsgezinden and Orangists attempt to mobilize the common man.

Further, factions of regents sometimes manipulated street riots as a lever to attain or increase their power vis-à-vis their rivals[17]. However, such riots had a strictly local character. Only, when political competition between parties was involved, did rioting take place on a national scale[18].

One may wonder to what extent social inequality and religious differences formed part of the Republic's segmented pluralism. Because the Republic was relatively modern for its time, one might well expect to find elements of an articulated class structure or signs of horizontal solidarity[19]. Dekker[20], who studied patterns of rioting in the Dutch Republic, argues that a class interpretation is too simple. Riots in which class contradictions played some role, like hunger-riots and tax-riots, had a local character and a limited scope. Enclosure riots were rare as the Netherlands had only very few enclosures. Strikes of a limited scope took place from 1600 to 1750, but only a very few are known after 1750. In general, only uprisings that contained goals supported by many segments of the population transcended the local level. These movements, however, were not limited to clearly identifiable classes. They drew their support from all strata. This holds especially for movements based on political and religious motives. In many cases such motives overlapped. Moreover, because of the conflicting interests between provinces, cities and factions, the ruling class was not united but fragmented. Thus in the Republic, class consciousness was almost absent and cannot be linked to segmented pluralism.

Although resistance against Roman-Catholicism had been a core element in the Revolt and Dutch Reformed Calvinism became the dominant religion in the beginning of the seventeenth century, the new Republic had a pluralist character. Around 1700, 50 per cent of the population belonged to the Dutch Reformed church, 10 per cent were Calvinist dissenters and about 40 per cent were Roman Catholics. These figures are rough estimates and differ from region to region[21]. Beside these main denominations, the spiritual life of Jews and Humanists flourished. In the Republic religious differences did not coincide with differences in social status, wealth or political power. The position of the Roman-Catholics is illustrative. Since 1572 Roman-Catholic churches and services had been officially banned. Roman-Catholics could not assume public office. The two southern Roman-Catholic regions Brabant and Limburg, conquered by the Republic during the rebellion, were ruled as colonies and economically exploited. However, unofficial freedom for Roman-Catholic aspirations was preserved. Roman-Catholics were allowed to conduct their services in secret[22]. In many cities Roman-Catholics formed part of or were related to the ruling oligarchy. There are many cases of Roman-Catholic merchants who reached positions of great wealth and prestige.

Religious contradictions that gave rise to political turmoil were seldom concerned with the relation between Roman-Catholics and Protestants, but rather with differences of opinion within the official Dutch Reformed church[23]. Because the House of Orange was traditionally identified with the Calvinist cause, Orangist movements often tried to promote their cause by appealing to anti-Catholic sentiments. As such, the religious aspect of the dispute was only a side-issue meant to mobilize public opinion. The hard core of the disputes between Staatsgezinden and Orangists had little to do with religious differences, but rather referred to the issue of federal versus de-centralised power.

Summarizing, our main conclusion with regard to the nature of segmented pluralism in the Dutch Republic is, that it was of a predominantly geographic nature. It concerned several issues. Firstly, the autonomy of the cities and their ruling elites.

Secondly, the often differing or even conflicting interests of cities and competing elite groups which caused the fragmentation of the ruling class. Thirdly, the balance of power between Holland and Zealand on the one hand and the other provinces on the other. Fourthly, throughout the history of the Republic, tensions arose between the Staatsgezinden and the Orangist parties. The latter represented the dynastic ambitions of the House of Orange and promoted centralism, the former favoured the interests of the regent class and the autonomy of cities and provinces. These four elements are strongly interwoven. Seventeenth and eighteenth century segmented pluralism did not concern the relation between classes and religious groups. Although used in public debates, arguments of class inequality and religious pluralism were subjected to overriding issues of centralism versus decentralism.

Mechanisms of Integration in the Republic

Structural Factors and Institutions

It is remarkable that the seven provinces remained integrated in the newly formed Republic after the success of the Revolt. In fact, they were only a hotch-potch of minorities without a significant central authority. One might well have expected disintegration in such a situation, but this did not happen. How can we explain the phenomenon of political unity in face of so many forces of disintegration, with such a highly dispersed power structure and so many often diverging interests, while, on top of it, its ruling class was so fragmented?

One of the arguments often advanced is that unity was fostered by the need for hydraulic management or a common struggle against the continuous threat of the sea. There certainly was an effort in the Republic in this field[24]. However, one may wonder whether this particular commitment to action was strong enough as a cultural element to foster unity in the social system as a whole. Land reclamation, the construction of canals and the building of dikes were in any case not organized on a national but on a local level. Land reclamation was often regarded as an investment opportunity for private entrepreneurs[25]. So the evidence in favour of hydraulic management as an integrative force in Dutch society is rather shaky[26].

A more powerful explanation is the consolidating function of external enemies. Faced with an external enemy, in the struggle for survival internal differences tended to be put aside. The Republic was founded by the seven provinces as an attempt to resist a common enemy. However, after gaining victory, the national territory and its commercial interests were constantly threatened by the French and the English. Although Queen Elizabeth of England supported the Dutch in their revolt against the Spaniards, in the seventeenth and eighteenth century the Netherlands and Britain were adversaries in a number of wars. In 1672 France and the German Bishop of Munster invaded the country. From its foundation the Republic was under constant pressure from abroad.

To a large extent this external pressure was a direct threat to the Republic's trade interests. As the position of the regent class in Holland and Zealand depended mainly on foreign trade, it was above all their interest to counter these threats effectively. This could only be done on condition that the federation remained integrated to a minimal extent[27]. As we have seen, Holland looked upon the other provinces as a buffer, a first line of defense against foreign invasion. Further, although Holland and Zealand together financed 67 per cent of the Republic's budget, one third was contributed by the other five provinces. Finally, army control was decentralized under the so-called repartition system by which each of the provinces was made directly responsible for financing the troops within its boundaries. Troop movements were subject to the approval of the provincial states concerned[28]. Thus, in times of military mobilization, the provinces had strong mutual interdependence.

The Republic was relatively small compared with the states surrounding it. Just as Holland and Zealand depended on the other provinces, the latter were also interested in maintaining the federation however loosely. On their own the means of each of them were too restricted to successfully counteract against strikes from abroad. Thus the best way to guarantee the continuity of autonomy was to accept, however unwillingly, the relative dominance of Holland and Zealand in the federation.

Besides this major structural factor, historical literature yields others that account for the maintenance of federal unity during the seventeenth and eighteenth century:

a. Firstly, the fact that the same language was spoken throughout the territories of the Republic fostered feelings of cultural unity.
b. A second integrating force was the Calvinist religion. The Calvinists had developed some rudimentary idea of a nation, e.g. they took the ideological position that theirs was the true faith that maintained the independence of the Republic. Further, although the Calvinist state church was organized on a provincial level, many supra-provincial relations existed and ministers could be called outside their provinces of origin[29]. Finally, throughout the Republic, all Calvinist churches used the same translation of the Holy Bible, the "Staten Vertaling".
c. A third integrative factor was the development of the regent class itself. Although the city remained the power base of this ruling class, many regents took political office on the provincial and federal level[30]. Further, the regent aristocracy gradually penetrated an increasing number of social sectors: the guilds, the citizen militia and even the church apparatus. As their power increased, so did the number of those who were dependent of them. A kind of patronage system grew with as clients personnel and those who acquired lower offices thanks to the protection of regents in higher places[31].
d. Finally, we noted that regent factions usually were organized on a family basis. In many cases such family ties extended over the borders of city or province. Thus, although the ruling class was fragmented and the social distance between regents and other citizen's increased in time, supra-city and supra-provincial interaction between regents took place and networks of interaction with lower strata were established by interlocking directorates and some kind of patronage systems.

Wansink[32] remarked that, although formally only a provincial official, the power of the Stadtholder as a force for federal integration was much bigger than the constitution of the Republic might make one believe. Firstly, the House of Orange had always hold the position of Stadtholder in more than one province at a time, with relatives occupying that office in the other provinces. Secondly, the Stadtholder had the right to nominate city magistrates from a select list. Some Stadtholders used this right ruthlessly to increase their power. Thirdly, the Stadtholder was commander in chief of army and fleet. Finally, because of its historical role in the Revolt, the House of Orange was very popular with the common people. Apart from their dynastic ambitions, the structural position of the Stadtholders made them to a binding force. However, during the Republic they never managed to overcome de-centralising tendencies completely. In fact they lived in almost constant conflict with the Staten-Generaal as to the limits of their authority.

Finally, as time passed by, institutions developed which enabled the Dutch to integrate or overcome the negative effects of a too strong commitment to group autonomy. Much decision making was informally prepared in the corridors of the Staten and the Staten-Generaal. From the mid seventeenth century onward, the "besogne" became a popular instrument to sort out political contradictions. Political issues too tricky to deal with were delegated to besognes or committees who worked behind closed doors and whose instructions usually were very vague or absent[33]. By means of such besognes, hot political issues were de-politicized and rival elite groups were able to compromise without loosing face. As we shall see, the institution of special committees played a major role in accomodating diverging interests in twentieth century pillarized Dutch society[34].

The Culture of "Living-Apart-Together"

The above description of how the Republic functioned leads to some analytical conclusions as to how it coped with segmented pluralism. As we demonstrated, seventeenth century geo-political and economic conditions of the seven provinces were the main structural factors that prevented the break-up of the Republic. In the long run these solved the dilemma between the necessity of central decision making and the provincial autonomy provided by the Republic's constitution.

Theoretically the dilemma could have been overcome by the complete political and military control of one of the provinces over the federation. In seventeenth century reality such a solution was not feasible for several reasons. The military was de-centralised over the provinces and its commanders in chief, the Oranges, had ambitions of their own which were contradictory to a potential supremacy of one of the provinces. Instead, they could gain from provinces that quarreled amongst each other. Secondly, financial power was concentrated in the hands of the merchant elites of Holland and Zealand. This ruling class was above all absorbed by the pursuit of its commercial interests. Thirdly, civil war would certainly have weakened the international position of the Republic which would have been detrimental to all. Thus the seven provinces were more or less sentenced to live together.

In the Republic principles of politics were therefore determined by three factors: the desire of each province to keep its autonomy at a maximum, the need to avoid the violation of the autonomy of other provinces and the requirement of some central decision making in order to safeguard the Republic's sovereignty in the interest of all. It is easy to see that in political reality these principles lead to conflicts. The defense of one of them may imply the violation of others. In other words, seventeenth and eighteenth century Dutch politicians had to cope with an optimum problem.

Of course, in solving political issues, optimal solutions were not found theoretically, but by trial and error and within the prevailing balance of power. As the latter had a dynamic nature, the solutions changed over time. The more Holland's power increased, the more the federation became dependent on it and the more it could influence central decision making in the Republic.

Thus, autonomy of the Republic's provinces has to be understood as a matter of relative autonomy. Consequently, political life was impressed by compromise as the only way to achieve the required national decisions. At the same time, the parties involved were anxious to maintain their autonomy to the highest possible degree. We have seen that institutions developed that smoothed the decision making process.

One could say that the political culture of the Dutch Republic under segmented pluralism resembles that of some of "modern" marriages where the partners live in separate houses with their own things, belongings and friends, because they believe that the sacrifice of their individuality will ultimately end in a therapist's consulting-room, a laywer's office, murder or suicide. It therefore seems appropriate to me to call the political culture of the Republic a culture or ethic of "living-apart-together".

By this I mean the institutional arrangement which enables mutually interdependent social and political groups to maintain their autonomy to a perceived optimum, within the frame of a national sovereignty. It ensures the integration of these groups to a minimal degree such as to prevent the jeopardizing of the national existence. A breakdown of the latter would be detrimental to the relative autonomy of all its constituting parts. Interactions between the groups involved are structured according to this institutional arrangement. This culture of, what I will call, "living-apart-together", has from the Revolt onward always been a core element in Dutch social and political development[35].

Modernization and the Change of Dutch Pluralism

National Unification

The dichotomization of the social order and the increase of social inequality in the second half of the eighteenth century provoked ideas about a better society to strive for. Around 1770 a new patriotic movement arose that was inspired by conceptions that in France led to revolution. In numerous pamphlets these ideas were spread among the population. They appealed to the creation of a better society in which the material conditions of life should be improved by a selective stimulation of economic life by the government. Further, such economic welfare policy should be supported by the moral and intellectual education of every member of society[36]. In 1782 the Patriots organized themselves on the national level. They drafted a secret program in which they included among their objectives the election of burgher committees and the establishment of free corps. Inspired by the American Declaration of Independence, the Patriots aimed for a democratic republic and thus attacked the traditional order. Reactions did not fail to come. In 1787 the threat to the Dutch establishment was removed with the help of Prussian troops and many Patriots fled to France.

Although the patriotic movement proved to be abortive, its ideas were the first signs of a change of climate. They addressed a national audience and were not focussed on particular cities or provinces. They had a limited, but democratic substance which identified with national interests and the removal of social misery.

In 1795 French troops occupied the Republic. Till 1813 the Netherlands remained under French influence. The French realized many ideas that the Patriots had hoped for: unification, centralization and democratization of the Netherlands. Napoleonic codification substituted the fragmented Republican legal system, which had evoluated from Roman-German legal principles. Because of the federative character of the Republic, each province had built on these a legal system of its own. As a result, in the Republic the administration of justice differed from province to province. Further, civil registration, civil marriage, a new tax system, a national administrative structure and land registry were introduced. The state bureaucracy was extended. Freedom of religion was now formally granted which brought relief to Roman-Catholics and Protestant dissenters who before had only been tolerated. Finally, in 1806 general public education was established by law.

Thus in the period 1795-1813, the dispersed power structure of the old Republic was largely overcome and the Kingdom of the Netherlands which started in 1813 was ruled by a truly central government in the Hague. However, as Van Dijk[37] has pointed out, many vestiges of the old power structure still remained. Because the civil service was relatively small, many decisions were executed by city governments. Although the latter were now chosen by an electorate, the circle of voters was so restricted as to enable many of the traditional elite to keep their positions of power. It would take till after the mid of the nineteenth century for this situation to change.

The Rise of Liberalism and the Constitutional Change of 1848

Around the mid of the nineteenth century Western Europe was affected by the desire for a political change to liberalism. In some countries this led to revolution in 1848. In Austria and Germany initial successes of the revolutionaries were overturned later on.

In the Netherlands, increasing social misery and the sharper stratification of the late seventeenth century led in the early eighteenth century to a social order containing two estates: the poor and the very rich. One could expect that such a dichotomization of society must have provoked social revolution. However, it did not, at least not in the sense of a sudden and violent change of the Dutch social and political order. The 1848 change of political climate came about rather smoothly, without revolution. Besides, the shift to liberalism proved to be more lasting than in Austria and Germany.

The reasons why the Netherlands did not experience revolution in the mid nineteenth century seem to be on the one hand a lack of class consciousness in the lower classes and, on the other hand, the flexible response of the Dutch monarchy to the prevailing political attitude.

Several factors contributed to the lack of class consciousness. First of all, although from an outside view Dutch society seemed sharply dichotomized as to life-style and social distance, within the lower classes modest possibilities for upward and downward mobility still existed. However, under the conditions of a stagnating economy such mobility was feasible only for individuals and not for whole social groups[38].

Secondly, like in many European societies education legitimized the existing social order. Primary education was differentiated as to structure, teaching programs and educational objectives. Education for the common man was based on religious ideas and directed at imprinting obedience and deference to the higher ranks. Although the creation of worthy and useful members of society was an educational goal, lower class education was certainly not meant to improve chances of social mobility[39].

Further, next to education, poor relief functioned as a powerful instrument of social control. In the first half of the nineteenth century religious denominations dominated poor relief. Next came private institutions administered by the well-to-do. Public institutions controlled by city governments played only a minor role. The view prevailed that poverty came from laziness, wastefulness and sin or that it was the will of God. In any case, the poor had themselves to blame and not society. Thus the poor had, above all, to be rehabilitated by teaching thrift, patience, fear of God and deference to those of higher social rank[40]. To be eligible for relief, the poor should commit themselves to the mercy of those above. So, besides income criteria based on means tests, social compliance was a prerequisite to obtain relief. As an example, weekly church attendance was quite normal a requirement to be eligible for poor relief.

Therefore, poor relief was a matter of morality. It justified the existing social order and was consciously aimed at preventing uprisings of the poor. It legitimized the ideology of a God given estate society. Furthermore, the different denominations used poor relief as a mechanism of social control to prevent their followers from defection to competing groups[41].

If there was any sense of class consciousness in the Netherlands around 1850, it was vested in parts of the middle and upper classes. Dutch economists were influenced by English liberal economic theories. These new ideas found roots because they were in line with the change in politico-economic conditions of the time. All kinds of protective measures were removed, forcing entrepreneurs to be more competitive. The world economy improved so that, in general, economic opportunities increased in the Netherlands too. The new economic liberals regarded entrepreneurs as a natural ruling class, the promotors of social progress by economic growth. One of their main themes was the freeing of education from traditional religious bonds. Instead, education should teach skills and abilities that would be the basis of economic progress and of improvement of lower class living conditions[42].

Liberalism as a political movement came to the fore around 1830. In 1813, by the Treaty of Vienna, the Dutch King William I had succeeded in merging the Republic, Belgium and Luxembourg into one nation state. For different reasons, this newly constructed state met with strong opposition from both Belgian and Dutch leading circles. Despite broad opposition in both parts of the kingdom, the king persevered, even waging a small scale war against the Belgian separatist movement. In 1839 he had to give in. Belgium became an independent state and Luxembourg was divided up between Belgium and the Dutch king.

The Belgium issue is an important factor explaining why the Dutch monarchy gave such a flexible response to the demands of reform which culminated in the constitutional change of 1848. Over the years it had weakened the position of king William I. Since 1813 he had gradually alienated himself from the leading circles and had lost support by clinging to the union. Related to the latter was the growth of public debt which became more and more problematic to the extent that it urged to the liquidation of postponed debt in 1841 and to monetary reform in 1844. Reacting to a public campaign against his intention to marry a Roman-Catholic Belgium countess, in 1840 William I abdicated in favour of his son William II. This course of events does not demonstrate a very powerful position of the Dutch monarchy in the mid of the nineteenth century.

In 1840 the constitution was changed as a logical result of the separation, but it only did to the extent required on account of the loss of Belgium. The new king, less liberal minded than his father, appointed Van Hall, a representative of the traditional Amsterdam aristocratic elite to Minister of the Crown, therewith removing the conservatives of Holland from the opposition camp. From that moment on one could discern a governmental axis between conservative merchant elites of Amsterdam and the Hague[43].

In fact this meant that the traditional supremacy of the merchant elites of Holland was partly restored. In that province the liberal opposition had little chance "to win over or replace ruling opponents strongly entrenched in traditional power. In some parts of the provinces, on the other hand, the liberal opposition was able to mobilize bigger forces and make a real break-through regionally. In this sense, therefore, liberalism, during the forties of the last century, became largely a provincial force, aimed at the Amsterdam - The Hague axis, more clearly so since 1842"[44].

It is clear that Dutch liberalism had become much more than a political ideology fostering political and economic freedom. In the Dutch situation these two core elements of the liberal ideology implied breaking the power of the traditional conservative elites in Holland and therewith the emancipation of the other provinces. So liberalism meant among others an attack on the vestiges of traditional Dutch geographic pluralism. How was this battle won?

In the constitution of 1814, suffrage on the basis of property qualifications was already introduced. However, the centre of government power rested with the king. The major change in the direction of constitutional monarchy came in 1848, a year in which in many European countries the democratic creed reached a peak. In the Netherlands, riots took place in Amsterdam and the Hague. The king, affected by these events and those abroad, on advise of leading politicians changed overnight from conservatism to liberalism. In November 1848 the constitution was changed. From now on, members of the Lower Chamber had to be directly elected by a district census system requiring an absolute majority on the first ballot with a second ballot between the two top candidates in case no candidate won in the first round[45]. The King's ministers became responsible only to parliament. Thorbecke, a leading liberal, became the first prime-minister under the new constitution.

Shifting Alliances After 1850

The change of the constitution did, however, not imply the complete victory of liberalism. In Holland the remnants of the republican oligarchies still dominated the local power structures. Although they did not fundamentally contest the idea of a unitary state, they longed for the autonomy they had in the glorious days of the Republic. Such feelings were often mingled with a sense of Protestantism as the true belief to restore the nation. However, these traditional elites were also affected by the democratic sweep of the 1840's, at least to the extent that they generally accepted the liberal constitution of 1848. For them this constitution meant a possibility to increase their influence on the national level without losing the power they had on the local level. The major source of conflict between conservatives and liberals, however, came from the fact that the former wanted to retain their positions of influence while the latter wished to bring new kinds of people into the traditional elites.

In 1850 the outcome of the elections was very favourable to the liberals. Out of the 68 members of the Lower Chamber they numbered 44. So the majority supported Thorbecke's government until the beginning of 1853 when the power base of local conservative bosses, whose support had substantially contributed to the liberal electoral success, was affected. When they withdrew their support, the liberal government collapsed[46]. The reasons behind this loss of support will be analysed below.

A second element of the liberal victory of 1850 was the fact that the Roman-Catholic provinces of Limburg and Brabant overwhelmingly voted liberal. During the period of French dominance the civil rights of Roman-Catholics had been formally restored. For the clergy in Brabant and Limburg supporting the liberals seemed to be the best guarantee to rebuild the Catholic church institutions, especially the schools. In return for their support, in 1853 the liberal government allowed the restoration of the Roman-Catholic episcopal hierarchy.

I agree with Van Holthoon[47] that the restoration of the episcopal hierarchy was an important factor contributing to pillarization. Since the Revolt in the sixteenth century, Dutch Roman-Catholics were directly controlled by Rome as a mission. Because of the distance between Rome and the Netherlands and the difficulties of transportation and communication inherent to that distance, Roman-Catholic leadership in the Netherlands had been weak. The restoration of the episcopal hierarchy meant the introduction, for the first time since the Revolt, of a strong national Roman-Catholic authority that could take decisions in the light of national needs. Secondly, it fostered within the Roman-Catholic community feelings of group consciousness. Further, as a reaction, it provoked the same sort of feelings within the competing Protestant groups, thus articulating the differences between Protestantism and Roman-Catholicism in the Netherlands.

The restoration of the episcopal hierarchy led to a wave of anti-papist feelings in the non-Catholic regions of the country. Many conservative Protestants, who had supported the liberals, withdrew their support. As the king showed signs of choosing the side of the protest movement, Thorbecke resigned.

At first glance, it seems as if the conservatives who had supported the liberals rebelled because they wanted to defend the Protestant position against the Roman-Catholics. However, the real cause of their revolt against Thorbecke was probably the attempt of the latter to change the power base of the local Protestant elites. Their defense was partly the mobilization of broad layers of the population by raising anti-papist sentiments. The mobilization of such sentiments had been a political strategy since the very beginning of the Republic. In what way then, did Thorbecke attack the power base of the local elites?

Since the end of the nineteenth century the poverty problem had increased enormously in the Netherlands and had become an issue of the highest priority. Poor relief was locally organized and gave local elites a powerful instrument of social control over the lower strata. As the number of those eligible for poor relief increased, the need for interference by the national government became more pressing. In 1800 a bill was introduced aimed at the centralization of poor relief at the national level. This bill met fierce resistance from the cities and the religious organizations. The constitutions of 1814 and 1815 did not prescribe poor relief arrangements. In 1818, however, the law on the "domicilie of onderstand" was enacted. It prescribed that the poor could only get relief from a public local institution in their place of birth or in the municipality where they had lived for four successive years and had paid taxes. This law led to a shifting around of the poor between municipalities trying to economize on the increasing burden of relief. The liberal constitution of 1848 was the first that prescribed poor relief arrangements. In 1851 Thorbecke introduced a bill that made poor relief the ultimate responsibility of central government. This implied a substantial weakening of local social control and aroused strong opposition, especially from the churches that regarded the bill as an unacceptable interference in their affairs[48].

When after the fall of Thorbecke's cabinet the conservatives took over, in 1854 the first poor relief act was enacted stating that the municipalities had to refrain from poor relief and only in cases of excessive need should public action be taken. Highest priority was given to the role of religious and private institutions.

Another factor contributing to the fall of the liberal government in 1853 may have been the introduction of the Local Government Act in 1851. This act restricted the autonomy of city administration and made local communities much more dependent on national government. Consequently it affected the power of local elites substantially.

Since the fall of Thorbecke's government the support of the Roman-Catholics for the liberals decreased gradually and ultimately led to a coalition of Roman-Catholics and Protestants against liberals. There are several reasons for these shifts in coalitions.

Firstly, before the introduction of the Local Government Act, the local Roman-Catholic elites in the south had ample opportunity to handle their affairs as they saw fit, even under the reign of a Staten-Generaal dominated by Protestants. The restriction of their autonomy at the local level, made Roman-Catholic organization at the national level more pressing in order to promote the Roman-Catholic interest.

Secondly, the restoration of the Roman-Catholic episcopal hierarchy and the resulting group consciousness of the Roman-Catholics created opportunities for Roman-Catholic social and political organizations. The Roman-Catholic hierarchy could formulate political policies that could be effectively pursued by the local clergy, whereas Dutch Roman-Catholics had previously operated from a position of second class citizenship. Moreover, as the primary aim of the Roman-Catholics - e.g. equal opportunity for their religious organizations - had been realized, the ideological differences between them and the liberals came more to the fore. In general, these differences can be summarized as the liberal promotion of modernization by secularization and the Roman-Catholic resistance to interference in affairs which they regarded as fundamentally the domain of the church. Their opposition to liberal attempts to secularize poor relief had been the first clear sign of these essential differences of opinion.

Further, perhaps as a result of the newly acquired status, within the Roman-Catholic camp a radicalization took place. Since 1853 an orthodox line became gradually dominant. It preached fierce resistance against any kind of modernism and demanded unconditional loyalty to Rome's authority[49]. For the time being, however, this enlarged Roman-Catholic consciousness did not promote militant opposition as the Roman-Catholics were cautious not to jeopardize their religious freedom.

Finally, as Daalder[50] has stated, the liberal outlook itself changed. It became more anticlerical. Partly because of events abroad such as the German Kulturkampf, partly because the need of modernization brought many liberal politicians in direct conflict with the different religious elites.

The Rise of Protestant Organization

The course of events during the nineteenth century demonstrated that the Protestant part of the nation was internally divided. The content of the conflicts changed over time. Because of their complicated nature, we cannot enter into detail here. However, in every conflict the competing parties were orthodox and non-orthodox. In the course of the nineteenth century two new Protestant churches were born out of the Dutch Reformed Church: the Christian Re-Reformed church and the Re-Reformed Church. It started in 1816 with the resistance of the orthodox against a new regulation for the Dutch Reformed Church. In 1834, internal conflict led to the "Afscheiding" (separation), but it was not until 1869 that a new Protestant Church, the Christian Re-reformed Church, was established.

A second important split took place in 1886, the "Doleantie". This orthodox reaction to the dominance of modernist theologians was led by Abraham Kuyper, who is generally considered the architect of the Protestant pillar. For Kuyper, modernism in theology and in daily life were two sides of the same coin. He believed that modernism could only be successfully fought if his Re-Reformed Church was able to defend and protect the identity of the Re-Reformed[51]. Instead of restricting himself to sermons, Kuyper's line of defense was the cultural isolation of his followers. By founding a political party, a press, schools and even a university, the Re-Reformed were "protected" against alien values, norms and ideas. At the same time, in this way feelings of group consciousness within the Re-Reformed community were fostered.

Kuyper's political actions were primarily aimed at the liberals who for him were the embodiment of atheism. Later on, Kuyper's founded the Anti-Revolutionary Party which fought liberalism and rising socialism. For Kuyper, political mobilization was a means to achieve governmental power. His record was impressive: he established a newspaper (1872), an Anti-School Law League (1872), a massive petition movement against a new Liberal School Bill (1878), the Anti-Revolutionary Party (1879) and the Protestant Free University in Amsterdam (1880).

> "If one considers that the proportion of the population eligible for mobilization cannot have been larger than 16 to 17 percent of the population, then this was a remarkable achievement"[52].

Kuyper's mobilization strategy succeeded. The Anti-Revolutionaries gradually replaced the Conservatives as the leading opposition party. Conservative seats in the lower house declined from 19 in 1868 to 1 in 1888. During the same years the Anti-Revolutionaries rose from 7 to 27[53]. Strange as it seems, this very success led to a split of the Protestant political force. The leader of the Anti-Revolutionary parliamentary group insisted on more freedom of action for the Anti-Revolutionary parliamentary representatives to be able to work out the necessary compromises. On this issue a break-up of the party took place which led in 1908 to the formal establishment of a second orthodox Protestant Party, the Christian Historical Union (CHU).

The strategies of Abraham Kuyper had profound influence on the shape of the new Dutch pluralism based on ideological differences. Firstly, Kuyper believed it the common duty of all religious people to fight modernism and atheism as represented by liberalism and, later on, by socialism. Although the theological differences between the churches should not be forgotten, they had a common enemy. Thus an ideological foundation was laid for the potential co-operation of different religious denominations while respecting the distinctions between the coalition partners.

Secondly, the acceptance of other denominations as battle companions included also the Roman-Catholic part of the nation. This was a new development in Dutch history. From the days of the Republic, the Roman-Catholics had repeatedly been used by the elites as scapegoats, a hidden enemy that could be used to mobilize public support. In the years after 1850 the Roman-Catholics were regarded by the conservative Protestants as the filthy helpers of the liberal anti-christ. Because of liberal policies, since the mid of the nineteenth century Roman-Catholic and orthodox Protestant interests coincided more and more. Although anti-papism did not wither away, the Roman-Catholics came to be the potential allies of the orthodox Protestants. As we shall see below, at the end of the nineteenth century this potential alliance was made effective and dominated Dutch politics till the 1960's.

Thirdly, already in the Republic, Dutch elites had found it useful to attract mass support as a means to sort out their disputes. Now Kuyper did so in a systematic and structured way. He did not mobilize the orthodox Protestants around temporary issues, but established organizations that were articulated to the strategic domains of social and political life, especially those regarding socialization. These organizations preserved public support over time and made it independent of the issues of the day.

Fourthly, the rise of the Anti-Revolutionaries should be seen as the rise of a new elite within the Protestant part of the nation. Contrary to the old elites, Kuyper's men stood for the protection of the common Protestant people against what they regarded the evils of modern times. Because of this appeal to the common people, they were able to erase the power of the traditional Protestant orthodox elite. However, in due time they themselves formed an oligarchy too. From his analysis of the Anti-Revolutionary elites, Kuiper[54] concluded that within this movement there was a hard core whose members took key positions in church, politics and universities. Sons and relatives had a strong chance of getting similar positions. Further, the organizations belonging to the Anti-Revolutionary movement were connected by interlocking directorates.

In the fifth place, although Kuyper's movement had started as an orthodox Protestant organization linked to the Re-Reformed church, his political party embraced not only the Re-Reformed but also many Dutch Reformed people. This was the logical consequence of his ideas on the acceptance of all religious allies against modernism. It implied that the political support and influence of the Anti-Revolutionaries went far beyond the number of Re-reformed people.

Finally, the formation of the Anti-Revolutionary movement and the success of its strategy later on acted as an example to other ideological groups in Dutch society, especially the Roman-Catholics.

The School Issue

Among historians, sociologist, and political scientists, the school issue is generally considered as giving momentum to the process of Dutch pillarization. We would like to emphasize two aspects of the school issue which influenced pillarization in the twentieth century and which are directly related to the social ethics of Dutch segmented pluralism: the principle of proportionality and the mechanisms by which this social problem was solved. Before plunging into the analysis on these points, we have to describe the content of the issue itself.

Before the middle of the nineteenth century, Dutch primary education was aimed to reinforce values that legitimized the prevailing social order. The estate society of the time was religiously sanctioned by the statement: "It was God's wish that there be estates". Thus education did not promote social mobility, it rather kept people in their proper places.

By the second half of the nineteenth century, it became obvious that the economy required manpower that had not only been indoctrinated by morality and obedience, but had also been taught functional skills. Besides, the economic changes of the mid nineteenth century inspired the striving of the lower middle class for the social advancement of their children. Their hope was for better and cheaper education provided by state schools. This demand was taken over by the liberals.

The orthodox Protestants of Abraham Kuyper thought to neutralize the effects of modernization by promoting special Protestant schools. Although the Roman-Catholics had won the freedom to organize their church, the liberals were not willing to provide the opportunities for Roman-Catholic education by congregational schools[55]. On the latter issue the orthodox Protestants and the Liberals stood united. This was certainly one of the factors that drove Roman-Catholics away from liberalism. Although the Protestants were in first instance not in favour of Roman-Catholic education, gradually it became clear that without the support of the Roman-Catholics the threat to Protestant education could not be countered.

The school issue concerned the relationship between church and state. In 1806 a law on education was enacted that gave public education an almost absolute prerogative over denominational schools. It was the logical result of the French domination. In 1857 a bill was introduced that gave better chances to denominational private schools. The establishment of the latter was now allowed, however, without any support of public money. Further, the act took a general Christian view as the basis of primary education. In first instance, the Roman-Catholics supported this act. It implied that, in the Roman-Catholic provinces of Brabant and Limburg education became in fact Roman-Catholic oriented. In the other provinces, however, the Roman-Catholics felt that they had to establish schools of a Roman-Catholic orientation. Roman-Catholic support eroded further with the papal encyclical "Syllabus Errorum" (1864) which inspired the Dutch bishops to reject all non-Catholic education. The Protestants, on the other hand, wanted public Protestant education. But that was unacceptable for the Roman-Catholics. Since the Protestants realized that they could never reach their aims without the help of the Roman-Catholics, they compromised by campaigning for fully state subsidized denominational education.

In 1878 the liberal Cabinet introduced a bill which stated that no public money should go to private schools. The bill further proposed an improvement of educational quality through increased government control. This put the spark to the finder. While the issue dragged on, the government made several concessions and a limited degree of public support for private schools was introduced. This did not satisfy Roman-Catholics and Protestants who mobilized the masses to stress their point. They organized a petition movement and gathered 300,000 signatures, which is considerable in relation to the total Dutch population and the numbers of Protestants and Roman-Catholics of the time. The petition asked the king not to sign the bill.

In 1889 limited public support for denominational schools was accomplished but only for one third of the total costs of these schools. Thus again the demands of Protestants and Roman-Catholics were not met. In 1904 and 1908 state subsidization was expanded to denominational secondary and higher education, but again total equalization of public and private education was denied.

Around 1910 the political situation was very tense. Next to the school issue, the social issue and the issue of suffrage divided the country. In 1913, on the eve of the First World War, the Cabinet appealed for a compromise between ideological differences. In December a committee was formed that embraced the leaders of all the political parties and their educational experts. Without any modification, the committee's proposals were taken over by parliament. It was decided that all schools, public and private, were to get public support in proportion to their enrollments. This principle of equalization was stated in the constitution of 1917[56].

The solution of the school issue was the start of the Dutch pacification policy. This concept is used by Lijphart[57] to explain the phenomenon of a pillarized democracy that can maintain its stability thanks to a set of specific rules. According to Lijphart these rules emerged around 1917. From then on, pillarization increased because the pillars sharpened their institutional differences. At the same time the overarching cooperation of the individual pillar elites was strengthened, reaching a peak at the end of the 1950's. In chapter V I shall analyse pillarization and its rules by proposing a model for the 1950's. For the moment I will present a preliminary analysis of the ways in which the school-issue was solved.

First of all, as Lijphart[58] remarked, the issue was solved entirely by the elites behind closed doors. To get the necessary leverage, however, the masses were mobilized in advance to create pressure. As we have seen, this was the way Dutch elites had settled their controversies in the Republic. The difference between the Republican era and the early twentieth century was the degree of institutionalization of the popular support. In the Republic it was more diffused and fluid depending on the issue at hand. The rise of religion in the nineteenth century as the new criterion for Dutch segmentation led to an institutionalization that was frozen along denominational lines. The equalization of public and private education enabled the denominations to install their norms and values into their offspring. By this, loyalty for the future was ensured. Thus, while educational segmentation in the Netherlands was the result of the process of pillarization as it took place in the late nineteenth century, at the same time it contributed to the progress of pillarization in the twentieth century.

Secondly, all parties were represented in the committee that settled the school issue, no party being excluded. As no party could reach a political majority the commitment of all was essential.

Thirdly, the essence of the settlement was the principle of proportionality. In 1917 this principle got a much wider application in Dutch society. In the constitution of that year the district system was changed into a system of proportional representation in parliament. In the twentieth century gradually every kind of social organization eligible for state subsidy was subsidized proportionally to the size of its client group. Further, representation of social organizations in overarching umbrella organizations was also determined proportionally. Thus political, social and cultural power became gradually allocated by the principle of proportionality. It follows that elites had a big stake in holding the loyalty of their clients as the latter directly affected their relative power versus competing elites.

The Rise of Roman-Catholic Organization

In his important work on pillarization Stuurman concluded that the school issue implied the start of an organizational process that molded the Anti-Revolutionaries into a pillar. According to Stuurman the contribution of the school issue to the pillarized organization of the Roman-Catholics has been much more limited, because at the time the Roman-Catholics had not yet a "modern" association or party and because many of its bourgeoisie had still a strong orientation to liberalism.

Among the Dutch Roman-Catholics involvement in politics was much more restricted to a small social and clerical elite[59].

There is a certain paradox in the development of Roman-Catholic organization. On the one hand, it started later than Protestant organization. On the other hand, once matured, it was more homogeneous and effective than that of other ideological blocks. In my view this has rather to do with the differences regarding the historical backgrounds of Protestants and Roman-Catholics than with the supposedly smaller importance of the school issue for the Dutch Roman-Catholics. As to the latter, it should be remembered that the papal encyclical "Quanta Cura" (1864) rejected neutral public education, a line of argument that was taken over by the Dutch bishops. Ideologically Protestants and Roman-Catholics were no doubt equally against the liberal attempts to reform education. However, they differed as to the extent in which each was capable expressing feelings of discontent.

Two conditions favoured the success of the Anti-Revolutionary mobilization strategy. The Anti-Revolutionaries had their roots in the Protestant elites that had ruled the Netherlands since the Republic. In first instance they differed from the conservative elites that, in their view, had become too latitudinarian. This Anti-Revolutionary grievance referred both to theological issues as well as to worldly affairs, especially with regard to the increased liberal influence in society. Thus the rise of the Anti-Revolutionaries was strongly related to the separatist movements within the Dutch Reformed church that developed in the course of the nineteenth century. It implied that as a separatist movement the Anti-Revolutionaries had available a religious organizational network from which further organization could be developed.

The Roman-Catholics on the other hand, had a recent past as second rate citizens. They were in the process of rebuilding their church and felt that they had to be careful not to provoke the traditional anti-papist sentiments that could frustrate their newly won religious and political freedom. So the school issue arose too early for the Catholics to develop their organizational capacities fully. Further, ideological differences within the Roman-Catholic community did not express themselves in organizational splits.

The second factor has to do with differences between Roman-Catholic and Protestant church organization in general and their respective traditions in the Netherlands in particular. Protestant church organization has a more democratic nature than Roman-Catholic which is more hierarchically structured. The Anti-Revolutionary movement was also rather heterogeneous, embracing different points of view within Protestantism. As a repressed minority the Roman-Catholics had developed a strong sense of group consciousness and a particular subculture[60]. Thus, once the Roman-Catholic elites had decided to organize their rank and file, their organizations were more coherent and effective than those of the Protestants.

Roman-Catholic organization matured in the first decades of the twentieth century, whereas the Protestants had reached that stage already at the end of the nineteenth century. Two alternative hypotheses explain the growth of Roman-Catholic organization. The first is that it was three hundred years of repression which provided the Roman-Catholics with the need to catch up with other groups in society. Goddijn[61] is the main representative of this argument. Van Heek[62], on the other hand, explained the success of Roman-Catholic organization by the urge to control and christianize society the Roman-Catholic way. A strive for dominance manifesting itself in an exceptional high birthrate.

I agree with Hendriks[63] that both hypotheses lack a time perspective. Hendriks discerns four phases of Roman-Catholic organizational development. The first phase lasted from about 1853 till the end of the nineteenth century. Then the Roman-Catholic aim was defensive - a realization of equal rights. Roman-Catholic elites tried to reach that aim by petionnements, addresses, etc., but not by the establishment of political or social organizations outside the domain of the church itself[64].

In the second phase Roman-Catholic aims changed and were directed at the creation of power. All kind of Roman-Catholic organizations were established, including education, politics, trade-unions, sport, the arts and, later, broadcasting. However, these organizations were inner-directed, providing the common Roman-Catholic with his own social and cultural world. Highlights in this phase were certainly the formation of a Roman-Catholic political party in 1896 and the foundation of a Roman-Catholic university at Nijmegen (1926).

The third phase covers the inter-war period. Roman-Catholic feelings of inferiority were substituted by feelings of superiority. Roman-Catholics had become the most self-conscious part of the nation using a well oiled organizational machine to articulate their interests. Finally, in the fourth phase, starting after the Second World War, a secularization of the Roman-Catholics took place leading to an erosion of their organizational structure.

Summarizing, regarding the issue of the differences between the Protestant and Roman-Catholic organizational networks, two factors of explanation can be identified. As to the difference in time-lag between both groups, in first instance the Roman-Catholics felt that they had to avoid any action that could be interpreted as a provocation. They had to manoeuvre from a position of second class citizenship. The Anti-Revolutionaries were much more closely linked to the dominant elites and culture.

Secondly, as to the difference in organizational coherence, the Roman-Catholics were in an advantageous position. Since 1853, the church had provided a powerful hierarchical authority structure. As to values and beliefs, the Roman-Catholics were more homogeneous than the Anti-Revolutionaries. Therefore, the Roman-Catholic elites could more easily control their rank and file by the threat of sanctions derived from the religious domain. In fact, there are many examples of the Dutch bishops threatening to withhold the sacraments to those disloyal to supposed Roman-Catholic interests.

The Rise of a Segmented Working Class

Working class conditions deteriorated in the second half of the nineteenth century. This promoted the rise or revival of ideas with regard to a better future. Van Tijn[65] discerns three complexes of ideas, or reflexes, two of which are important for understanding the segmentation of the Dutch working class. The first reflex, the socialist one, did not accept the capitalist order. Its adherents believed that the human dignity of labour could only be achieved by the organization of the working class aimed at the creation of a new social an economic order. Although Van Tijn labels these ideas as socialist, many differences of opinion, ranging from anarchism and syndicalism to social-democracy can be distinguished within this field[66].

The second reflex was progressive liberal. Modern capitalist development is accepted as inevitable, only the negative side-effects have to be compensated for by the cooperation of all social classes based on mutual respect. This liberal oriented ideology prooved abortive in the Dutch situation[67].

The third reflex is traditional and reactionary. Labour should be revalued, but within the framework of an estate society according to God's will. The respect to which each estate is entitled should be achieved by a re-christianization of society. As Van Tijn[68] notes, this was a protest against a rational money economy, a nostalgia to a past that had never existed. It is an anti-modern working class reaction based on the religious ideas of both Protestantism and Roman-Catholicism.

How did these ideas affect the way Dutch working class organization developed? The phenomenon of the trade union in its contemporary meaning arose in the Netherlands only about the beginning of the twentieth century. There are several reasons why working class organization did not start earlier. In the Napoleonic era the French introduced a ban on the coalition of employees to organize for higher wages. In 1872 this coalition ban was lifted. Secondly, the enlargement of scale as an aspect of modern capitalist development set in relatively late. Until the latter half of the nineteenth century most production took place in small units where employer and employee directly interacted in a patron-client oriented way. Finally, throughout the nineteenth century religious legitimation of the estate society prevented Protestant workers from developing some sense of class consciousness. For the Protestant part of the population in the first instance the issue was rather a struggle against modernism. In this battle employers and employees stood united, thus blurring the contradiction between the classes.

In the middle of the nineteenth century some labour organizations already existed. However, they were aimed at the provision of mutual insurance and the organization of festivities. Management usually visited the gatherings as respected guests[69], which proves that these early local unions cannot be regarded as an expression of class struggle. They were rather manifestations of the estate ideology that prescribed mutual respect and responsibility within and between the estates. As far as class contradiction came to the fore, it was in the few large scale industries and in the countryside. Although in the latter some strike tradition had settled, countryside strikes had most of the time a wildcat character[70].

As both the Protestant and the Roman-Catholic ideology were inimical to working class organization, it is not surprising that the first pioneers of national trade-unionism were socialists and atheists. In 1871 the Algemeen Nederlandsch Werklieden Verbond (ANWV) was founded. It had a social-liberal signature, but also included members of the first International. In 1876 the Protestant Amsterdam members left the ANWV and founded the Protestant local union Patrimonium that became national in 1880. Its aims were to preserve Sunday observance and to prevent neutral state education and social legislation because the latter was believed to affect the rights of employers and parents. How strongly Patrimonium was connected with the Protestant denomination became manifest in 1886. In that year the separation movement within the Protestant church led to the split of Patrimonium into the Re-Reformed Patrimonium and the Dutch Reformed oriented Christelijk Nationale Werkmansbond.

In the meantime the non-denominational ANWV experienced all kinds of internal conflicts regarding differences of opinion between syndicalists and those in favour of a more social-democratic attitude. Finally, in 1906 the trade-union federation Nederlandsch Verbond van Vakvereenigingen (NVV) was founded. It wanted to be ideologically and politically neutral and open to all kind of workers. In fact, however, the NVV that became the largest non-denominational trade-union federation, has always been closely related to the Dutch social-democratic movement.

Similar to Patrimonium, the influence of religious values and attitudes is clearly recognizable in nineteenth century Dutch Roman-Catholic labour organization. The Nederlandsch Roomsch-Katholieke Volksbond (founded 1888) had formulated as one of its major aims the safeguarding of the labour estate and petty bourgeoisie from the socialist errors of the time. This organization was led by the Roman-Catholic upper class including many employers. In Enschede, the center of textile industry, the success of the general unions stimulated the Roman-Catholic clergy to establish a Roman-Catholic Union aimed "to promote the religious and social interests of the Roman-Catholic labourers and to safeguard them from anti-Christian socialist influences"[71]. The organization did not succeed in attracting many Roman-Catholic textile workers. In 1895 Unitas was founded, an interdenominational trade union for textile workers with separate departments for Protestant and Roman-Catholic workers. Similar initiatives developed in the mining areas of Limburg.

The Roman-Catholic clergy was now confronted with the question whether or not it wanted Roman-Catholic workers associated with Protestants. The lower clergy favoured interdenominational labour organization. In 1906, however, the Dutch bishops took the official point of view that Roman-Catholic workers should be organized in Roman-Catholic organizations. This implied the end of the interdenominational labour movement[72]. After the turn of the century national Roman-Catholic unions arose. To control the religious life of the working class families, the clergy insisted on double membership. Besides, a spiritual councilor, whose will was law, participated in all governing bodies, both on the local and on the national level[73].

The history of Roman-Catholic trade-unionism in the Netherlands shows the great influence of the higher clergy and the obedience of the rank and file. Windmuller[74] explains this phenomenon by pointing at Roman-Catholic isolation and discrimination since the seventeenth century, which gave the Roman-Catholic clergy disproportional authority in worldly affairs, at least compared with other West European countries. Besides, the presence of militant Calvinism had eventually promoted among Roman-Catholics a militant frontier mentality. As a result the Roman-Catholics in the Netherlands had become more Roman-Catholic than those in Rome.

In 1903 the already existing cleavage within the working class was sharpened. In that year the railway workers went on a wildcat strike. The socialist union movement had to take over as it did not succeed in dampening the strike. Surprisingly, the management of the two railway companies gave in very soon which was regarded as a loss of face for the socialists and as a victory for the syndicalists who had supported the strike from its very beginning. The latter now came into a revolutionary mood. The government, in which the Anti-Revolutionary Party participated, felt the victory of the strikers as a threat to the established social order and introduced a bill that banned strikes from public utilities and the railways.

As a reaction socialists and syndicalists founded a Committee of Defense. The syndicalists wanted an immediate general strike on the national level. The socialists, on the other hand, were afraid to provoke the government. Eventually, both groups compromised. It was decided to proclaim a railway strike at a favourable moment, but a general strike was to be organized only if it turned out to be a prerequisite to the success of the railway strike.

To support the government, the Protestants founded two railway unions, one for every company. The Roman-Catholics installed a committee to fight revolutionary agitation. On April 2nd the bill banning strikes was accepted by the government. On April 8th a general strike was proclaimed but it very soon turned out to be a complete failure.

As Windmuller[75] writes, the railway strikes of 1903 had a profound influence in moulding twentieth century Dutch labour relations. They sharpened the cleavage between syndicalists and socialists, and between non-denominational unions on the hand and denominational unions on the other hand. The latter now had proof that their distrust of the socialists as disloyal to throne, nation and the national institutions had been justified. But the railway strikes made clear something else. The denominational unions were not able to act independently from the political parties to whom they were connected. They had proved to be applicable as a powerful tool to control large parts of the Dutch working class.

In the concluding chapter of his thesis on the origins of pillarization Stuurman[76] writes:

> "The thesis that pillarization was a counterattack of the churches and the confessional bourgeoisie against the progressive movements of the last quarter of the nineteenth century, is easy to defend. Confessional labour organization was seriously attempted in those periods and in those regions where socialists developed some strength. And the fight against the socialists was practically an official goal of the denominational organizations".

Something should be added to Stuurman's thesis. The first Protestant unions were a defense against modernity in general. Patrimonium had as one of its primary aims the prevention of neutral state education. According to the Protestant view, the interests of the working class could be served best by a re-integration of the estate society. In first instance, the liberals were the supposed enemies to this ideal. The shift to socialism as the main enemy came indeed about only in the last decades of the nineteenth century when socialism rose as a reaction to the same modernism that was reacted against by Protestantism. For the Roman-Catholic labour movement the picture is different. As Roman-Catholic organization in general started later than Protestant organization, the Roman-Catholic action was directly confronted with the rise of socialism. In any case, the social and political developments of the late nineteenth century led to a segmentation of the Dutch working class which has not been ended even in our days.

Conclusions

This chapter analysed the relationship between modernization of the Netherlands and the associated institutional change from the sixteenth to the twentieth century. The conclusions can be broadly summarized as follows.

1. Because in the middle ages the Low Countries were only loosely integrated within the German empire, a strong sense of autonomy developed. The revolt against the Spanish Habsburg monarchy, which marks the origin of the Dutch state, was above all a conservative reaction to a modern Spanish policy of centralization. From this one can understand that the Revolt produced a Republic characterized by a highly de-centralized power structure. Till the nineteenth century Dutch pluralism was primarily of a geographic nature. Class and religious solidarity played only a minor role. The citizens of the Republic identified themselves above all with their cities and provinces.

2. The negative effects of geographic pluralism could easily have endangered the very existence of the Dutch state. They were overcome by internal stabilizing forces - the role of the House of Orange, the fact that the Calvinist church was organized on the national level, blood relationships between members of the elite and the interchange of offices - , by the development of a culture of "living-apart-together", including rules which made compromise possible, and by external factors. The latter had to do with the geo-political conditions of the Netherlands. They led to a situation in which it was in the interest of all the constituent parts of the federal Republic to maintain unity vis-à-vis the outside world.
3. As long as entrepôt trade flourished, geographic pluralism yielded enough flexibility for the Dutch economy to prosper. When, at the end of the seventeenth century, entrepôt trade started to decline, the de-centralized power structure prevented institutional innovations necessary for the pursuit of new economic activities that perhaps could have compensated for the weakening of the Replublic's economic position. Only in the beginning of the nineteenth century was geographic pluralism partly overcome due to the French occupation of the Netherlands.
4. When in 1813 the Netherlands had become a kingdom, liberalism presented itself as a modernizing force. It further weakened the power position of the traditional elites of Holland, and promoted institutional innovation and legal reform. The latter implied increasing centralization. Because of a flexible response by the Dutch monarchy which was weakened by the Belgium affair, in 1848 liberal constitutional reform could be enacted without revolution.
5. In the second half of the nineteenth century liberalism was fought by those Protestant elements which perceived it as the main representative of evil modernism. The former allies of the liberals, the Roman-Catholics, later joined this resistance movement. Their motives had to do with internal developments within the Roman-Catholic world church and with the loss of autonomy of the Roman-Catholic elites in the south due to legal reform promoted by the liberals.
6. The Protestant Anti-Revolutionaries, who were the pioneers of the nineteenth century anti-modernist movement, used methods to create leverage which originated from the Republican days. However, the mobilization of the masses was now structured and embedded in organizations. Thus as in the second half of the nineteenth century social differentiation took place as part of the more general process of Dutch modernization, newly established organizations were brought under the influence of religious groups, notably the Protestants and the Roman-Catholics. In this way, a triple structure of organizations could arise. For many social functions to be fulfilled, different organizations with different ideological outlooks were founded: public schools, Protestant schools, Roman-Catholic schools and so on.
7. The result of these developments was that Dutch society became divided in ideological blocks along vertical lines of solidarity. Each of these blocks claimed a high degree of autonomy or *sovereignty in its own circle*. Thus at the end of the nineteenth century traditional Dutch geographic pluralism had changed into ideological pluralism. Reviewing this historical process, two aspects are striking. Firstly, both forms of pluralism implied a high degree of de-centralism or autonomy of categories of the Dutch population. Secondly, in both cases the negative effects of pluralism were overcome by a culture of what I call "living-apart-togeth-

er". Around the school issue, rules arose that made compromise between the different ideological blocks possible. It is remarkable that those rules show such a close resemblance to those that were used in the Dutch Republic to solve issues of national importance.
8. The division of Dutch society in ideological blocks along vertical lines of solidarity implied a corresponding segmentation of the working class. Two further factors contributed to this process. Firstly, denominational organization was earlier than social-democratic. Thus, when at the end of the nineteenth century the socialist movement attempted to organize labour, large parts of the working class were already incorporated in the Protestant and Roman-Catholic blocks. Secondly, at the end of the nineteenth century and the beginning of the twentieth century the Dutch socialist movement was internally highly divided.
9. Although at the end of the nineteenth century Dutch religious forces turned against socialism, the organization of Protestants and Roman-Catholics came about primarily as resistance against the secularization of society which can be associated with modernization and which in the second half of that century was represented by liberalism.

CHAPTER V

DUTCH PILLARIZATION: THEORY AND PROCESS

Introduction

The previous chapter analysed long term changes in the plural character of Dutch society. According to the theoretical model of modernization presented, the changes of the Dutch social structure that came about in the second half of the nineteenth century, as well as the related shift from geographic to institutionalised ideological pluralism should be regarded as integral parts of the process of modernization in the Netherlands. The resulting segmentation of Dutch society in a Roman-Catholic, a Protestant, a Socialist and a Liberal bloc has become known as pillarization.

This chapter consists of two parts. Part one reviews the main definitions and explanations of pillarization as known in sociology and political science. It ends with the presentation of an ideal typical model of Dutch pillarization. Such a model is required to enable the measurement of pillarization as an empirical phenomenon.

In part two pillarization will be monitored by the analysis of time series of secondary data. A disadvantage of the choice for the construction of time series is that the fit between the theoretical and the empirical parts of the chapter is not perfect. Because of the inadequacy of available sources not all variables incorporated in the theory could be measured. I therefore decided to construct a limited number of time series that come as close as possible to the theoretical statements involved. I attempted to compensate the remaining lacks by the use of less systematic qualitative material. As a result the empirical part of the chapter is not a test of hypotheses in the strict methodological meaning. I therefore prefer to speak of *monitoring* pillarization.

The Concepts of Pillar and Pillarization

In the second half of the nineteenth century traditional Dutch pluralism took a new shape. Since the Republican era the pattern of social and political organization had changed from highly dispersed and fragmented to a society divided in four large blocs representing four ideologies: Protestantism, Catholicism, liberalism and socialism. Within each bloc subdivisions existed ranging from right to left wing, from orthodox to modernist.

Especially the Protestant bloc was rather heterogeneous. It embraced the two major Protestant denominations, the Dutch Reformed and the Re-Reformed. Even within these two separate categories adherents to somewhat different lines of theological interpretation could be distinguished. Next to the two major Protestant denominations, a few smaller independent congregations are usually reckoned as being part of the Protestant bloc.

In some cases these ideological subdivisions within Protestantism led to separate and independent political and social organization. Thus within Protestant education, schools could be oriented towards orthodox thinking or express a more liberal view. In parliament right wing Protestantism was represented by two (and since a few years by three) small parties that each had at most three members of parliament out of a total of 150. There always were subtle ideological differences between newspapers belonging to the same bloc.

Despite greater or lesser political and cultural heterogeneity within all four blocs, there seems to be a broad consensus among Dutch social and political scholars that they gradually came to represent the main lines of cleavage in Dutch society. As neither bloc could achieve a majority position, at the political level some cooperation was required in order to govern by coalition cabinets. Because of this combination of competition between the blocs and overarching cooperation at the national level they are called *pillars* (in Dutch: zuilen).

In 1940 the idea that Dutch society could be understood as divided into four pillars was first ventilated in a Dutch newspaper[1]. Only fifteen years later the phenomenon drew the attention of Dutch sociologists. They emphasized the connection between ideology, politics and social organization as characteristic of each bloc. In Dutch society every social function was thought to be exerted by ideologically diverse organizations or pillars[2]. Thus each of the four blocs had its own political party, socio-economic organizations, schools, broadcasting association, youth association and so on. These were formally or informally linked by interlocking directorates and could be regarded as more or less independent subsystems within the larger system of Dutch society. In this way all social activities were embedded in an ideological framework of organizations. Or, to put it differently, each of the four ideological blocs had its own organization for every social activity. The concept of pillar then referred to the interrelated network of organizations based on a shared ideology.

Table 5.1. Factual or Preferred Ratio of Friendship Relations within Ones Own Denominational Group as a Percentage of the Total Number of Friends

Author:	Catholic	Dutch Reformed	Re-Reformed	Others	No Denomination
Constandse[*]:					
factual	72	58	69		41
Gadourek[**]:					
preferred	59	19	35	7	0
Kruijt[***]:					
factual	20	16	26	n.a.	n.a.
preferred	31	16	31	n.a.	n.a.
Kuiper[****]:					
factual	68	68	71	36	46

Source : This table has been taken over from G. Dekker, 1965, p. 110. The sources used by Dekker are: [*]A.K. Constandse, 1958-1959, p. 72, [**]I. Gadourek, 1956, pp. 112-113, [***]J.P. Kruijt, 1957, p. 61, [****]G. Kuiper, 1954, pp. 62-63.

Pillarization also heavily affected the level of interaction in Dutch society. At the basis of Dutch society social distance between the pillars was considerable. According to researchers who investigated preference patterns of individual interaction, the Dutch had a strong inclination to interact with people belonging to their own ideological groups (see table 5.1.). This emphasizes the relationship between politics, social organization and individual identification. It is this very combination that makes up the essence of the concept of pillar.

One should keep in mind that the idea of Dutch society being a pillarized social system is a model that does not equal the realities of the empirical world. First of all, not every social function was pillarized. Some large organizations like the Dutch touring club ANWB and the consumer organizations were never related to any pillar. Secondly, I mentioned that the pillars were not homogeneous to the extent the model suggests. And third, not all Dutch citizens felt themselves affiliated to one of the pillars. Especially large parts of the intelligentsia and the artists took an independent position.

The fact that pillarization is a matter of degree would make it useful to have one indicator by which the extent of pillarization could be expressed. This would enable us to compare levels and degrees of pillarization in time and between countries. Such a measure has not yet been developed, probably because of the enormous amount of work and difficulties its construction would involve. For every year it would require a complete inventory of all organizations, voting behaviour as well as a measurement of the extent to which individual behaviour is guided by pillars' ideologies. To my knowledge Kruijt and Goddijn[3] are the only researchers who undertook such an attempt. They restricted themselves to organizations and investigated these for one specific year. Even then, they had to abstain from a complete review of all organizations because of difficulties in obtaining the data required.

Many Dutch sociologists focussed their analyses on the functioning of separate pillars instead of on the mechanisms underlying the social system as a whole. For many, the peculiarity of Dutch pillarization lay in the organizational relations between denominations and profane social activities[4]. In the early days of theorizing on pillarization J.P. Kruijt[5] presented a model consisting of concentric circles. The church is positioned in the inner circle. Functions closely related to the church, like welfare and education, are located in the second circle. In the outer circle more worldly activities like interest articulation are ordered. According to Kruijt the strain between social activity and its legitimation grows as the distance to the nucleus of the set of circles increases[6].

In the 1960's political scientists started to engage in pillarization research. Their interests, however, differed from those of the sociologists. Political science concentrated not so much on particular pillars, but rather on the political mechanisms underlying pillarized society as a system. In 1968 Arend Lijphart set the tone with his book *The Politics of Accomodation: Pluralism and Democracy in the Netherlands*. His main problem was to explain how ideologically divided Dutch society achieved political stability under conditions of democracy. Later Lijphart and his successors broadened the theory on Dutch society to plural societies in general. This line of thought has become known as the consociational democracy school.

The definition of the concept of pillar as a network of organizations centered around a religious denomination gives rise to the question how to evaluate the position of social-democrats and liberals. In the preceding chapter I described the history of the process of pillarization in the Netherlands. We observed a time order in this process: the Anti-Revolutionaries of Abraham Kuyper were the first to set up an organizational network to mobilize the rank and file and to prevent them to defect to rising socialism. A few years later the Roman-Catholics started to pursue the same strategies. Around the turn of the century the social-democrats responded in the same way. Thus, although not organized around a denomination, from about 1900 till the 1970's Dutch social-democratic organization looked pretty similar to that of Protestants and Catholics. From this point of view the social-democrats could be considered a separate pillar.

On the other hand, the Protestant and Roman-Catholic pillar both recruited their members from all social classes, while the followers of the social-democratic pillar were primarily working class. According to a current definition of pillarization we preliminarily presented before, pillarization is the process by which society is divided into vertical columns rather than into horizontal layers. As such, the concept of pillarized society is opposed to that of class society. One of the meanings of denominational pillarization was to counteract class antagonism by appealing to a common denominational background. On the other hand, before the 1920's Social-Democratic organization regarded class struggle as its primary aim[7] and was thus opposed to denominational pillarization[8]. So at first sight it does not seem logical to regard social-democratic organization as a pillar. Following this argument, Stuurman[9] rejects the idea of a Socialist pillar because it is based on a class interest. But then the whole idea of pillarized society becomes doubtful since this very idea comprises society as a whole and not just as two parts of it.

There is still another difference between Protestant and Roman-Catholic ideologies on the one hand and those of social-democrats and liberals on the other. The latter are both forces originating from the Enlightenment. They stand for a society that does not discriminate as to ideological backgrounds. From this point of view, one could imagine that social-democrats and liberals would have resisted pillarization.

Some find a solution by amalgamating socialism and liberalism into one general secular pillar. Dutch society can then be subdivided in three pillars: Protestant, Roman-Catholic and secular. This position is taken by Lijphart[10]. He gives two arguments for this classification. Firstly, all three pillars have their historical roots in Dutch history. Both liberalism and socialism can be considered as offspring of the humanist tradition. Secondly, all three pillars lived *in worlds of their own, separated from others*[11].

Lijphart's first argument does not seem valid to me. If one suggests to add up socialism and liberalism to one general pillar because of their common relationship to humanism, one could likewise propose to merge Protestantism and Roman-Catholicism since they are both part of the Christian tradition. However, the distinction between a Protestant and a Roman-Catholic pillar is not predominantly made on the basis of a cultural criterion, but rather on social and political criteria anchored in the history of Dutch society. As we have demonstrated before, these historical circumstances eventually led to the organization of Dutch Roman-Catholics and Protestants into two separate blocs. In the end it is the fact that Roman-Catholics and Protestants have separate organizational networks that makes them two different pillars.

Lijphart's second argument seems true enough for Roman-Catholics and Protestants, but fails when it is applied to the so-called general pillar. Within this pillar the social-democrats certainly lived in a "world of their own", but this world did not include the liberals. In the twentieth century the Dutch liberals defended the status quo and the interests of the bourgeois class. They propagated the ideas of individualism and laissez-faire, while the social-democrats were strongly in favour of state intervention in social and economic policy. An alliance between social-democrats and liberals proved difficult to form. Since 1922 liberals and social-democrats together with Roman-Catholics and Protestants were represented in the same cabinets only from August 1948 to September 1952. However, this happened under exceptional postwar circumstances in which there was a widely felt urge to unity in order to rebuild a devastated country. The fact that liberals and social-democrats recruited their voters from opposing classes made them natural enemies and weakened the position of the general pillar vis-à-vis the Roman-Catholic and the Protestant pillar.

Because of these fundamental contradictions one may question the idea of a general pillar itself. It does not apply to the political and social positions of liberals and social-democrats. Although they had parallel points of view regarding several political issues like public education and morals, these domains of agreement were constantly overshadowed by conflicting ideas on economic policy. The liberal idea of economic policy was more consistent with that of Catholics and Protestants. Therefore, it is not a surprise that in the political history of the Netherlands we see the liberals so frequently making coalitions with these parties and not with the social-democrats. Besides, because of the different recruitment bases of liberals and social-democrats, respectively bourgeoisie and working class voters, they did not belong to the same organizational network.

From the above it may be clear that if religion is made the nucleus of the concept of pillar, one runs into serious difficulties because the concept then acquires a too specific meaning. On the other hand, if we drop the religious criterion as a mean to distinguish pillars, we run the risk that we create a catch-all concept which can be applied to all plural societies in which different subcultures show some form of social and political organization. In that way even tribal African societies could be labelled as pillarized[12].

To solve this theoretical dilemma one could detach the ideological aspect of the pillar concept from the religious content it has got in Dutch history. Thus a pillar may be organized around any ideology, e.g. ethnic, linguistic as well as religious. However, it is crucial to see that whatever ideological organization, the latter can only be conceived as a pillar if relevant ideology, social organization and individual identification are closely tied together because the concept of pillar stands for the very relationship between those three elements.

Further, the concept of pillar contains the metaphor of the ancient temple. Each pillar is a piling up of stratification segments of unequal rank. As a collection of such segments, pillars are standing on their own in vertical direction but they are also related as they are integrated in the same overall construction.

With regard to the first element of this metaphor, the model of pillarization does not fit to the reality of twentieth century Dutch society, because only the denominational groups are pillars that embrace segments of all social classes. They can be conceived as cross cutting society rather vertically than horizontally. The social-democrats and the liberals do not comply to the pillar image as their respective power bases are predominantly shaped by parts of one social class. Therefore, if one would stick to the definitory element of vertical lines of social division, about half of the Dutch population would be left out if one takes voting behaviour as a criterion.

Denominational pillarization started to counteract secularization brought about by accelerating modernization. It was aimed at uniting all people irrespective of social class. Thus the social-democrats failed to be the exclusive representatives of the working class, because large parts of the latter were incorporated in denominational organizations. Their only chance to successful competition was the use of strategies of social control fairly identical to those of the denominational pillars. In that way their methods of organization became to look very similar to those of the already existing denominational pillars. The social-democrats too formed gradually an independent subsystem that linked ideology, political power, social organization and individual identification.

Although the history of the Dutch liberals differs from that of the social-democrats, at the political level they formed a separate political party with traditions and styles clearly distinct from others. At the meso level many liberal oriented organizations mutually related by interlocking directorates can be pointed at: sporting clubs, cultural associations, newspapers, employers organizations, a broadcasting association. However, in most cases these organizations did not advertise themselves as "liberal", but took a profile "voor alle gezindten" (for all persuasions). It is probably due to the liberal tradition of individualism, that the liberal bloc abstained from powerful organization that implied strong social control.

The liberal bloc is the most difficult to label as a pillar because it is the most diffuse. Unlike the social-democrats, the Protestants and the Roman-Catholics the boundaries of the liberal subsystem are not always clearly recognizable. Although liberals and social-democrats had many interests in common vis-à-vis the denominational parties - the promotion of public education, a more liberal legislation on moral issues - both their respective power bases and their organizations were different. For that reason they can not be considered to form together one pillar.

If one abstains from the vertical component in the pillar concept and instead emphasizes the element of pillars being subsystems within the larger society that link ideology, political behaviour, social organization and individual identification, many of the above problems are solved. We are then able to distinguish a Protestant, a Roman-Catholic, a social-democratic and a liberal pillar in Dutch society, be it that there are differences to the extent in which these subsystems show cohesion[13]. In this study I will use the term pillar in that sense, although I realize that it deviates from the original meaning in the literature on pillarization that uses to stress the component of vertical social divisions.

The second element in the metaphor is that pillars are related because they are all part of the same construction. In other words, a pillar can only exist in a pillarized society. Logically then, pillarized society consists of more than one pillar. It is a social system in which ideological and political heterogeneity is structured by separate subsystems that link ideology, political behaviour, social organization and individual identification.

Although mutually competitive, pillars must cooperate at least to some minimal degree to sustain the social order. It is here that the before mentioned concept of the culture of "living-apart-together" comes into play. If ideological and political heterogeneity are organized in a pillarized way and there is no minimal cooperation between the pillars involved, the social order itself is challenged.

By the requirement of minimal cooperation the pillar concept is certainly related to the study of social order. This, however, does not imply that social dynamics are neglected beforehand. Pillarized society is the result of the relative power positions of the social groups involved and the existence of a culture of "living-apart-together". It is an equilibrium that may be disturbed by structural changes in the relative power positions or cultural change. As outcome of such processes pillarization may erode and new power configurations may arise.

To summarize, I consider as a pillar a subsystem in society that links political power, social organization and individual behaviour and which is aimed to promote, in competition as well as in cooperation with other social and political groups, goals inspired by a common ideology shared by its members for whom the pillar and its ideology is the main locus of social identification. Pillarization then is the historical process by which a society becomes divided in pillars.

It should be emphasized that the above definition describes a pillar as an ideal typical model. In reality pillarized configurations deviate. In chapter I we gave a preliminary definition of pillarization which is current in the literature: the process by which society becomes divided along ideologically based vertical lines rather than along horizontal cleavages. From the above it may be clear that I regard this definition as too far away from empirical reality. We have seen that it produces major analytical difficulties. To make the analysis and explanation of pillarization fruitful, it therefore has to be rejected. My definition of the pillar-concept draws the attention to the fact that pillars can be shaped around different ideological concepts, that they are more or less autonomous subsystems and that they should be studied as interdependent entities.

Theories of Pillarization

Different scholars have pointed to different factors in the search for an explanation of pillarization. An unambiguous classification of prevailing theories is difficult to make as most authors use a mixture of explanatory factors in their theories. Thus Daalder discerned four categories: the emancipation hypothesis, the social control explanation, consociational theory and, what he called, "traditional pluralism, early constitutional relations, slow democratization and pillarization"[14]. In a study on Roman-Catholic pillarization in Europe Righart presents three main lines: emancipation, protection of faith and political mobilization[15].

I propose a classification that is slightly different from the above: the emancipation hypothesis, the protection hypothesis, the social control approach and consociational theory. It is based on a dichotomy between emancipation and protection on the one hand and social control and the consociational theory on the other. The former hypotheses explain pillarization primarily from the actions of churches and religious groups. Their main focus is the individual pillar. The latter are much broader and attempt to explain changes in society as a whole. In most theories reference to processes of modernization is made to a larger or lesser extent.

The Emancipation Hypothesis

The emancipation hypothesis explains pillarization as the effect of a process of emancipation of Roman-Catholics, the Protestant petty bourgeoisie and socialist workers against a conservative liberal upper class[16]. In the last quarter of the nineteenth century these groups resisted discrimination, paternalism and economic deprivation by organization and ideological exclusiveness. In course of time competing movements grew into pillarized structures. The pillars functioned as the roads to social esteem and equality. After these objectives had been reached, de-pillarization could set in.

The emancipation hypothesis has several weaknesses. First of all, it cannot make clear why Dutch Roman-Catholic pillarization came about only in the second half of the nineteenth century. As we mentioned before, the Roman-Catholics were a repressed minority since the end of the sixteenth century. Besides, the Netherlands was not the only country with a repressed minority[17].

Secondly, the three groups differed as to the moment in which they organized. Protestant Anti-Revolutionary organization came first, followed by the Roman-Catholics. The Social-Democrats organized latest. By acceptance of the emancipation hypothesis, the question remains why it is that especially the emancipation movement of the Protestant petty bourgeoisie resulted ultimately in a pillarization of the whole structure of Dutch society. Daalder comments:

> "It is not clear whether these emancipation movements developed simultaneously or whether the emancipation of the one blocked the emancipation of the other"[18].

Thirdly, when the Dutch Roman-Catholics pillarized they had already achieved substantial emancipation. They had gained equal civil rights. Freedom of religion was granted. They were permitted to have educational institutions of their own and the episcopal hierarchy had been restored. So Roman-Catholic pillarization came about after the Roman-Catholics emancipated and not before. Equally, the Anti-Revolutionary and Roman-Catholic emancipation was completed long before the Second World War. According to the emancipation hypothesis, de-pillarization should set in after emancipation. Yet, the 1950's are generally regarded as a peak of Dutch pillarization.

Fourthly, it remains obscure what the subject of emancipation was. As Hans Righart has put the question:

> "Have Roman-Catholic workers been pillarized as workers because of their emancipation as Roman-Catholics? Is that not a contradiction? Were their social, or class interests, not better served by unity instead of fragmentation?"[19].

It is evident that in the case of Roman-Catholic and Protestant workers such questions can only be answered from an ideological point of view.

The Protection Hypothesis

The second hypothesis states that the development of denominational organizations has to be seen as an attempt to protect the purity of faith and the autonomy of churches against the ideas of Enlightenment, liberalism and rationalism. In general, as a defensive strategy against the forces of secularization and modernization[20]. The faithful had to be protected against the evils of modernity that threatened the power of the churches: secularization, materialism and class struggle. This could be done by the creation of subcultures which would indulge the individual completely.

Righart concludes at the end of his study on Roman-Catholic pillarization in Austria, Belgium, the Netherlands and Switzerland:

> "In all four countries pillarization...was a church strategy against secularization. It was the lower clergy, which during the last decades of the nineteenth century understood that in order to survive the modernization of society the church had to adapt to the development of society. It was protection by adaptation"[21].

Righart's study is limited to the organization of Roman-Catholics. He does not analyse pillarized societies as a whole. For that reason he easily admits that the protection hypothesis does not explain non-denominational pillarization[22]. Although strongly related to modernization, its explanatory power is therefore limited.

A further argument can be put forward against the protection hypothesis. It is similar to that against the emancipation hypothesis. This criticism refers to the fact that in most modernizing countries denominational groups did not pillarize. The problem of denominational groups confronted with worldly ideologies is not an exclusively Dutch problem neither is it restricted to Austria, Belgium and

Switzerland. The protection hypothesis cannot answer the question why some countries showed pillarized structures while others did not.

The Social Control Approach

The social control approach regards pillarization as an institutional pattern by which elites exert social control[23]. This thesis is the antithesis of the emancipation theory. Thus Van Schendelen[24] understands Dutch pillarization as an attempt of elites to acquire and defend power positions by consciously sharpening social cleavage. Pillarization is the product of elite behaviour. Social control is exerted by the exchange of material rewards and punishments (health care, schooling, social assistance) and by a strengthening of ideology by pillarized socialization[25].

Within the social control approach a marxist subcategory can be distinguished. According to marxist theorists, pillarization is a deliberate attempt to cope with class contradictions and to repress class struggle[26]. So Stuurman[27] regards pillarization as a Dutch variant of the more general process of capitalist development, a reaction to the rise of socialism. This could be inferred from the objectives stated in the programs of the denominational organizations as well as from the fact that in those areas where socialism rose, denominational organization was strongly promoted. Especially within the Roman-Catholic denomination religious solidarity was emphasized in an attempt to mitigate class contradictions[28].

The social control approach cannot explain either why denominational organization took the shape of pillarized structures. The marxist variant raises the question why it was so easy for conservative elites to deal with "fundamental" class contradictions by manipulating "superficial" heterogeneities, and why it was so difficult for progressive leadership to neutralize such conservative politics[29].

Finally, the social control approach seems to overlook the fact that the Anti-Revolutionary and Roman-Catholic elites of the second half of the nineteenth century did not defend existing power positions. They were coming men who challenged the traditional Dutch establishment by the mobilization of the masses in order to achieve power. So originally, at the turn of the century the success of denominational pillarization implied a change of the Dutch power elites instead of a continuation of the traditional Dutch oligarchy.

The Consociational Democracy Theory

The most well known political science study of Dutch pillarization is Lijphart's. His work dominated the field for nearly two decades. Only recently have his ideas been criticized[30]. Lijphart's approach is different from those we discussed before. It was not Lijphart's aim to give an explanation of Dutch pillarization as such, but rather to use it as a case-study for a more fundamental theoretical problem: how could a divided society maintain internal peace and reach effective decision making? As a result, in Lijphart's work the rules of internal equilibrium and decision making are central elements. Because of its importance, we will discuss his contribution to the understanding of plural societies more extensively than the approaches treated before.

Lijphart's starting point is the observation that the Netherlands is divided by social cleavage, or pillars, while at the same it shows a stable and effective democracy[31]. At first sight this seems to be a paradox that needs an explanation.

Lijphart distinguishes in Dutch society four distributions based on religion and class. These result in three types of social cleavage. The first is the one between religious and non-religious. The second is within the religious part of the nation, i.e. between Protestant and Roman-Catholic. Finally, the third one involves the differences between Social-Democrats and Liberals within the non-religious category. These four distributions correspond to four different networks of social and political organization or pillars: Roman-Catholic, Protestant, social-democratic and liberal.

For Lijphart pillarization started around 1917. Then the tension in the relationship between the four blocs reached a peak because of three related issues: the school issue, the social question (unemployment, poverty and disability) and the introduction of general suffrage. The solution of these social controversies took place in 1917. From then on a set of informal political rules was accepted by which minimal consensus at the top of society could be achieved. In this respect Lijphart speaks of the Dutch "Pacification Democracy" and the "Policy of Accomodation". Although analytically distinct, these two are indissolubly linked to Dutch pillarization.

Lijphart explains the stability of this political system by two factors: 1) a basic sense of nationalism among the members of all four pillars reinforced by a few national symbols; 2) the cross-cutting of religious and class cleavage. The first factor promotes unity. The second mitigates sharp social divisions[32]. The indirect explanation of the stability of the Dutch political system is the elitist character of political leadership and the passivity of the masses. The core factor is "prudent leadership" manifested in the policy of accomodation. Accomodation means the "settlement of divisive issues and conflicts, where only a minimal consensus exists"[33]. In order to maintain the system as a whole, social cleavage at the basis of society should be respected, while the elites must be capable and willing to compromise at the top. For that purpose a number of rules are required: a business-like approach of politics, an agreement to disagree, summit diplomacy among the elites, proportionality between the elites, de-politicization of issues, secrecy and the acceptance of the government's right to govern.

> "Here the value of the metaphor of pillarization becomes apparent: the four blocs are like pillars, each standing on its own, but they are overarched by elite accomodation .....Prudent leadership, then, saved the system"[34].

Reacting to criticism, in a recent publication Lijphart emphasizes that his hypotheses are of a probabilistic nature. Under favourable conditions there is a high probability that a pacification democracy will keep social peace and maintain the democratic order. Such favourable conditions are: 1) all social segments are minorities; 2) external threats strengthen the country's unity; 3) there is a high degree of socio-economic equality between the social segments; 4) the elites have a strong tradition of tolerance[35].

Lijphart's study on Dutch society is a case-study aimed to the formulation of a more general theory on plural societies. The latter are those which are characterized by segmental cleavage of whatever nature[36]. By that Lijphart's work is related to Dahl's[37].

In 1977 Lijphart summarized his ideas derived from comparative research on segmented plural democracies. Consociational democracy is characterized foremost by coalition government in which the political elites of all the significant segments of plural society participate. Further, there is mutual veto, proportionality and a high degree of autonomy for each segment[38].

In Lijphart's model social pluralism is synonymous with segmental cleavage. However, it is not clear when a social division has to be considered as cleavage and when as segmental cleavage[39]. As every society always shows cultural heterogeneity, the specific form of cultural segmentation needs definition. This is not done by Lijphart. He uses a mixture of elements like outward isolation of groups and inward cohesion. Likewise, Lijphart fails to give criteria to decide when overarching loyalties and cross-cutting cleavage transform segmentation into universal diversity[40]. Further, cross-cutting cleavage - one of the elements guaranteeing political stability according to Lijphart - can also endanger the political order[41].

Responding to his critics, Lijphart proposes four criteria to measure the degree of societal pluralism: (1) the degree to which the different segments of plural society can be clearly discerned; (2) whether or not the size of each segment can be measured exactly; (3) whether or not there is parallelism between on the one hand the borderlines of the different segments and on the other hand the boundaries of social, economic and political organization; (4) whether or not the political parties based on social segments enjoy a stable support of those segments[42]. These criteria, however, do not solve the problem of the difference between pluralism and segmental cleavages. Actually they strengthen the suggestion that segmental cleavage is characteristic of societies with a high degree of organized pluralism.

Regarding the Netherlands, the above criticism is directly related to the empirical level of analysis. Roman-Catholics and Protestants cannot be regarded as homogeneous categories. Lijphart fails to define the segmented nature of Dutch social cleavage because his model is predominantly oriented towards politics. His preoccupation is the observed political stability and the functioning of the Dutch democratic system[43]. Thus the strong intertwining between ideology, political power, social organization and individual identification is not a central element in Lijphart's model. It is the very interdependence of these which sets pillarization apart from plural societies in general.

A major criticism to Lijphart refers to the fact that his model of Dutch pillarized politics is built on the premises of a spirit of prudent elite behaviour, accomodation and co-operation. But, because the Dutch multi-party system results from the structure of the electoral system (extreme proportional representation, free entry of new parties, no threshold for mini-parties), co-operation is inevitable to create a majority. "So, more important than any spirit is the objective need to co-operate"[44]. However, as Daudt remarked, instead of showing a spirit of accomodation and co-operation, Dutch elites treated each other as unequal competitors[45]. Within the frame of the necessity to cooperate, Dutch elite relations were competitive. Therefore, the Dutch pillarized society represents the same type of dead-lock situation as that in the Republic.

As to the supposed unique character of the rules of accomodation policy, such informal rules are not unfamiliar in other plural democracies as well[46]. We have seen that similar rules were applied by the Republican oligarchy too. In other words, they are not representative for the era of pillarization in Dutch history.

Within the group of consociational democracy theories a subcategory can be distinguished which has been labelled by Daalder as *traditional pluralism, early constitutional relations, slow democratization and pillarization*[47]. This school of thought is closely related to the works of Stein Rokkan[48] and Robert A. Dahl[49]. It is a comparative approach that tries to answer the question how in Western societies lines of political division developed during the nineteenth and twentieth century. This question differs from Lijphart's. The latter refers to the mechanisms underlying equilibrium in plural democracies. In correspondence with Lijphart, however, the interest of this group of scholars is primarily of a political science nature[50]. As to Dutch society their main thesis is that because of a long tradition of mutual accomodation the coming of mass society transformed social divisions into pillarized structures.

In the Netherlands Daalder is the main representative of this approach. He himself feels highly affiliated to Lijphart's work: "What have been the factors that have prevented the social system from flying apart?"[51]. His explanation, however, is more of an historical nature. Daalder points to the following factors[52]:

a. "political elites historically were in a strong position so that they had little fear for the danger of total challenge or replacement and hence showed a degree of permissiveness...;
b. throughout Dutch history elite groups were never completely closed or homogeneous...;
c. a tradition of pluralism caused differences of opinion to generate relatively little heat...there was widespread awareness that at most power might be shared rather than conquered;
d. older traditions of compromise were transferred to newcomers into the political process;
e. against the emancipationist pressures of Protestants, Roman-Catholics and socialists...certain government institutions (the Cabinet, the judiciary and the bureaucracy) retained some degree of unity above the groups;
f. the need for coalition government forced groups to enter into transactions on matters of common concern. The bureaucracy, for all its increasing diversity of recruitment, retained a common tradition - facilitating it to play a brokerage function;
g. one might speak of an effect of accumulating experience, a learning process suggesting that a recognition of claims for autonomy need not conflict with practical cooperation among groups".

Like Lijphart, Daalder, as a political scientist, is mainly interested in the rules of equilibrium. Next to the structural factor of the need for coalition government, he introduces a new explanatory variable: the stabilizing function of the Dutch bureaucracy[53]. Further, Daalder mainly emphasizes the country's cultural heritage. However, he does not elaborate on the change of Dutch pluralism in the nineteenth century and its effects on the process of pillarization. This implies that his theory cannot explain why pillarization came about at the end of that century and not some hundred years before.

Modernization and Pillarization

None of the theories presented above yields an explanation of pillarization which does not evoke further questions. They all have their pros and cons. Thus, it cannot be denied that denominational and socialist pillarization had an emancipating element. The purity of faith was certainly protected by orthodox Protestants and Roman-Catholics. The latter experienced also strong impetus from the Vatican. In all pillars the social control dimension is clearly recognizable. So is the fact that pillarized elite behaviour had its roots in the Dutch political tradition.

It seems to me, however, that we are left with three crucial problems: 1) why is it that non-denominational groups pillarized too?; 2) Why did Dutch pillarization take place at the end of the nineteenth century?; 3) Why did pillarization develop in the Netherlands and not in all countries?

Before we saw that Dutch political history is characterized by a delicate equilibrium between the constituting parts of a fragmented ruling class. Mainly because of external threats a culture of living-apart-together could settle. It prevented the country from falling into pieces. In the course of time the informal rules of elite interaction became institutionalised. Daalder puts forward a strong argument when he remarks:

"Older traditions of compromise were transferred to newcomers into the political process" and that "one might speak of an effect of accumulating experience, a learning process suggesting that a recognition of claims for autonomy need not conflict with practical cooperation among groups"[54].

Such phenomena are part of the cumulative effect of history on the change or formation of - what Eisenstadt called - social ethics.

Therefore, the behaviour of new elites - whether their goals were emancipation or the protection of the purity of faith - was structured along the lines of the prevailing cultural traditions. This holds for denominational as well as non-denominational elites. This conclusion, however, does not necessarily imply that elite actions had to produce pillarization as a logical result. Given the Dutch cultural heritage one may expect that the new nineteenth century elites strongly valued autonomy. For that sake they could have used the traditional ways of incidentally mobilizing supporters to promote their cases as was common practice with the Republican elite factions. So we may wonder why they did not and why their mobilization efforts became gradually structured in a pillarized way.

I will propose two factors that, in combination with the cultural tradition of living-apart-together, make Dutch pillarization understandable: the extension of franchise and the course of the process of modernization as it developed in the Netherlands in the second half of the nineteenth century. As we have seen in chapter IV the coming of constitutional monarchy in 1848 was heavily influenced by political and social events abroad. These made the extension of franchise an urgent problem.

In the constitution of 1814 suffrage on the basis of property qualifications was introduced. By that of 1848, members of parliament were elected directly, but suffrage was restricted to the well-to-do. To be enfranchised one had to pay a specified amount in taxes. In view of the fact that direct tax rates were low, the franchise was a privilege of the upper class. In the parliamentary election of May 1880, only 13.1 per cent of all men above 25 years of age were enfranchised. All women were excluded. In 1887 suffrage was extended to men satisfying minimal educational requirements. In the election of 1890 the percentage of enfranchised men had risen to 28.6. By the electoral reform of 1896, several classes in the electorate were distinguished according to living situation, level of income, level of education, etc. In 1900 the proportion of persons entitled to vote increased to 49 per cent of the male population above 25 years of age. In the elections of 1910 the figure was 63.2 percent[55]. The 1917 amendment to the constitution introduced universal and compulsory male suffrage and that of 1919 extended the suffrage to women.

Due to the extension of franchise a new situation arose. The Republican oligarchy once could maintain equilibrium and rule without systematically being held responsible by the rank and file. Only in cases of intense conflict did the latter need to be mobilized. Like in other countries, the extension of franchise caused the elites to be much more responsive to public opinion and increased their dependency on their rank and file. In a democratic system electoral support is generally dependent on the content of the policies promoted and executed. Therefore, after 1848 the autonomy of the Dutch elites became more and more limited. By this, the very equilibrium of the system was challenged.

The most appropriate way to counteract such tendencies is the broadening of social control so that electoral support is made relatively independent of the policies pursued. By the linking of the social coordinates of individuals directly to ideology, long lasting loyalty can be obtained independently of political conviction regarding current issues. This can be achieved if the individual identifies primarily with the ideological group involved and if his pattern of interaction is restricted to fellow-believers.

The availability of an organizational network creating a "society of its own" and having a straight and unambiguous ideological outlook, supports the exigencies at the level of individual interaction to a major extent. Taking both the organizational and the political level into account, in pillarized society a change of party preference would confront the individual with a loss of social identity and would deprive him of significant social relations. Therefore, pillarization is a powerful mechanism of social control which fits remarkably well into the Dutch tradition of sovereign elites.

The second factor that makes Dutch pillarization understandable is the course of the modernization process itself. The extension of franchise is related to modernization as it stands for an increase of political participation. However, other dimensions of modernization too evoked the need for a broadening of social control.

At the end of chapter III we concluded that: 1) economic growth accelerated from 1870 to 1920; 2) urbanization started around 1850 and accelerated from 1880 to 1890; 3) the growth of public expenditures increased substantially from 1870 to 1890; 4) industry underwent a process of enlargement of scale from 1880 to 1910; 5) social relations became more formalized after 1873; 6) standardization of information progressed after 1860 and 7) social mobility of the lower strata started to increase after 1870.

The acceleration of modernization in the last decades of the nineteenth century implied that for large parts of the population traditional social ties became looser. Further, the growth of education and information presented competing ideas to the common men, but, at the same time increased the power of information emitting agencies. Finally, the establishment in 1813 of the Dutch centralized state had shifted the gravity point of power from city and province to the level of national government. Standardization induced by national laws and the increase of public expenditures further promoted elites to compete for power at the national level. At the same time, however, these very developments made the traditional ways of social control - e.g. by the allocation of offices at the local level - obsolete, while keeping the rank and file in line became more than ever a requirement to get hold on the distribution and allocation of national scarce goods.

To summarize , Dutch pillarization has to be seen not as a reaction to, but as an integral part of the process of Dutch modernization. The structural dimensions of modernization - mobility, enlargement of scale, standardization of information, increasing political participation, formalization and growing public expenditures distributed at the national level - gradually produced an incompatibility with the traditional ways of power allocation. Without any changes of the system, this might well have led to outburst of social and political conflict detrimental to the progress of modernization itself. Enabled by technological developments - improved means of transportation and communication - the observed incompatibility was levelled by the rise of a new method of social control by which individual identification with ideology was secured. So loyalty to elites was made independent of current political issues. Elite equilibrium was preserved and prevented the jeopardizing of Dutch society as a configuration of minorities. Pillarization was a modification of the traditional culture of living-apart-together as well as its structural concomitants brought about by the forces of modernization.

We can now conclude by formulating a model of Dutch pillarized society. It can be regarded as a summary of the main characteristics of Dutch pillarization that we described in the foregoing pages. One should, however, keep in mind that any model is an exaggeration to the extremes and as such it does not equal reality. Further, this model only refers to the Dutch variant of pillarization which can be discerned from other forms by its high degree of social and political stability. If one or more of the prerequisites stated in the model are not fulfilled, a high degree of instability may be the result.

We will discern three levels of analysis. Firstly, the political level that refers to the distribution of power in society at large. Secondly, the organizational level. Thirdly, the level that deals with the interactions between individuals. It should be quite clear that these three levels can not be regarded independently of each other, but that they are logically interrelated. For example, the division of society in pillars has as its logical result a pillarized organization structure and leads to a high degree of social distance between individuals belonging to different pillars. The model of Dutch pillarized society can now be presented by fifteen statements:

1. Society is divided into pillars based on competing ideologies;
2. These pillars are in a state of equilibrium: no pillar dominates any other;
3. Because none of the pillars has a political majority, political power is exerted by coalitions of two or more pillars;
4. Scarce means are divided among pillars according to the principal of proportionality. This principle is applied to all pillars, whether they are in power or not;
5. Conflict management and the safeguarding of the national interest is realised by the application of a set of informal rules which regulate the political process. All pillars adhere to these rules;
6. There is a strong relation between ideology on the one hand and political and social organization on the other hand;
7. While the function of ideology is to separate members of different pillars, pillar elites have frequent and often institutionalised consultations on matters of mutual interest that can only be arranged on the national level.
8. Every pillar has an organizational network of its own. For every social function every pillar has its own social organization;
9. Within each pillar these organizations are connected by interlocking directorates. The political party is the centre of the organizational network;
10. Because of pillarization every social function is exerted by several organizations of different pillars. Organizations differing as to ideological background but

performing the same social function, are overarched by an umbrella organization which enables the leaders of the member organizations to reach cooperation if necessary;
11. Pillarized organizations are characterized by a low degree of internal democracy.
12. For the individual the main locus of social identification is the pillar to which he belongs;
13. There exists a large social distance between individuals of different pillars. Every individual interacts exclusively with others that belong to the same pillar;
14. The individual experiences normative social control to a very high degree. In cases of deviation the ultimate sanction is the loss of social contacts and therewith the loss of social identity. For this reason social control is very effective;
15. In all situations of interaction the individual is confronted with the same set of consistent norms derived from the ideology of his pillar.

Monitoring Pillarization

Politics

As a mechanism of social control, pillarization is not restricted to the political domain. The strength of this mobilization strategy is that the political behaviour of the individual is made dependent on the often lifelong identification with ideological groups outside the political arena. In pillarization the political party is indirectly and strongly related to social life. Furthermore, pillarization is a historical process and not a timeless momentum: I.e. degrees of pillarization can be discerned.

According to the census of 1947 42.6 percent of the Dutch population considered themselves as Protestants[56], 38.5 percent as Roman-Catholics, 1.7 percent belonged to other denominations and 17.1 percent was without denomination. For 1960 these figures were 40.4, 39.1, 2.0 and 18.4 percent respectively[57]. If political pillarization is complete, every Roman-Catholic, Protestant and so on should vote for the political party representing his denomination. As we one can see in table 5.2., Dutch pillarization never reached this ideal typical form.

The core of pillarization is the political domain. Pillarization is the strategy by which political elites ensure themselves of stable constituencies. How effective has this strategy been since the introduction of general suffrage? Has it stabilized the power relations between the different political elites successfully?

There are two angles from which we can approach the issue. The first one is that of the stability of the proportion of votes each political party got in national elections over time. Table 5.2. presents the percentages of votes by political parties in national elections. The table regards only the major parties. Small parties are left out. The column "Protestant" is the sum of both Anti-Revolutionaries and the more orthodox CHU (till 1977) as they both together formed the Protestant pillar. From the elections of 1977 onwards, the Roman-Catholic party, the Anti-Revolutionaries and the CHU participated in one united Christian-Democratic party, the CDA. For reasons to enable comparison in time, I added up Roman-Catholic and Protestant vote from 1922 onwards (column 1+2). In table 5.2. Roman-Catholic and Protestant votes are also given separately. In figure 5.1. only the CDA-votes are presented.

Table 5.2. Percentages of Votes in National Elections for Roman-Catholics, Protestants, Social-Democrats and Liberals as a Percentage of All Votes Cast, 1922-1986

Year	Catholics (1)	Protestants (2)	(1+2)	Social-Democrats (3)	Liberals (4)
1922	29.9	24.6	54.5	19.4	9.3
1925	28.6	22.1	50.7	22.9	8.7
1929	29.6	22.1	51.7	23.8	7.4
1933	27.9	22.5	50.4	21.5	7.0
1937	28.8	23.9	52.7	21.9	3.9
1946	30.8	20.7	51.5	28.3	6.4
1948	31.0	22.4	53.4	25.6	7.9
1952	28.7	20.2	48.9	29.0	8.8
1956	31.7	18.3	50.0	32.7	8.8
1959	31.6	17.5	49.1	30.4	12.2
1963	31.9	17.3	49.2	28.0	10.3
1967	26.5	18.0	44.5	23.6	10.7
1971	21.8	14.9	36.7	24.6	10.3
1972	17.7	13.6	31.3	27.3	14.4
1977			31.9	33.8	17.9
1981			30.8	28.3	17.3
1982			29.4	30.4	23.1
1986			34.6	33.3	17.4
average:					
- 1922-1956	29.4	22.3	51.7	24.1	7.4
- 1956-1986			38.8	29.2	14.2
stand.dev.:					
- 1922-1956	1.2	1.8	1.7	3.1	1.6
- 1956-1986			8.1	3.3	4.3
coefficients of variability:					
- 1922-1956	0.04	0.08	0.03	0.13	0.22
- 1956-1986			0.21	0.11	0.30

Sources: Figures for elections in the period 1922 till and including 1972 were derived from Peter Flora, 1983, pp. 133-134; figures on 1977 and 1981 from CBS, 1983, p. 10; figures on 1982 and 1986 from CBS, 1987, p. 8.

If we take the coefficient of variability as a measure of stability of the constituencies involved, we can conclude that from 1922 till and including 1956 the parties based on a religious ideology show a very stable pattern over the years. Sharp fluctuations in their electoral support are non-existent. Even in the thirties, when they were in power and took a number of unpopular decisions, they did not experience a substantial downfall in the number of their adherents. The constituencies of the parties of liberals and social-democrats turned out to be far less stable. After the 1950's this pattern changed.

A second conclusion to be drawn is that, with the exception of 1952, till 1956 the denominational parties together formed a political majority in parliament. After then, this position crumbled down to about one third of all votes cast. However, because of the antagonisms between social-democrats and liberals, the Christian parties could still dominate the political arena. From 1917 onwards they always participated in the respective cabinets.

Figure 5.1. Percentages of Votes in National Elections for Christian-Democrats (Roman-Catholics + Protestants) (a), Social-Democrats (b) and Liberals (c) as a Percentage of All Votes Cast, 1922-1986

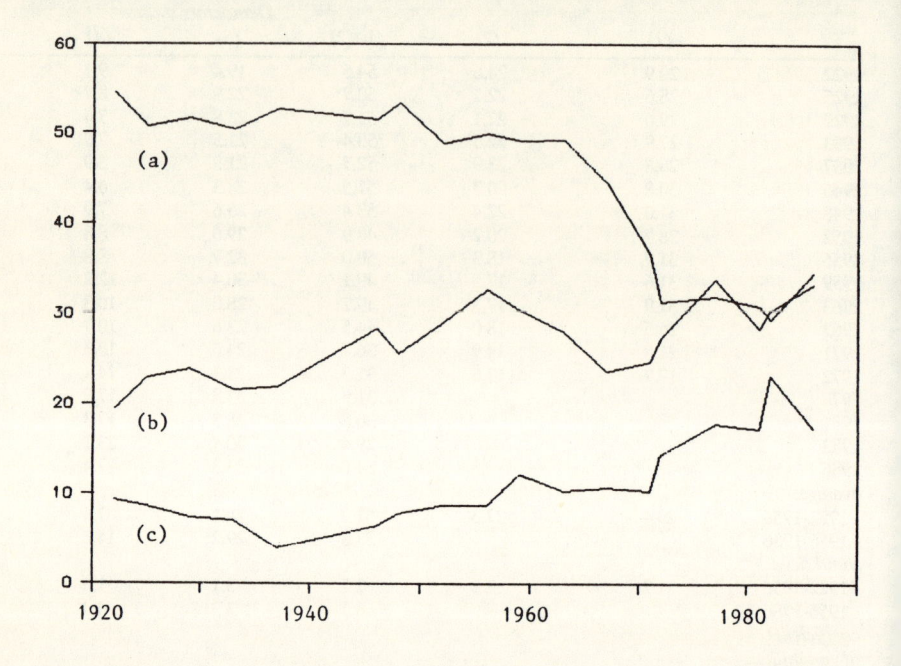

Source : Table 5.2.

A second approach to the stability of political power regards the stability of government indicated by the number of cabinet changes. Stable electoral relations are advantageous to that. Much depends also on the ways elites deal with each other. Because of the culture of "living-apart-together" we may expect that in the Dutch pillarized political system government stability is relatively high. If one compares the Netherlands to some other West-European democracies it responds to this expectation rather well (see table 5.3.).

Table 5.3. *Number of Government Changes in Austria, Belgium, France, Germany and the Netherlands, 1918-1984*

	Austria	Belgium	France	Germany	Netherlands
1918-1925	10	6	8	11	2
1925-1935	10	11	24	10	4
1935-1940	n.a.[a]	6	9	n.a.[b]	4
1945-1955	3	11	20	2[c]	5
1955-1965	6	4	8	3	4
1965-1975	3	6	8	5	5
1975-1984	3	8	5	4	3
1918-1984	35	52	82	35	27

Notes : a) Parliamentary rule was suspended in March 1933. The last pre-war Austrian cabinet was that of Prime-minister Dolfuss established in May 1932; b) Since January 30th 1933 presidential cabinet of Hitler; c) The first post-war parliamentary cabinet was that of Konrad Adenauer established September 20th 1949. See further appendix B.1.
Source : See Appendix B.1.

Organizational Pillarization

To measure pillarization at the organizational level, Kruijt and Goddijn tried to answer the question what percentage of the members of some significant social organizations participated in denominational organizations[58]. For these authors pillarization is exclusively related to religious organization. They discerned a "pillarized quotum" (Roman-Catholic and Protestant) and a "general quotum" (social-democrats and liberals). Both quota together add up to hundred per cent. Further, as they measured organizational pillarization for only one year, their approach is rather static[59].

I attempted to dynamize the method used by Kruijt and Goddijn by the construction of time series. Because of lack of data this was not possible for all types of organizations presented by Kruijt and Goddijn. In the following, data on farmers' organizations and newspaper readers are therefore missing[60].

— Education

Education is a major agent of socialization, reinforcing and internalizing values and norms of social groups. If education performs its function of socialization effectively, it is a powerful instrument to the continuation of the status quo. Pillarization was aimed at the creation of stable constituencies. The linking of the school system to the pillars ensured the feelings of political loyalty to be handed over from parents to children and thus paved the way for the stability of the electorate to come. Pillarization of the school system was a crucial step in the process of extending social control by the elites.

From the bottom to the top the Dutch school system was - and for a large part still is - pillarized. At the age of four Dutch children entered a nursery school of their parents' denomination. In all cases of primary and secondary schools, professional and even academic training[61] parents could choose between the various denominational institutions and the public non-denominational educational system.

Public non-denominational schools are administered by the municipality or by central government. For denominational schools authority lies in the hands of private boards generally consisting of respected parish members and representatives of the clergy. All schools are financed by public means. Because of its financing power, the Dutch state can formulate the requirements concerning the cognitive content of the teaching program. From the elementary level onwards, the state sets the terms for the final examinations so that there is no difference between graduates of public and denominational schools.

Since the cognitive content of the curriculum is controlled by the state, in this domain the meaning of the term "freedom of education" is rather restricted. Consequently it refers to the ideological dimensions of teaching. In denominational schools prayers are said and Bible interpretation is part of the curriculum. Besides mathematics the youngsters are taught to be good Protestants or Catholics and to live as such. Their view of the world is moulded according to the values of their pillar. In fact, the difference between public and denominational school is expressed in the ideological construction of reality[62].

It is, however, quite clear that in parts of the curriculum ideology and the transmission of knowledge are intertwined. In his study of the community of Sassenheim, Gadourek[63] gives a striking example. He examined the textbooks in use at the three types of elementary schools in the community. For Roman-Catholic children the Inquisition was pictured as a tribunal of the church, its members being wise and pious bishops and priests. Calvinist children were told that the Inquisition was merciless, murdering and torturing innocent people. Dutch-Reformed children learned that "everybody hated the Inquisition. William of Orange, while still being a Catholic, hated it too, together with many other Catholics". As socializing agents schools acted as instruments, making young members of the pillar aware of the fact that members of other pillars belonged to a different world. Schools gave pupils that kind of "we feeling" so important for the rest of their lives in pillarized society.

Table 5.4. presents the number of pupils of Roman-Catholic, Protestant, public non-denominational and "other" elementary schools as a percentage of all elementary school pupils. The column "others" is of a hybrid nature. It contains both non-denominational special schools and schools of small religious groups who do not form a pillar. I added up public non-denominational and "others" to a non-denominational quotum and Protestant and Roman-Catholic pupils to a denominational quotum. Both quota are represented in figure 5.2.

Table 5.4. The Number of Pupils of Elementary Schools by Denomination as a Percentage of All Pupils of Elementary Schools, 1850-1984

Year	Public (1)	Protestant (2)	Catholic (3)	Others (4)	Denomin. Quotum (5)	Non-Denom. Quotum (6)
1850	n.a.	n.a.	n.a.	n.a.	23.0	77.0
1870	n.a.	n.a.	n.a.	n.a.	23.0	77.0
1890	n.a.	n.a.	n.a.	n.a.	29.0	71.0
1910	n.a.	n.a.	n.a.	n.a.	38.0	62.0
1930	37.7	25.0	35.8	1.5	60.7	39.3
1950	27.4	27.5	43.3	1.9	70.8	29.2
1960	26.8	27.1	44.2	1.8	71.3	28.7
1970	27.5	27.5	43.0	2.0	70.5	29.5
1980	31.7	28.2	37.3	2.8	65.5	34.5
1984	31.6	28.8	35.5	4.1	64.3	35.7

Source : Appendix B, table B.2.II.

Figure 5.2. demonstrates the educational dimension of Dutch pillarization rather well. It clearly shows the effect of the solution of the school issue (see chapter IV) around the turn of the century. The relative growth of denominational education ends in the 1950's. The denominational quotum then contains about two thirds of all elementary school pupils. After 1960 the denominational quotum slightly decreases.

Table 5.5. reveals that the denominational division of elementary school pupils is very dissimilar to that of university students. Although the denominational quota increase over time, in 1983 public non-denominational state universities still control the field. There are two possible causes for the small denominational quota and their relative growth after World War II. Firstly, in 1575 the oldest Dutch university, that of Leiden, was established. In the seventeenth century others followed. However, the first Protestant university was founded only in 1880 in Amsterdam. The initiative was taken by Abraham Kuyper and was part of his strategy of "sovereignty in one's own circle". In 1923 the Roman-Catholic university of Nijmegen was opened and four years later that of Tilburg[64].

Thus, the denominational universities came relatively late and had to start from scratch. Because of the expensive nature of academic training and research not all

Figure 5.2. Denominational (a) and Non-Denominational (b) Quotum of Pupils of Elementary Schools as a Percentage of All Pupils of Elementary Schools, 1850-1984

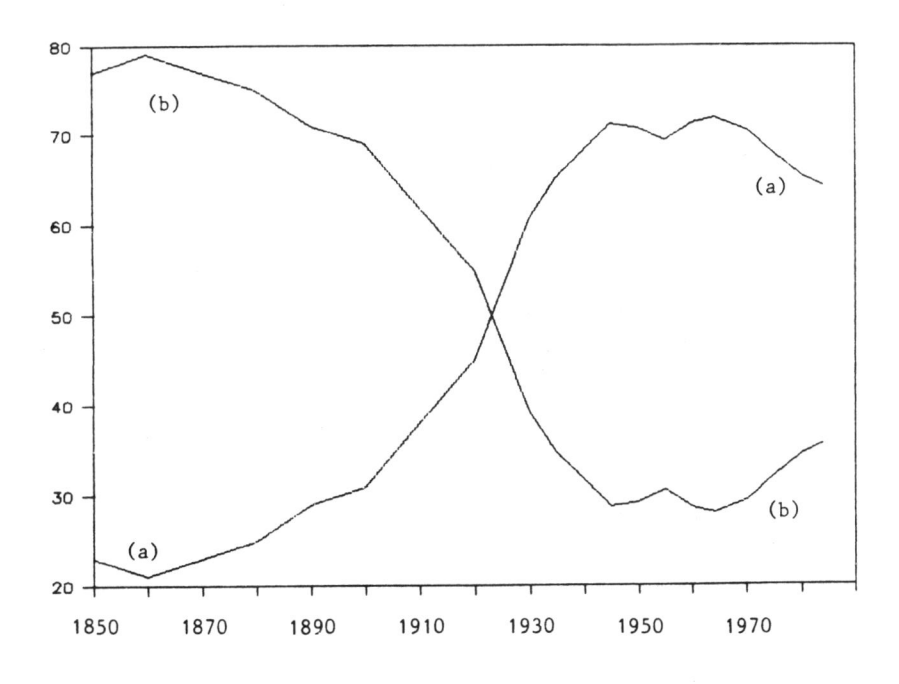

Source : Appendix B, table B.2.II.

sciences could be covered by denominational universities. Consequently many a Protestant and Roman-Catholic student had to enroll to a public state university.

Secondly, the Roman-Catholic universities of Nijmegen and Tilburg are located in the South of the Netherlands. Till 1975, when the state university of Maastricht was opened, no public university was available in that part of the country. So the Netherlands showed a geographic disequilibrium as to the dispersion of denominational and non-denominational state universities. Urged by the increasing demand for academic training since the 1960's, the government granted the universities of Nijmegen and Tilburg financial support to establish new faculties. As a result the number of the enrollments of these universities increased relatively. This, however, cannot be considered a product of pillarization.

Table 5.5. Number of University Students by Denomination of the University as a Percentage of All University Students, 1920-1983

Year	Protestant quotum	Catholic quotum	Non-Denominational quotum
1920	2.6	0.0	97.4
1925	3.7	3.0	93.3
1930	4.1	4.4	91.5
1935	4.9	5.3	89.8
1937	4.9	5.3	89.8
1945	4.3	5.6	90.1
1950	4.9	6.3	88.9
1955	7.0	8.0	85.0
1960	7.7	9.9	82.5
1963	7.5	10.9	81.6
1968	8.3	12.4	79.3
1974	9.3	13.9	76.8
1979	9.0	13.5	77.5
1983	8.1	13.4	78.5

Source : Appendix B, table B.2.III.

— Health Service

In the domain of health service Kruijt and Goddijn[65] present figures for the home nursing services and the control of hospital beds for respectively 1954 and 1957. In the Netherlands home nursing services are voluntary private organizations that provide their members with medical aids (wheelchairs, special beds, crutches, etc.) and district nursing. Traditionally there is a Roman-Catholic, a Protestant and a general non-denominational home nursing service. Members have to pay a small subscription fee. Consequently the services are heavily subsidized by the state.

Table 5.6. presents the members of the three home nursing services as a percentage of the total number of members. Over the years, the figures are relatively constant. Probably this can be explained by the fact that membership is of a rather passive nature. It is generally considered to be an insurance against the incidence of sickness and disability.

Table 5.6. Members of Home Nursing Services by Denomination of Home Nursing Service as a Percentage of All Members of Home Nursing Services, 1951-1972

Year	Non-Denomin.	Catholic	Protestant
1951	55.5	29.8	14.8
1958	56.8	30.7	12.5
1963	56.4	31.5	12.1
1967	55.8	31.1	13.1
1970	56.1	31.9	12.0
1972	56.3	32.1	11.5

Source : Appendix B, table B.3.I.

In the Netherlands hospitals may be private organizations administered by a private foundation, public institutions controlled by local authorities or academic hospitals. Private hospitals may be denominational or non-denominational. In all cases hospitals are financed by government regulations as well as by private and public sickness insurance funds. Table 5.7. presents the number of hospital beds by denomination as a percentage of all hospital beds.

Table 5.7. Number of Hospital Beds by Denomination, 1950-1985

Year	Catholic	Protestant	Non-Denominat.
1950	45.0	15.2	39.8
1955	43.8	16.6	39.6
1960	42.3	17.3	40.4
1965	43.8	18.9	37.3
1970	43.4	20.6	36.0
1975	43.2	17.3	39.5
1977	41.5	20.3	38.1
1985	36.2	19.2	44.6

Source : Appendix B, table B.3.II.

Regarding health services, one can observe a high degree of denominational influence. This is not surprisingly, as care for the sick has always been a core element of the Christian tradition. Centuries ago the first hospitals were founded by the Roman-Catholic church. In the Netherlands, after the Reformation the Protestant poor-relief boards took over. The fact that after the 1950's the distribution of members of home nursing services and the control of hospital beds between the Roman-Catholic, Protestant and non-denominational agencies is relatively stable (with the exception of a relative downfall of Roman-Catholic hospital beds) may be explained by the perception of these institutions as relative value-free provisions.

— Broadcasting

In the world of broadcasting the Dutch take an exceptional position. The Netherlands do not have a state directed broadcasting system neither are radio and TV the realm of private enterprise. Instead, voluntary private associations are responsible for the programs. The government only sets the frame by the allocation of public means, the determination of the total number of broadcasting hours and the number of broadcasting hours per broadcasting association. The share of each association in the total number of available hours depends on the number of its members.

Since the start of Dutch broadcasting in the first decades of this century, the Netherlands had two Protestant associations (NCRV and VPRO), one Roman-Catholic (KRO), one Social-Democratic (VARA) and one "general" association (AVRO). The latter always was clearly oriented to the liberal tradition. Thus Dutch pillarization was reflected in its broadcasting system.

Table 5.8. Members of Denominational and Non-Denominational Broadcasting Associations as a Percentage of All Members of Broadcasting Associations, 1950-1986

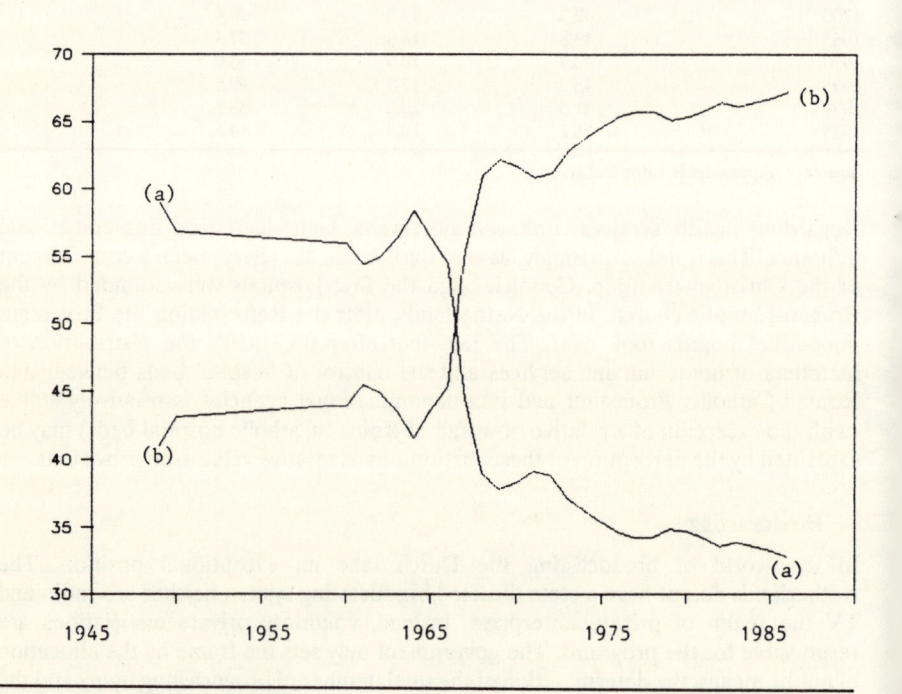

Year	Non-Denominational	Denominational
1950	43.1	56.9
1960	44.0	56.0
1965	43.6	56.4
1970	61.7	38.3
1975	64.7	35.3
1980	65.3	34.7
1986	67.2	32.8

Source : Appendix B, table B.4.II.

Figure 5.3. Members of Denominational (a) and Non-Denominational (b) Broadcasting Associations as a Percentage of All Members of Broadcasting Associations, 1949-1986

Source : Appendix B, table B.4.II.

Till the 1970's the Netherlands had only two radio stations and one TV channel. Till 1965 all three were completely covered by the broadcasting associations mentioned. The latter openly used their programs to reinforce the norms and values of the pillar to which they were connected by ideology and interlocking directorates. As an example, pictures with some reminiscence to the erotic were not broadcasted by the denominational associations. In general, the idea prevailed that broadcasting should enlighten the public instead of providing mere entertainment. The kind of enlightenment was strongly made dependent on the ideology involved.

Since 1965 the pluralism of Dutch broadcasting changed. Not only as to the number of broadcasting agencies but also with regard to the contents of the programs. A newcomer, the TROS, entered the field. It proclaimed not to be involved in ideology and to serve the public by providing entertainment. Ten years later the TROS was followed by a former radio pirate, VOO, who managed to get a legitimate status. In fact, VOO had the same objectives as the TROS, be it that the former directed itself more to the younger generation.

The appearance of the TROS affected the whole system. To hold their relative position within the system, the traditional associations now had to compete for the public's favours. A shift took place from ideology to entertainment with the effect that the programs of the different associations became more to look alike. With reference to the TROS who disturbed the equilibrium of the system, this process has become known as *vertrossing*. Especially in orthodox Protestant circles it led to dissatisfaction with the functioning of the NCRV that was accused of becoming too worldly. In 1970, a new association, the orthodox Protestant EO, came to the fore.

Table 5.8. and figure 5.3. present the membership of denominational and non-denominational broadcasting associations as a percentage of the total number of all broadcasting associations. The spectacular decline of the share of denominational broadcasting strikes the eye. This decline is even greater than depicted in the figure because of the *vertrossing*-effect. It also affected the Social-Democratic VARA which is included in the non-denominational category.

To conclude, till the mid of the 1960's Dutch broadcasting was heavily pillarized. It served to reinforce the values and norms of the pillars involved. After 1965 the relative power of pillar connected associations decreased.

— Economic Life and Industrial Relations

Although Dutch pillarization led to political stability, some of its characteristics could be detrimental for the economic process to develop freely. Here we allude to the high degree of social distance between the religiously different parts of the population and the exclusiveness these social categories felt towards each other. In its extremes such social barriers may limit economic exchange and the optimal seizing of opportunities. There are no signs that point to a systematic relation between Dutch pillarization and economic stagnation. In the seventeenth century, when the foundations of pillarization were laid, Dutch economic performance was outstanding in the world. After stagnation in the eighteenth century, the Dutch economy saw unprecedented growth after World War II, an era when the degree of pillarization was high. Consequently, the hypothesis that Dutch pillarization limited economic performance does not seem feasible.

Did pillarization create economic barriers in the Netherlands? Some authors think it did[66]. They have the impression that in business relationships, like employer-employee and buyer and seller, members of the same pillar had a strong preference to each other, especially in villages and little towns. However, the Dutch economy never has been split in segments corresponding to pillars, neither have economic contacts between pillars been strictly regulated the way it was in, for example, Indian caste society. Formally no regulations were ever issued forbidding economic transactions with members of other pillars. An exception to this has been the attitude of the Roman-Catholic bishops. In 1958 they issued[67]:

> "For the planning of buildings, the tendering and engineering of building-works and the supply of materials, the parochial church- and public assistance committees and all the other Catholic institutions that are accountable to the Episcopacy, have to assign, respectively: solely Catholic architects or building experts, Roman-Catholic organized building contractors, sub-contractors and surveyors, as many as possible Dutchmen who supply Dutch manufacture, and as many as possible Roman-Catholic organized workers...We sincerely trust that also the boards of Catholic institutions that are not accountable to us, will observe the regulations stated."

Illustrative for the spirit of pillarization as these kinds of regulations may be, they never affected the economic process seriously. As one of the major functions of pillarization since the beginning of the twentieth century was the protection of the working class to the evils of socialism, pillarization influenced the domain of industrial relations to a much larger extent.

Roman-Catholic and Protestant leaders saw the foundation of denominational trade unions as a defense against rising socialism. Gradually three trade-union federations emerged. Employers organized themselves in liberal, Roman-Catholic and Protestant associations. By means of interlocking directorates unions and employers' organizations were closely connected with the respective political parties. So the structure of industrial relations was part of the larger pillarized social system.

Although a certain mutual understanding and cooperation existed between the leaders of the three trade-union federations, religious oriented workers used to be put under pressure by the clergy to join a union of their denomination. Since 1918 the church did not allow Roman-Catholics to be members of the Social-Democratic party nor of the Social-Democratic trade-union federation. By their notorious *Mandement* of 1954 the bishops confirmed their position: Roman-Catholic members of the Communist Party and the by Communists dominated trade-union federation Eenheidsvakcentrale[68] were threatened with excommunication. Roman-Catholic members of the Social-Democratic trade-union federation NVV were no longer allowed to receive the holy sacraments and were denied a Roman-Catholic burial[69]. Although the Mandement raised strong emotional feelings among social-democrats and the more liberal Catholics and Protestants, its effects were negligible. It was outdated because its aggressiveness did not fit into the post-war inclination to harmony. The elites of the trade-union federations consequently continued to cooperate according to the rules of pillarization.

With respect to the cooperation of union's elites the period of the German occupation from 1940 to 1945 was of special significance. During this period union leaders met secretly and plans for a post-war unified labour movement matured. However, the denominational federations could only agree to unification if it would have a religious basis. This turned out to be totally unacceptable for the leaders of the Social-Democratic NVV. The employers, being afraid of a unified working class, supported the denominational opposition to it. The ultimate outcome of this process was the foundation of a Council of the Trade-Union Federations (in Dutch: Raad van Vakcentrales) in which the leaders of the federations met each other to discuss matters of mutual interest. In 1954 this Council was abolished because the NVV left it as a protest against the Mandement of the Roman-Catholic bishops.

Table 5.9. *Members of Social-Democratic (NVV), Roman-Catholic (NKV), Protestant (CNV) and Other Trade-Union Federations as a Percentage of the Total Number of Trade-Union Federations, 1914-1980*

Year	NVV	NKV	CNV	Others
1914	31.7	10.9	4.1	53.3
1920	36.2	20.6	9.8	33.3
1926	38.5	18.3	9.8	33.3
1933	40.6	23.2	13.9	22.3
1937	39.2	23.2	15.0	22.7
1940	40.0	23.3	14.9	21.8
1947	38.0	28.4	15.0	18.6
1950	38.3	29.7	15.6	16.4
1952	38.5	29.4	16.0	16.2
1961	36.2	29.4	16.0	18.3
1964	36.0	28.6	15.6	19.9
1967	36.2	27.6	15.7	20.4
1971	38.8	25.6	15.1	20.5
1975	40.0	21.1	13.3	25.6
1980	42.0	18.2	17.0	22.8

Source : Appendix B, Table B.5.I.

Also during the occupation, unions, employers' associations, political parties and high ranking civil servants of the Ministry of the Interior had consultations on the structure of post-war industrial relations. These meetings resulted in the foundation of the Council of Labour (in Dutch: Stichting van de Arbeid), an institutionalization of contacts between employers and employees.

The NVV, which had the strongest desire for a more unified labour class, had to accept the compromise of the mere institutionalization of contacts between trade-union federations and a closer cooperation between employers and employees. By this the characteristics of pillarization in industrial relations were in fact maintained after the war.

Many who strived for unity of the working class and who wished a change of the social structure after the war, were disappointed by the outcome of the negotiations that started in wartime. By an appeal to the general feeling that the social barriers of the past should be erased, they hoped to mobilize many workers into a new organization that could make a fresh start. They founded the Eenheidsvakcentrale (EVC) that discerned itself from the other trade-union federations, not only by its striving to a unified working class, but also because it refused to give priority to the national interest over the interests of the workers. The traditional organizations aimed at social harmony as a condition to rebuild the country as soon as possible. The EVC on the other hand rejected the rules of pillarized society and propagated a conflict perspective. In the first years after the war the EVC organized some major strikes of a mainly political character. Because of its aggressiveness and its ideology of working class solidarity, at the outset the EVC succeeded in mobilizing a fairly large number of adherents. However, as the organization became more and more dominated by communists it lost support[70] and after 1950 it was no longer of importance in Dutch industrial relations.

The history of the rise and fall of the EVC demonstrates that, despite the post-war national desire for unity, at the end the forces of pillarization proved to be stronger. The working class maintained its segmented character. By the Mandement of the Roman-Catholic bishops the traditional divisions were confirmed again.

Not only the segmentation of the working class but also a tendency to corporatism had been a peculiar trait of the structure of Dutch economic relations of this century. The essence of corporatism is the denial of class conflict and the conviction that employers and employees have shared responsibilities and should therefore live in social harmony. Originating from the Roman-Catholic tradition, it is not surprising that corporatism goes well along with pillarization. In both the idea that the basic division in society is that between social classes, is rejected.

Already in the beginning of this century Roman-Catholic intellectuals got the idea of reforming the capitalist Dutch economy in a more corporatist direction. This far reaching proposal has become known as Corporate Industrial Organization (in Dutch: Publiekrechtelijke Bedrijfsorganisatie, abbreviated: PBO). Every branch of industry would be administered by an industrial council consisting of representatives of employers and employees. The government should delegate authority to the industrial councils so that these autonomously could take decisions concerning production, distribution and industrial relations. In the 1930's these ideas became accepted by all pillars with the exception of the liberals who feared too many restrictions for the freedom of entrepreneurs.

In 1950 the PBO was introduced. As an umbrella organization of all industrial councils the Social Economic Council (SER) was established. With the exception of the agricultural sector, the system of Corporate Industrial Organization turned out to be a failure. The SER on the other hand acquired much prestige as an advisory council for the Dutch government and for a long time it has been the cornerstone in the formulation of social and economic policy. It consists of 45 members: 15 representatives of the trade-union federations, 15 representatives of the employers and 15 independent members appointed by the Crown. The last category does not function as a governmental delegation, but is considered as completely independent expertise composed of leading professors from the economics and law faculties. Within the fraction of employers and employees, members representing their organizations are appointed according to the principle of proportionality.

If we now summarize the organization of the Dutch economy in the 1950's, we can discern two main characteristics. Firstly, employers and employees were organized in pillarized organizations. In industrial relations the main locus of identification for the individual was not his social class, but the pillar to which he belonged. As a result their was no united working class. Secondly, consultations between leaders of different interest groups had nowhere been institutionalised to such a high degree as in the economy. The SER was the apex of this institutional framework.

These two characteristics are in line with the trend that set in at the beginning of the century. It was strengthened by the experience of the German occupation. After the war the striving to unity of the working class did not succeed in erasing its traditional segmentation. Representing the aim for class consciousness, the EVC only caused a temporary turbulence in the industrial relations and receded at the end of the forties. The tendency to corporatism, so well in line with some of the fundamentals of pillarization, in combination with the necessity to rebuild a devastated country, led to a far reaching institutionalization of consultations between the different economic elites and a shared responsibility for the social and economic policy of the Netherlands.

We can now return to the question whether pillarization led to stable industrial relations. The relations between the economic elites, as we described it, and the institutionalization of their consultations do not fit very well in a type of society dominated by class conflict. Although pillarization is not the only explanatory variable, it certainly stimulated the segmentation of the working class and offered a fruitful soil for corporatistic features to develop. Relative to other countries we may thus expect Dutch strike activity to be low. We will investigate this hypothesis by comparing the twentieth century Dutch strike activity to that of other industrialized countries. Because Dutch industrial relations institutionalised after World War II, it is reasonable to expect a dissimilarity between the interbellum and the period after World War II.

Strike activity usually is measured by the strike volume being the product of the size of strike (i.e. the number of workers involved per strike), the average duration of strike (i.e. man-days lost per worker involved) and strike frequency (i.e. the number of strikes per thousand workers)[71]. The strike volume equals the amount of working days lost by industrial disputes related to the labour force. As the strike volume is the product of three variables, two similar strike volumes may be the effect of different sizes, durations and frequencies.

For our own computations we used a somewhat different indicator for strike activity: the number of working days lost by industrial disputes as a percentage of the demand for labour. We prefer this measure to that of the strike volume because ours corrects for the economically non-active part of the labour force and for changes and fluctuations in working hours between countries and time periods.

A comparative analysis of the whole period from 1920 till 1979 is not possible, because for most countries data on the first decades of the twentieth century are either lacking or incomplete. For the Netherlands, however, both average strike activity and the standard deviation are higher before World War II (respectively .105 and .107, see appendix B, table B.6.I.) than after.

Table 5.10. Arithmetic Means and Standard Deviations of Strike Activity as a Percentage of the Demand of Labour for Twelve Industrialized Countries, 1950-1979

Country	Arith. Mean	Stand. Dev.
Austria	.009	.016
Belgium	.084	.087
Denmark	.052	.137
France	.072	.051
Germany	.013	.017
Italy	.274	.171
Japan	.038	.024
Netherlands	.010	.013
Norway	.029	.056
Sweden	.015	.025
Switzerland	.001	.003
UK	.119	.115

Source : Appendix B, table B.6.I.

For the Netherlands I depicted the strike activity as a percentage of the demand for labour as well as the average strike activity for the whole period in figure 5.5. The latter is represented by the horizontal line. With regard to the Interbellum, the center of gravity of the instability of Dutch industrial relations lies rather in the early 1920's than in the 1930's. In the post-war period the low average level of Dutch strike activity is only beaten by Austria and Switzerland (See table 5.10. and figure 5.4.)

If the Dutch strike volumes of the inter-war and the post-war periods are desaggregated to size, duration and frequency, one can observe remarkable differences[72]:
- in the inter-war period the size of strikes is small. In the post-war perid the size is large, mainly because of the predominantly national character of strikes;
- in contrast to the post-war period strikes in the interbellum had a long duration;
- in the inter-war period strike frequency was high; after World War II it was low;
- contrary to the Interbellum, in the post-war period the number of wildat strikes systematically exceeds the number of official stries.

These differences in patterns of strike activity are largely explained by differences in the degree of pillarization of Dutch society between the inter-war and the post-war

Figure 5.4. Arithmetic Means and Standard Deviations of Strike Activity as a Percentage of the Demand for Labour for Switzerland (1), Austria (2), Netherlands (3), Germany (4), Sweden (5), Norway (6), Japan (7), Denmark (8), France (9), Belgium (10), UK (11) and Italy (12), 1950-1980

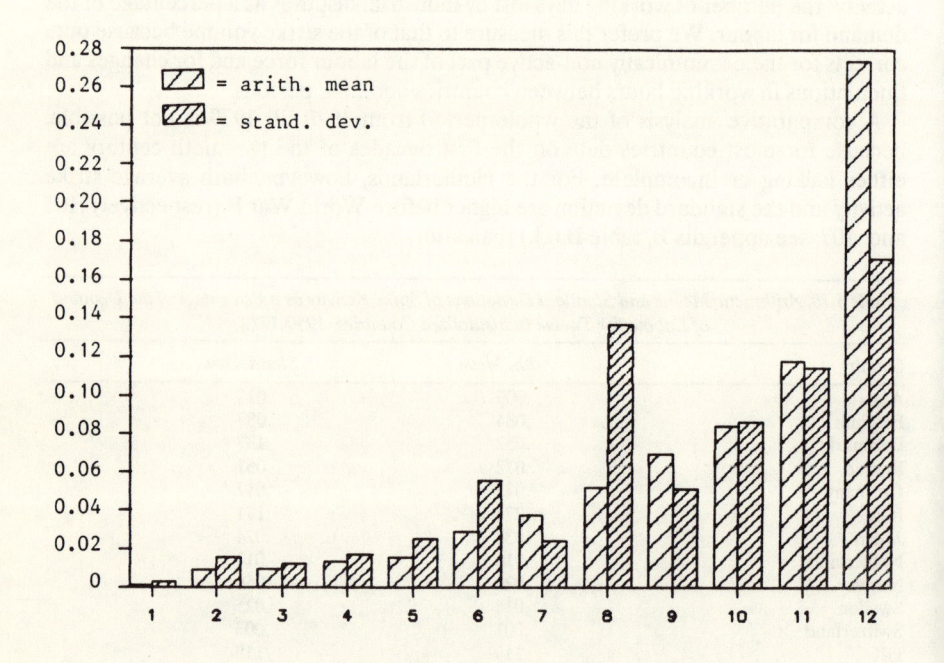

Source : table 5.10.

periods. In his major work on industrial relations in the Netherlands, Windmuller points at the special character of the Dutch nation which is to his opinion the basis of the low level of strike activity[73]. Albeda is more specific. He mentions the negative attitude of Dutch employees towards strike activity as the result of the conscious indoctrination by the Christian-Democratic unions under whose wings were gathered more than fifty per cent of all organized Dutch employees[74]. Both authors seem to imply that the Dutch social structure as well as the Dutch values have been extremely fruitful for the institutionalization of industrial relations: a highly centralized structure of negotiations, stable relations between unions and strong trade-union federations, a relatively low communist influence on unions, a mutual recognition of employers' and employee's organizations as negotiating partners along with a certain will to cooperate, institutionalised participation of employers' and employee's organizations in the formulation of national socio-economic policy, participation in government of the labour party and a great involvement of government in industrial relations. These factors, stressed by Ross and Hartman as basic to the stability of industrial relations[75], are all promoted by the pillarized social structure as we

Figure 5.5. Strike Activity as a Percentage of the Demand for Labour in the Period 1920-1979 and the Arithmetic Mean, The Netherlands

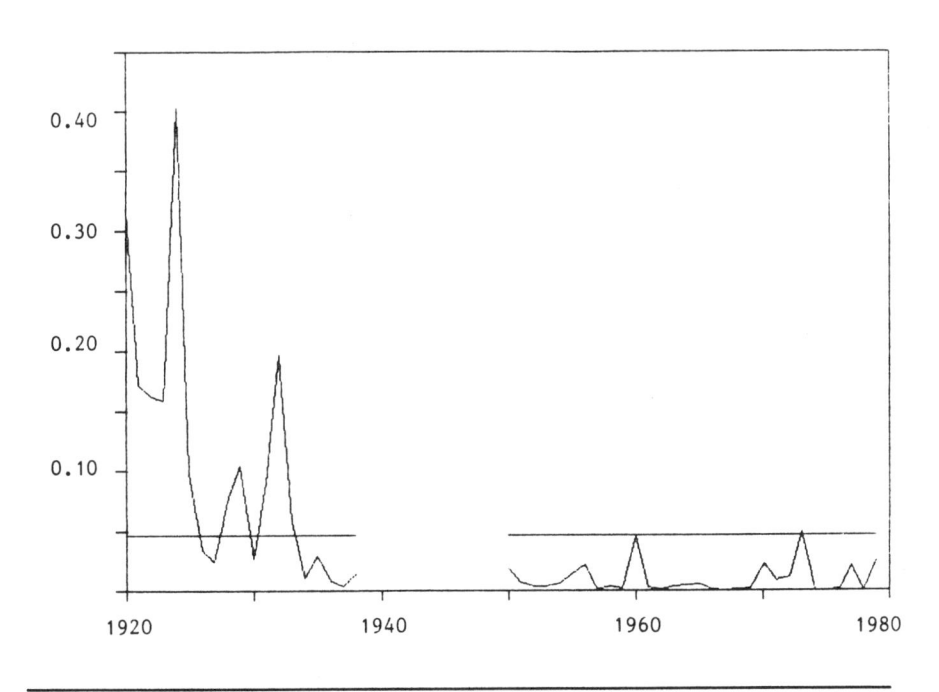

Source : Table 5.10. and appendix B, table B.6.I.

described it for the Netherlands. The differences in patterns of strike activity before and after World War II can be related to differences in the degree of pillarization.

In the 1920's, when pillarization came about as an institutional framework, consultations on industrial relations between employers, employees and government were not yet institutionalised to such a degree as after World War II. Negotiations did not take place on a centralized level. Union's policy and structure were not yet strongly centralized on the national level. What is more, unions were often mutually competing for the favours of the workers, a factor promoting their engagements in industrial disputes.

Employer's organizations and unions became more and more integrated in pillarized structures as time went by. The denominational unions became part of the very same political structures as their natural counterparts, the denominational employers' organizations, which was a favourable condition for the institutionalization of consultations to develop. Next to this, the integration of unions into the pillarized system meant that the unions' elites took over the culture of that system. Although the unions still presented themselves as rivals competing for the workers' favours, their elites gradually started to follow the rule of compromising. This made the realization of a certain coordination of unions' policies possible, while the inclination was suppressed to start strike activities for reasons of gaining workers' favours.

In the 1950's the institutional framework of industrial relations matured further to a tripartite relationship between unions, employers and government. Besides, in the 1950's the Social-Democratic Party participated in government. Many issues of dispute between employers and employees were removed to the political arena[76]. As the unions' elites institutionally were so connected with government and employers, their attitudes towards strike activity were negative. It explains why the wildcat strike was so dominant during those years[77]. So far as the official unions organized strikes, they mostly did as a reaction to wildcat strikes with the aim of preventing the loss of members. The wildcat strikes after World War II indicate the tensions between union's elites and rank and file caused by the strong commitment of unions' policies to national socio-economic policy. The characteristics of the system of industrial relations after the war - integrated into the pillarized structure and highly centralised - as well as its built-in tensions, explain why the strike activity in that period was of a relatively short duration and manifested itself at the national level.

Individual Behaviour: Social Distance

In a pillarized social system the social distance between the members of different pillars will be large. In daily life people prefer to interact with those who adhere to the same values and norms. This is a natural tendency of the human kind. Everywhere people form networks, be it on the basis of either class or status divisions or both. Although in pillarized society status and class are still relevant concepts in defining social categories, the main criterion determining the chance of interaction is membership of the pillar. So Protestant workers prefer to interact with other Protestants rather than with Roman-Catholic workers or socialists. The same can be said with concern to the middle and upper classes. This does not mean that within the pillar social distance between classes or statusgroups is small. What I want to stress is the fact that in pillarized society for the individual the main locus of identification lies within the pillar rather than within a broader peergroup which shares the same economic and social conditions outside the pillar.

An indicator for measuring the degree of social distance between social categories is the connubium, i.e. the degree to which people marry within their own social category. As a variable the connubium is expressed by the number of the mixed or homogeneously married. The mixed married are defined as those who are married to a partner adhering to a different denomination. So, for example, in case a Roman-Catholic is married to a Protestant, both of them are regarded as being mixed married. Since we regard the qualification "no denomination" as a subcategory of the variable "denomination", the same holds true for e.g. a Roman-Catholic married to a partner who adheres to no denomination at all. If both partners belong to the same denomination, they are considered to be homogeneously married.

Table 5.11. *Homogeneously and Mixed Married in the Netherlands per Denomination as a Percentage of All Married per Denomination in 1947 and 1960*

	Roman-Catholic 1947	Roman-Catholic 1960	Protestant 1947	Protestant 1960	Others 1947	Others 1960
homogeneously	93.6	94.7	90.3	92.2	60.7	71.5
mixed	6.4	5.3	9.7	7.8	39.3	28.5
total (x1000)	1,305	1,814	1,710	2,007	171	204

	No-Denomination 1947	No-Denomination 1960	All Denominations 1947	All Denominations 1960
homogeneously	81.9	87.1	88.5	91.3
mixed	18.1	12.9	11.5	8.7
total (x1000)	750	1,013	3,936	5,038

Source : Figures on 1947 derived from CBS, 1951, table 7, p. 108; figures on 1960 from the national census 1960 computed on the bases of G. Dekker, 1965, tables 7 and 8, pp. 98,99.

Table 5.11. presents the homogeneously and mixed married per denomination as a percentage of all married per denomination. At first glance, the percentage homogeneously married seems pretty high, especially for the Roman-Catholics and the Protestants. For a more precise evaluation we need a frame of reference with which we can compare the data of table 5.11. Therefore we accounted the probabilities for men and women to marry a spouse of the same denomination as if the choice of a partner proceeded completely at random. These probabilities are given in table 5.12. Of course, this way of comparing the real situation with a hypothetical one is not perfect. It is a well known fact that in general people tend to meet and relate more frequently with their fellowmen that belong to the same social class or statusgroup and adhere to the same ideology. So even in non-pillarized societies we may expect the number of marriages mixed by class, status and ideology, to be lower than their probabilities as a result of a random process. Further, the degree of geographical dispersion of a denomination affects the probabilities on homogeneous and mixed marriages.

Table 5.12. *Probabilities to Be Homogeneously Married for Men and Women per Denomination in 1947 and 1960*

	Catholic 1947	Catholic 1960	Protestant 1947	Protestant 1960	Others 1947	Others 1960	No-Denomination 1947	No-Denomination 1960
men	33.5	36.4	44.0	40.1	4.6	4.3	18.0	19.2
women	32.8	35.6	42.9	39.6	4.1	3.8	20.2	21.0

Source : See table 5.11. For each denominational category we accounted the probabilities to be homogeneously married separately for men and women. The probability to be homogeneously married for men of a certain denomination is the number of married women of that denomination divided by all married women. The probability to be homogeneously married for women is the number of married men of the denomination divided by all married men. N.B.: Singles are left out. The probabilities refer only to existing marriages.

Table 5.13. *Homogeneously Married in Current Year by Denomination as a Percentage of All Married in Current Year by Denomination, 1936-1980*

Year	Catholic	Protestant	Others	No-Denom.	Total
1936	86.8	86.2	85.2	59.6	83.3
1940	87.0	83.4	53.7	61.5	80.2
1946	87.6	83.8	37.7	62.8	80.5
1950	88.7	83.2	40.7	62.0	81.1
1955	90.4	84.0	44.8	68.0	82.9
1960	89.7	82.5	44.0	67.4	81.9
1965	88.4	80.2	41.7	66.7	80.4
1970	82.9	74.1	39.7	66.3	75.3
1975	78.5	68.6	43.7	64.6	71.2
1980	77.6	68.4	50.3	66.9	71.1

Source : Appendix B, table B.7.II.

Taking these disadvantages into consideration, however, the comparison of the actual with the hypothetical situation reveals the important role of ideological frames of reference in the daily lives of Dutchmen at the end of the 1950's. The deviation between the actual percentages and the probabilities of the homogeneously married is considerable for all groups. Using connubium as an indicator, we may conclude that in the 1950's in the Netherlands the social distance between denominational groups was large indeed.

The tables 5.11. and 5.12. relate to existing marriages. Marriages that lasted for years and just married couples have been taken together. Social factors dating long before the time of measurement may therefore express themselves in these tables and bias our conclusions with respect to the 1950's. It may well be that for young couples at the time the choice of a marriage partner was governed by other considerations than those in earlier generations. To get an estimate of this bias we have to look to the marriages that were contracted in the year of measurement. Table 5.13. gives the percentages of the homogeneously married in current years from 1936 to 1980. With the exception of the column "Total" the same figures are also presented in figure 5.6. These percentages are considerably lower than those of the homogeneously married.

Two conflicting explanations for the difference between the percentages of the homogeneously married and the homogeneously married in current year can be put forward. The first states that among the married many have contracted their marriages in a period in which the mixed marriage was taboo and that new couples are more inclined to a mixed marriage[78]. From table 5.13. and figure 5.6. one can see that only after 1960 the percentage of homogeneously married in current year decreases steadily. This means that not earlier as the 1960's the married in current year show a greater inclination to mixed marriages.

The second explanation is known as the hypothesis of the tendency to homogeneity[79]. It says that once married, in a mixed marriage one of the partners tends to change denomination in due course. The result is that a contracted mixed marriage becomes an existing homogeneous one.

In view of our own data the first hypothesis seems less plausible than the second. If the hypothesis of the tendency to homogeneity is not too far removed from the truth, it illuminates some other aspect of the issue of social distance in pillarized society. The Dutch churches worried a lot about the phenomenon of the mixed marriage. It

Figure 5.6. Homogeneously Married in Current Year by Denomination as a Percentage of All Married in Current Year by Denomination, 1936-1980 (RC(a); Prot(b); All(c); No(d); Others(e))

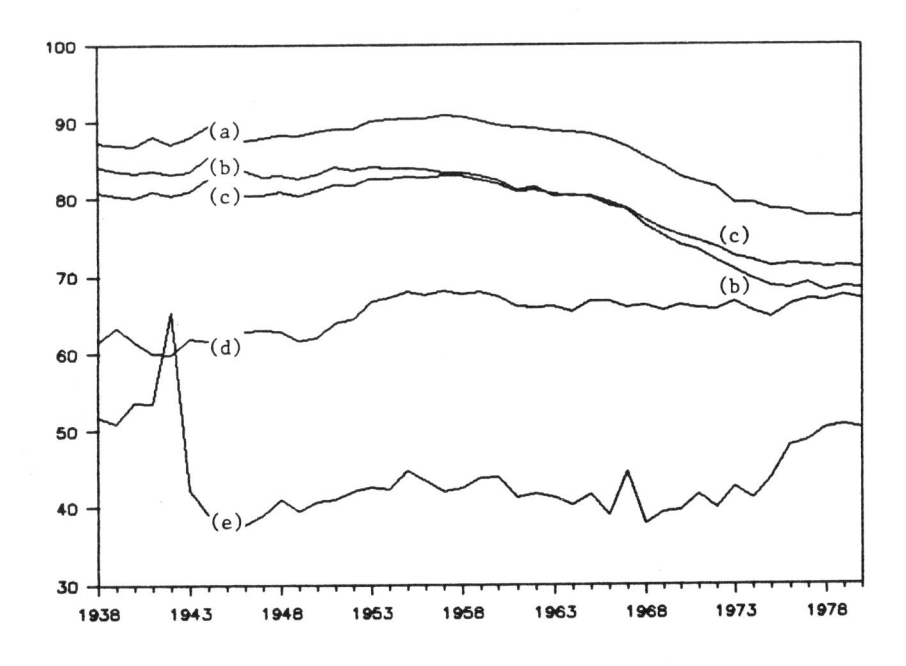

Source : Appendix B, Table B.7.II.

was generally considered a threat to the purity of faith. Parents used to strongly advise their children to look for a spouse of their own denomination. If the couple to be married turned out to be stubborn and successfully resisted social control, during their marriage they had to follow explicit rules concerning the religious education of their children. In other words, before and after contracting, a mixed relationship was subjected to severe social control and evaluated negatively by both the denominations and the families of the partners. The transition of one of the partners to the denomination of the other was a logical solution. The hypothesis of homogeneity then points to the effectiveness of social control in holding out the separation between the denominations.

Similar to existing marriages, for the married in current year I also computed the probabilities of a homogeneous marriage for both men and women (see appendix B, table B.7.III). Consequently I subtracted these probabilities from the real percentages of homogeneously in current year married men and women per denomination. The figures 5.7. till and including 5.10. present the results of the latter operations. As the curves approach the X-axis, the more the observed percentage of

Figure 5.7. Differences Between in Current Year Homogeneously Married Roman-Catholics (Men (a);Women (b)) as a Percentage of All in Current Year Married Roman-Catholics (Men resp. Women) and Their Probabilities of a Homogeneous Marriage, 1936-1980

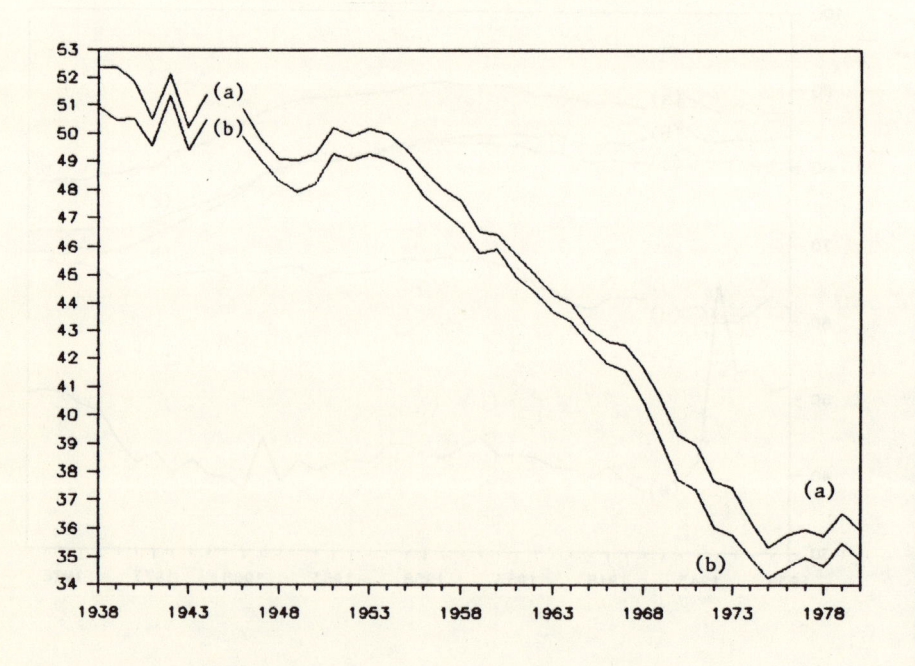

Source : Appendix B, tables B.7.V.a. and B.7.V.b.

homogeneously married men respectively women approximates the corresponding probabilities. In other words, the higher the percentage difference between actual percentage and probability the larger the social distance between the denominational group involved and others.

Figure 5.7. shows for both Roman-Catholic men and women a decreasing tendency to engage in homogeneous marriages. This development set in as early as 1953. For 1936 and 1937 the differences between Roman-Catholic homogeneously in current year married men respectively women are even higher than after 1937 (see appendix B, tables B.7.V.a. and V.b.). This elicits the conclusion that at least in the second half of the 1930's the social distance between Roman-Catholics and other denominational groups was higher than after World War II. In contrast, before the War the Protestants had a lower difference-rate between the percentage homogeneously married in current year and the corresponding probabilities.

It is difficult to interpret the Roman-Catholic pattern. Due to a history of discrimination, once emancipated, Roman-Catholic social control may have been

Figure 5.8. Differences Between in Current Year Homogeneously Married Protestants (Men (a); Women (b)) as a Percentage of All in Current Year Married Protestants (Men resp. Women) and Their Probabilities of a Homogeneous Marriage, 1936-1980

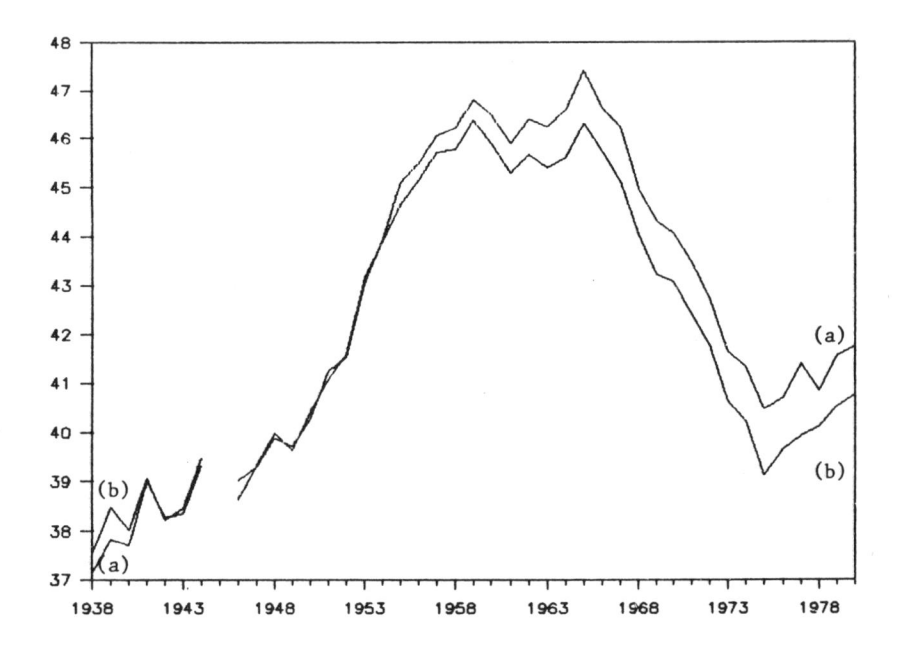

Source : Appendix B, tables B.7.V.a. and B.7.V.b.

stronger than that of other groups. However, without more detailed research we cannot be sure.

The Protestant pattern (figure 5.8.) is almost contrary to that of the Roman-Catholics. From the Interbellum onwards the differences between factual homogeneously in current year married Protestant men and women and their probabilities of a partner of the same denomination increase steadily till 1953 and decline after 1965.

With the exception of a peak in 1942, the curve of the category "other denominations" (figure 5.9.) is rather flat. Before World War II the figures on which the figure is based, include only Jews. The 1942 peak may be caused by the exceptional social circumstances related to prosecution by the German Nazis which implied that for Jews the availability of non-Jewish partners decreased.

After World War II the figures of figure 5.9. include other small religious groups as well. Some of these belong to orthodox traditions characterized by high degrees of group consciousness and social isolation. Others are more liberal oriented. The hybrid nature of the category "other denominations" explains why it does not show a

Figure 5.9. Differences Between in Current Year Homogeneously Married "Others" (Men (a); Women (b)) as a Percentage of All in Current Year Married "Others" (Men resp. Women) and Their Probabilities of a Homogeneous Marriage, 1936-1980

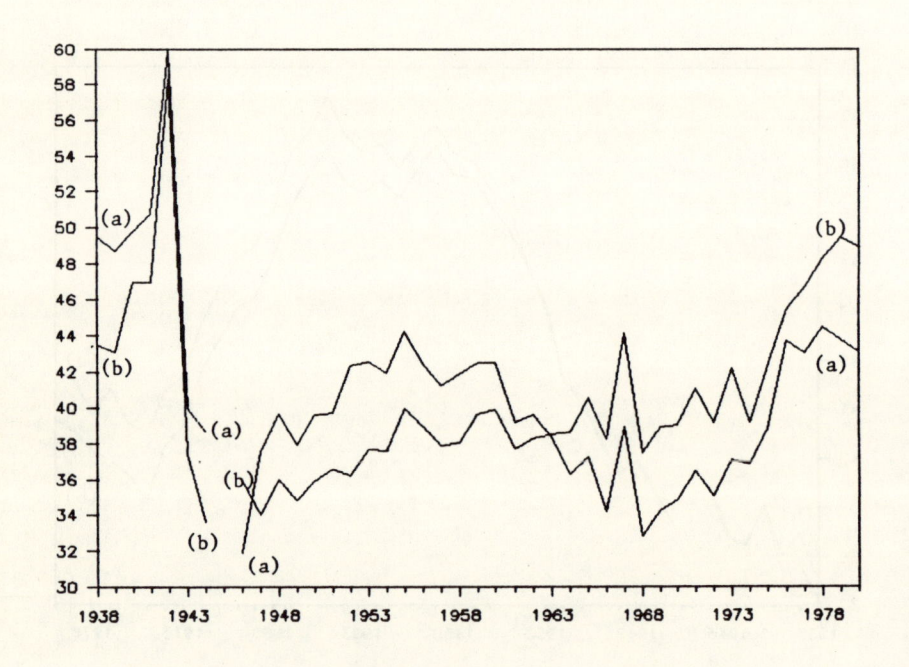

Source : Appendix B, tables B.7.V.a. and B.7.V.b.

clear unambiguous pattern. Because of its religious melange it cannot be considered a pillar.

The curve of the category "no denomination" (figure 5.10.) shows a peak in the second half of the 1950's and declines after 1958.

Our method of comparing reality to probability does not enable us to draw conclusions with regard to the differences in levels between the figures. We did only measure the extent to which the choices of men and women for a marriage partner were affected by the availability of possible marriage partners belonging to other denominations. The figures clearly show that it fluctuated in time and that the patterns of the different denominations are not similar. The above analysis makes clear that at the level of individual interaction one cannot speak of the 1950's as a peak of Dutch pillarization. The picture is much more complicated. While in those years the Protestants became more inclusive as a group, the Roman-Catholics became more open to adherents of other ideologies.

Before ending the analysis of the mixed and homogeneous marriages in the 1950's, I would like to make some remarks on the effect of regional variability. If

Figure 5.10. Differences Between in Current Year Homogeneously Married "No Denomination" (Men (a); Women (b)) as a Percentage of All in Current Year Married "No Denomination" (Men resp. Women) and Their Probabilities of a Homogeneous Marriage, 1936-1980

Source : Appendix B, tables B.7.V.a. and B.7.V.b.

denominational groups are exclusively concentrated in certain regions the probability of mixed marriage in those regions will be low and that of homogeneous marriage high. In other words, besides the cultural factors mentioned before, the regional concentration of denominational groups is also one of the determinants of this probability.

In his book on mixed marriages Dekker gives figures of the mixed married in all the provinces and big cities of the Netherlands for 1947 and 1960[80]. Firstly, in the Southern provinces of Brabant and Limburg the percentage of mixed married is below the national average. This is explained by the fact that these provinces were largely dominated by the Roman-Catholic denomination. Secondly, in the provinces of Utrecht, Noord-Holland and Zuid-Holland the percentages are above the national average, probably because the big cities with their liberal climate of cultural heterogeneity are located in these regions. Thirdly, in every province and for each denomination there is a decrease in the percentage of mixed married from 1947 to 1960, which corresponds to the national trend. Fourthly, as on the national level, in all provinces the percentage of mixed married is lower than that of mixed married in current year.

Due to regional differences in denominational concentration rates and urbanization, a regional variability in the percentages of mixed marriage can be observed for the 1950's. On the other hand, national trends are reflected in all provinces: a decrease in the percentage of mixed marriages and a trend to homogeneity after contracting marriage. Because of the reflection of the national trends in the regional data it is unlikely that regional variability in the percentages of mixed marriage makes the foregoing analysis invalid.

Concluding Remarks

Pillarization is commonly described as the process by which society becomes divided along ideologically based vertical lines rather than along horizontal cleavage. In part one of this chapter I theoretically analysed the concept of pillar and the related pillarization process. The outcome was that the emphasis on the religious nature of pillars explains why there has been so much attention to the vertical division of Dutch society. At the same time it turns pillarization into an almost uniquely Dutch phenomenon that is restricted to Roman-Catholics and Protestants. The prevailing pillar concept leads to serious theoretical and empirical problems.

Instead I regarded a pillar as a subsystem in society that links political power, social organization and individual behaviour and which is aimed to promote, in competition as well as in cooperation with other social and political groups, goals inspired by a common ideology shared by its members for whom the pillar and its ideology is the main locus of social identification. According to this definition different pillars may be organized around different types of ideology. This implies that a vertical division of society is not a necessary effect of pillarization, but becomes dependent on other factors as well.

The definition emphasizes that a pillarized society consists of more mutually related pillars. It draws the analysis away from the individual pillar to the relations between the pillars as part of a social system.

I further stressed that pillars link power allocation, social organization and individual behaviour, a point that is often neglected in studies on pillarization. It is this very linkage which sets pillarization apart from other forms of pluralism.

The definition presented led to the formulation of a fifteen statement ideal typical model. As can be noticed from the second part of this chapter reality often deviates from theory.

A review of major theories on pillarization yielded the conclusion that all have their pros and cons and that all leave us with questions. However, Daalder's historically oriented theory combined with the social control thesis is the most powerful. It is substantially strengthened if it is enriched with modernization theory. Pillarization then results from the incompatibilities between modernization and traditional ways of power allocation. It is a way of social control embedded in the history of Dutch society.

Part two of this chapter monitored pillarization as an historical process. The introduction of the time factor brought out some interesting phenomena.

The qualitative analysis of nineteenth century pillarization in chapter IV led to the conclusion that the social groups involved pillarized at different moments in time. The figures on connubium presented as an indicator of social distance between members of different pillars also revealed diverse patterns for the various pillars.

Further, the three levels of pillarization did not show a synchronized development. The denominational vote started to decline as early as the first half of the 1950's. The downfall of pillarized organization was far less evident and in some cases even absent. Especially in the field of industrial relations, attempts to neutralize pillarization shortly after World War II failed. The exception to this meso pattern is broadcasting. Denominational broadcasting declined relatively after 1965. However, this rather spectacular development took place a decade after the decrease of denominational vote set in. At the level of individual behaviour group inclusiveness decreased for Roman-Catholics and people of no denomination since about 1958 and for Protestants since about 1965.

From the analysis of time series we learned that pillarization has never been a broad general phenomenon which applies to all sectors of society to the same degree at the same moment in time. One cannot properly speak of the 1950's as the peak of Dutch pillarization. It is not that simple. A specification is needed as to what social groups as well as to what kinds of social behaviour the statement is directed.

CHAPTER VI

DUTCH DE-PILLARIZATION: THEORY AND PROCESS

Introduction

Chapter V explained Dutch pillarization in terms of the acceleration of modernization that took place in the second half of the nineteenth century. Chapter III concluded that Dutch modernization advanced further in the 1960's. According to the theory presented so far, we might thus expect the degree of pillarization to increase after the 1960's. However, in the literature on modern Dutch society it is generally assumed that in the 1960's de-pillarization set in[1]. Many of the time series previously presented indeed show a break in that decade. How can we explain that modernization produced pillarization in the late nineteenth century and is associated with de-pillarization in the second half of the twentieth century?

I will take the position that pillarization as a strategy of social control leveled incompatibilities within the nineteenth century process of modernization. However, as modernization accelerated again in the 1960's, existing pillarization became incompatible with new technologies and the associated social requirements. In other words, this chapter attempts to demonstrate that both pillarization and de-pillarization are related to qualitatively different periods[2] in the modernization of Dutch society.

Before plunging into theoretical discussions, we have to review the data collected to monitor pillarization again, but now from the perspective of de-pillarization. This implies frequent reference to tables and figures presented in chapter V.

Monitoring De-pillarization

Politics

From a survey of the Dutch electorate it was concluded that since the beginning of the 1960's the percentage of religious voters has decreased steadily and that the intensity of religious belief measured by frequency of church attendance has declined. Both changes were the effect of a changed composition of the electorate: the inflow of new cohorts of young voters and the disappearance of old cohorts. An average decrease in the intensity of religion, however, can be found among all age-groups[3].

In chapter V (table 5.2.) the percentages of votes in national elections for the Roman-Catholic party, the two Protestant parties, the Social-Democratic party and the Liberal party were given for the years from 1922 to 1986. Since 1977 the Roman-Catholic and the Protestants came with a combined list, the Christian-Democratic Appeal (CDA). From table 5.2. we can learn that the combined electoral support for the denominational parties (column 1 + 2) dropped from 50 per cent in 1956 to 29.4 per cent in 1982; that is a negative average annual compound growth rate of -2.0 per cent! This drop in public support for the denominational parties set in after the elections of 1963. If we compare electoral support for the Roman-Catholics with that for Protestants, the latter suffered less than the former in the period 1956-1972. For those years the negative average annual compound growth rates for the Roman-Catholic party and the two Protestants parties are respectively -3.6 per cent and -1.9 per cent.

From 1982 to 1986 electoral support for the CDA rose substantially from 29.4 per cent to 34.6 per cent. When the 1986 election results became known, many comments in the mass media expressed the opinion that the CDA success could be explained by an increase of CDA attractiveness to those voters not affiliated to church or religion.

According to a survey of the 1986 electorate, only 5 per cent of all CDA voters were traditionally non-church members, 11.3 per cent were former church members, 47.4 per cent were church members who attended church incidentally and 36.4 per cent were regular church members[4]. So in the 1986 elections the proportion of non-church CDA voters was rather small (16.3 per cent) whereas those who voted CDA and had no religious background whatsoever were only 5 per cent of all CDA voters. Therefore, the conclusion that the 1986 rise of the CDA vote can be explained by substantial support of non-religious voters seems doubtful.

On the other hand, in 1986 almost half of the CDA electorate consisted of people born and educated as Christians and still considering themselves believers but only incidentally attending church. This may reflect a *vertrossings*-effect in Dutch politics, i.e. a gradual secularization of the content of Christian-Democratic politics may have broadened its electoral base, in particular to those who by tradition feel affiliated to Christianity. I would suggest that the 1986 increase of CDA votes is an expression of progressive secularization of Christian-Democratic politics and not of a restoration of political pillarization.

If we compare the years from 1922 to 1956 with those from 1956 onwards (see table 5.2.) the weakening of the position of the denominational parties becomes the more clear. Firstly, taking the coefficient of variability as a measure of the stability of constituencies, the instability of the denominational constituency after 1956 increased, the coefficient of variability being .03 for the period 1922-1956 and .21 for the years from 1956 to 1986. So, after 1956 pillarization lost its effectiveness as a political strategy to guarantee stable constituencies.

Secondly, as a result of the decrease of electoral support after 1956, the denominational parties together were no longer able to form a political majority. This fact, however, did not weaken their power position substantially as the country could still only be governed by coalitions in which Christian-Democrats participated. Thirdly, in the seventies the percentage of liberal votes increased, which explains why the liberal coefficient of variability is higher after 1956 than before.

In table 5.2. only figures for the major parties are presented. Beside these parties the Dutch system of proportional representation has implied the existence of many small parties. Some of these are orthodox denominational parties, others belong to the extreme left or right without a denominational base. Although since the introduction of general suffrage only the major parties listed in table 5.2. have been of importance for the division of power in society, the trend towards political de-pillarization becomes more evident if we take into account the smaller parties as well. In table 6.1. and figure 6.1. all votes cast are dichotomised into votes for denominational based parties versus votes for non-denominational parties.

The material presented in this paragraph suggests that after the 1960's the organization of democratic power in the Netherlands changed. If religion became less important in Dutch society, it is understandable that it lost its effectiveness as a way of binding constituencies to denominational parties. Dutch politics de-pillarized. The remarkable loss of electoral support for the denominational parties, however,

Figure 6.1. *Votes Cast for Denominational (a) and Non-Denominational Parties (b) in National Parliamentary Elections (2nd Chamber) as a Percentage of All Votes Cast, 1946-1986*

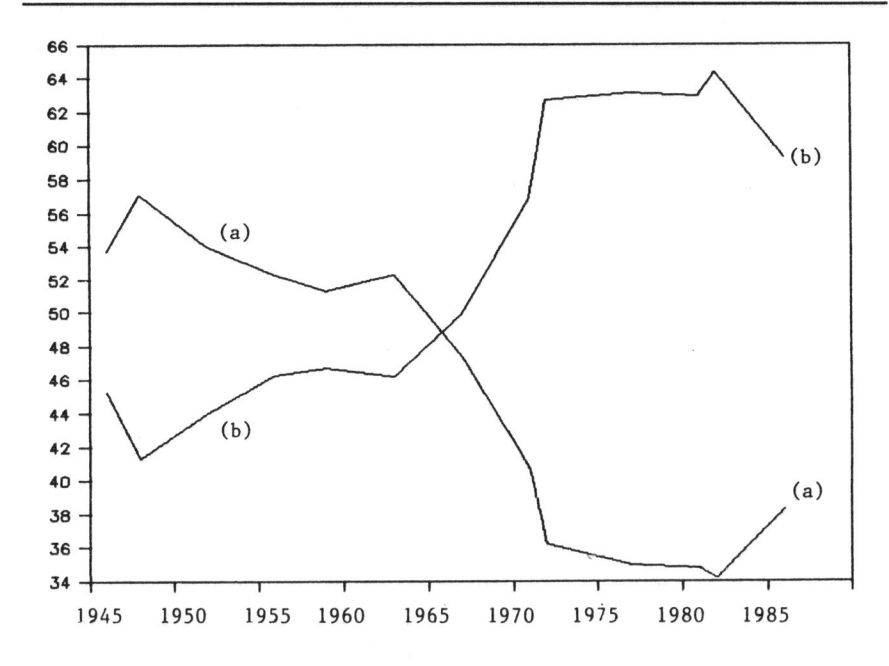

Source : Table 6.1.

did not invoke a loss of their political power to a comparable degree. Since the late sixties, social-democrats and liberals have mutually excluded each other as coalition partners. As a result both have had to compete for the favour of the Christian-Democrats in order to rule. This kept the latter in the centre of power.

De-pillarization is a matter of degrees. In the Netherlands of to-day the older part of the religious electorate still feels a need to vote for a denominational party because of religious motives. In the denominational constituencies the elderly are over-represented which, under unchanged conditions, represents a threat to the future of these parties. Thus, as the significance of religion as a basis of political power decreases, denominational parties are urged more and more to win votes on the basis of political issues. The favours of the religious electorate can no longer be taken for granted, but have to be competed for with non-denominational parties like the social-democrats and the liberals.

In contrast with the past, loyalty to the Christian-Democratic party became dependent on the choices made in every-day political life. This has confronted the Christian-Democrats with a left-right dilemma. In the old days denominational parties had succeeded in binding left wing as well as right wing voters. Nowadays, right wing favoured decision-making by these provokes the left wing of their electorate to desert to the social-democrats, while right wing decisions elicit electoral loss to the liberals. Thus a decreasing importance of religiously oriented power in Dutch politics brings the class dimension to the fore.

Table 6.1. *Votes Cast for Denominational and Non-Denominational Parties in National Parliamentary Elections (2nd Chamber) as a Percentage of All Votes Cast, 1946-1986*

Year	Denominational	Non-Denominational
1946	53.7	45.3
1948	57.1	41.3
1952	54.0	44.0
1956	52.3	46.3
1959	51.3	46.7
1963	52.3	46.2
1967	47.4	49.9
1971	40.7	56.9
1972	36.2	62.7
1977	35.0	63.1
1981	34.8	62.9
1982	34.2	64.3
1986	38.3	59.3

Note : Percentages do not add up to 100 percent because of the votes for parties that did not get a seat in parliament.
Source : 1946-1986 : C. van der Eijk and B. Niemöller, 1983, table 1, p. 171.
          1986 : *Keesing Historisch Archief*, 1986, p. 354

The merger of the three major denominational parties into one united Christian-Democratic party is as such a sign of de-pillarization. It implies the levelling of political barriers between Roman-Catholics and Protestants. It was the Roman-Catholics in particular who, having lost the most electoral support after the 1963 elections, strived for this merger.

The equilibrium between the different ideological elites that was characteristic for the Dutch pillarized political system, was menaced by a more general trend in the 1960's: the democratization of society. As we stated before, the Dutch political system based on ideological cleavage at the basis of society could only achieve stability because at the national level the political game was played according to certain rules. The democratization of society led to a democratization of internal party procedures that made the observance of these rules almost impossible. Two of these rules, the de-politicising of hot issues and bargaining behind closed doors, could only exist thanks to a certain degree of passivity of the constituencies and a lack of internal party democracy which gave party elites room to manoeuvre.

As in many other Western democracies, the student revolt of 1968 was one of the first manifest signs of a growing tendency towards democratization. Soon it had its effects on nearly all domains of social life. Everywhere authority came under attack. Most of the time because the attackers wanted a change of things, but sometimes the attack was launched against the very phenomenon of authority itself. In this atmosphere authority became suspect.

In the Netherlands of that time the Provo movement, representing the counter culture, was aimed at provoking authority wherever it could and for its own sake. In the political domain, however, as early as 1966 a new political party was established that aimed at changing the Dutch political system into a more democratic direction, i.e. at giving the electorate a more direct influence on the process of government formation. The intellectuals, mainly from liberal and social-democratic origins, that were gathered in this new party D'66, took the British voting system as their example. As far as politics was concerned, for them the principle of proportionality was to be abandoned. They favoured the American way of campaigning, i.e. a direct and open approach to the public which stood in sharp contrast to the traditional Dutch political style. In a short time the ideas of D'66 became very popular. In the elections of 1967, the first in which they participated, the party got 4,5 per cent of all votes; for the elections of 1971 the percentage was 6.8.

The impact of D'66 was greater than these percentages do suggest. The traditional parties, although not taking over the specific reform proposals of D'66, incorporated the spirit of the new ideas. This holds true especially for the Social-Democratic party, the PvdA. So, the least one can say is that D'66 transferred the general trend towards democratization to the political field and that the democratic attitude of this party influenced the other parties to a substantial degree.

This all resulted in a growing resistance to the traditional power game. Loyalty to leadership was no longer self-evident. Instead, party elites were held responsible for the decisions they took. In dealing with the leaders of competing parties, they were forced to let their parties approve the results of the negotiations. Due to this rather fundamental change of mentality the formation of a government proved to be more and more difficult and time consuming. The results of the negotiations at the national level were in some cases judged by special party congresses and rejected so that the negotiations had to start all over again. Agreement between the national elites was reached with much more difficulty than in the fifties. In 1977 a record was established: the formation of a new government required 208 days!

Of course, the impact of the new democratic ideology was a matter of degree and differed from party to party. The left wing parties were influenced to the highest degree, the impact on the right wing parties was slightest. It is significant that even the Dutch Communist Party (CPN), although small in its support and never playing a role in national politics, had to give up the practice of "democratic centralism".

Not only national issues were politicised by democratization, but also the differences of opinion that existed within every political party. In some cases this led to separation and the establishment of new, mostly small, parties. So, the new culture of democratization effected a change in the party system itself: in the period from 1946 to 1967 only five parties played a national role, the Roman-Catholic party, two Protestant parties, the Social-Democratic party and the Liberal party. Along with these five, several small parties existed that were oriented either to the more radical left or to the more orthodox right. But all these parties showed ideological consistency and electoral stability to a high degree.

The new parties that arose in the late sixties, however, had no historically determined ideological roots and were formed to promote the solution of social problems like reforming the political system, solving the environmental problem and solidarity with the Third World. Ex post one could say that for a great deal they represented the fashions of the era. Their constituencies consisted largely of a floating vote and were therefore unstable.

Two parties, however, had a somewhat different outlook and must be mentioned by name. The first is the Farmers Party (Boerenpartij) which assembled under its wings all those who were, for one reason or another, dissatisfied with and frustrated by the state of the nation. This right wing and autocratic party had no clear ideology and can best be compared with the French Poujadist party. In the elections of 1967 it reached the peak of its electoral success with 4.7 per cent of all votes. In the elections that followed it faded away. In 1981 it gathered only 0.2 per cent.

The second party to be mentioned is DS'70 (Democratic-Socialists 1970). This was a right wing split of the Social-Democratic PvdA. DS'70 gathered parts of the traditional PvdA elite which could not agree with the radicalization of their party of origin. They wanted to go back to the traditional style of social-democratic leadership and hoped to assemble all those who were against the New Left. For one year, 1972-1973, DS'70 had governmental responsibility. After forcing a cabinet crisis in 1973 its political power, already small, crumbled.

De-pillarization and democratization led to an increase of the floating vote[5]. In the combat for this floating vote politicians were forced to play more than ever for the public, to outline their mutual differences sharply and to present their opinion to the electorate whenever possible by the use of the mass media. In the 1970's this had as an unavoidable effect political polarization and it undermined the traditions of the pillarization and accommodation era further. At the end of the seventies a new development set in. The influence of the smaller parties diminished and only three parties remained of importance for the division of political power: CDA, PvdA and the Liberal party VVD.

History never repeats itself. Although the old parties had come to dominate the political arena again, the rules that determined their interrelationships had changed profoundly. Because the system of pillarization had broke down to a substantial degree, the phenomenon of the floating vote remained a menace to each of them. As a result the elites of the three parties had to continue polarization and to accept the influence of party rank and file on which they remained highly dependent. Therefore, the old rules seemed to have passed away forever.

Another effect of the trend towards the democratization of society, was the rise of action groups as an alternative to the traditional pillar connected interest groups. Although both must be considered as pressure groups, their outlook, structure and way of promoting their goals is completely different. Action groups are pressure groups that are loosely organized to promote a limited goal by recruiting those who have an interest in that goal irrespective of ideological background. Action groups promote their goals usually by extra-parliamentary and often even by illegal means: blocking roads to promote the provision of traffic lights, blocking motorways by tractors to influence agricultural policy, occupation of universities by students to enforce changes in education and refusal to pay taxes as a protest against nuclear armament. It is by the illegal nature of their actions that these groups try to attract the attention of the media to their cause. The public attention they raise in this way has proven to be a powerful weapon to influence the political establishment.

The traditional interest groups of pillarized society are, on the other hand, pressure groups linked to political parties by interlocking directorates. They try to promote their objectives by putting pressure on politics via this political party. Interest groups are therefore an integral part of the political system. They have a formal organizational structure and they recruit as their members those who have the same interests as well as a similar ideological background. Examples are trade unions, farmers unions, employers organizations, etc. In the economic domain, these pillarized interest groups fitted perfectly in the institutionalised system of consultation as described above. It follows that interest groups only use accepted legal means. Besides the means they use, the crucial difference between the action group and the interest group is that the latter is ideologically more homogeneous and strongly connected with the political system, while the former is not. Action groups do not fit into pillarized social structures, since they do not integrate people on the basis of a common ideology as well as a common interest, but only on the basis of an often short term common commitment to a particular issue.

The first action groups appeared on the Dutch scene in the late sixties. Nowadays they are a more or less accepted phenomenon to which politicians are accustomed. Their significance lies in the fact that they give the political arena a double structure: on the one hand the political parties with their interest groups, on the other hand the action groups that try to influence these parties, parliament and government. On the one side of this double structure the organizing principle is still ideology - although to a far lesser extent than in the times of pillarization -, on the other side the organizing principle is a common interest promoted by people that may be dissimilar in ideological background.

To complicate the matter further, it may well be that people adhering to a certain action group are also members of particular interest groups. Further, that relations exist between action groups and well established interest groups. An example of the latter is the relationship between action groups of social assistance beneficiaries and the unions.

The most important conclusion that can be drawn is, however, that the rise of action groups and the enormous spread this phenomenon has taken since the late sixties, indicate the trend that ideological cleavage at the basis of Dutch society has become less important than before and that, in the political domain, people tend to organize around issues instead of ideologies.

Organizational De-pillarization

From 1966 onwards the Dutch Social and Cultural Planning Agency (Sociaal en Cultureel Planbureau, SCP) has measured public attitudes on pillarization by surveys of nationally representative samples. One of the instruments the SCP uses is a *pillarization-index* based on a number of items indicating the preferred relation between religion and social organization[6]. Over the years the index was measured for both church members and all respondents. In its report on 1986 the SCP concludes that "in the period 1966-1970 an important change took place among church members into the direction of heterodoxy and de-pillarization"[7]. The same trend to de-pillarization can also be observed for all respondents together. After 1970 these trends stabilize[8].

The SCP pillarization-index focuses on attitudes to organizational pillarization. Our concept is somewhat broader. Firstly, we try to measure actual behaviour as well. Secondly, our concept of pillarization includes political trends and individual behaviour as well. With regard to organizational de-pillarization we shall now monitor this phenomenon by the same indicators we used in chapter V.

— Education

From table B.2.II. (appendix B) and figure 5.2. it becomes clear that since 1970 the percentage of pupils attending denominational elementary education has decreased. Regarding university students we can see a reverse trend (table 5.5.), but as explained before, we cannot reach valid conclusions from the latter as it is biased because of increased supply of academic facilities.

The shift from denominational to non-denominational elementary education is also reflected in the surveys of the Dutch Social and Cultural Planning Agency. As an item of the above mentioned pillarization-index nationally representative samples were asked a question regarding the preference for non-denominational or denominational elementary education (table 6.2. below). What strikes the eye is not so much a rise of the preference for public non-denominational schools as the increase of the category "no preference" and a decrease of the category "preference for denominational schools". This trend holds for all respondents as well as for the subcategory church members. Thus the distinction between non-denominational and denominational elementary education has weakened in the minds of large parts of the Dutch population in a period of twenty years. The choice of a particular school is more dominated by profane factors like the quality of education and the distance from home to school.

Table 6.2. *Public Opinion of All Respondents (a) and Church Members (b) on Preferences for Public Non-Denominational and Denominational Schools (percentages), 1966-1986*

	1966 (a) (b)	1970 (a) (b)	1980 (a) (b)	1986 (a) (b)
public school	39 17	35 16	36 12	39 12
no preference	6 5	24 25	26 24	25 26
denominational school	55 78	41 59	39 64	36 62

Source : *Sociaal en Cultureel Rapport 1986*, 1986, pp. 373-374, annex table 11.29.

From both the change in the number of denominational enrollments as well as from the shift in public opinion we may conclude that the impact of denominational schools as socializing agents for the Christian pillars has lost importance. The latter has been strengthened by the fact that the ideological content of the curriculum of many a denominational school has become restricted over the years. In many cases the Christian character of a denominational school is only clearly recognizable around Christmas time. In the West of the country one can easily find Christian schools attended by Islamic youth. Thus the curricula of pillar administered and public schools have grown more similar over the years.

Under conditions of progressive secularization of society and a falling of the birth rate, pillar administered schools could only compete with other publicly administered types of education by decreasing the ideological component of their curriculum. Inevitably, by this they lost importance as an agent of socialization for the pillar, the more, as this development implied that the population of pupils had increased in heterogeneity. Nowadays, many Protestant children go to Roman-Catholic schools, just as many Catholics attend Protestant education; a situation unthinkable in the 1950's.

In the 1980's, Dutch educational policy seems to be hindered by the historical remnants of the pillarized school system. The principle of "freedom of education" is sacred for Christian-Democratic politicians. A recent example is the School for Journalists. There was a public School for Journalists which could easily satisfy the demands of the labour market, but the principle of proportionality, so strongly connected with pillarization, required the establishment of a formally Protestant and a formally Roman-Catholic School for Journalism as well. Thus, while for demographic reasons the total number of pupils is decreasing and while the curricula of pillarized and public schools are becoming more and more similar, the pillarized school system is maintained and, in some cases, hinders a rational planning of educational facilities

– Broadcasting and Health Service

As shown in chapter V, de-pillarization can be clearly traced in the domain of broadcasting. Since 1965 the relative shares of denominational and non-denominational broadcasting association membership have been reversed (figure 5.3.). I further pointed to the so-called *vertrossings-effect*, i.e. the phenomenon that in the competition with non-denominational broadcasting associations the denominational associations softened the ideological tones in their radio and TV programs so that the broadcastings of the diverse associations became more similar.

In the domain of health services there has traditionally been a substantial control by denominations. If we compare them with other indicators of organizational pillarization, they have been relatively stable. Thus, in the membership of home nursing services, the denominational quota was almost constant from 1951 till 1972 (table 5.6.). Because figures are only available till 1972, our time horizon is necessarily restricted. However, in politics and broadcasting de-pillarization evolved in the second half of the 1960's. So if the home nursing services de-pillarized at all, they did so after 1972. At that time many local home nursing services merged. Such mergers between denominational and non-denominational home nursing services in itself point to de-pillarization.

With regard to control over hospital beds, the picture is more difficult to interpret (see table 5.7.). Since 1975 the number of beds in Roman-Catholic hospitals decreased by 7 per cent. The percentages of Protestant beds increased from 1950 to 1970 by 5.4 per cent, then declined till 1975 by 3.3 per cent, increased again by 3 per cent in 1977 and decreased by 1.1 per cent in 1985.

As the Protestant and non-denominational quotas fluctuate so strongly over the years, we cannot draw conclusions with regard to a de-pillarization in the field of hospital bed control.

– Economic Life and Industrial Relations

Chapter V illustrated the Dutch system of industrial relations of the 1950's. We observed that pillarization had resulted in a highly segmented labour class. Industrial relations had become an integral part of the larger pillarized social system characterized by a high degree of centralism. It was argued that these institutional traits could largely be seen as a specification of pillarization in the domain of industrial relations. Although the necessity of rebuilding a devastated country certainly encouraged social stability, we concluded that the major factor explaining the extremely low strike activity of those years, was the pillarized character of the industrial relations system. If pillarization dominated the industrial relations in the 1950's to such a degree, we must expect that de-pillarization of Dutch society had consequences in this field.

Most authors agree on the relation between the state of the economy and industrial relations. They differ, however, to the degree to which the two are interdependent[9]. As we do not want to suggest that changes in Dutch industrial relations can only be understood from the viewpoint of the de-pillarization process, we shall summarize the major economic developments since World War II. Where needed, we shall link these economic developments to the changes of industrial relations.

The Dutch economic attitude of the 1950's was dominated by four problems. Firstly, during the German occupation the economic infrastructure had largely been devastated and it had to be rebuilt as soon as possible. Chapter III notes that this coalesced with a process of rationalization by enlargement of scale. Secondly, the loss of Indonesia as a rich colony was felt as an economic deprivation that had to be compensated for economically. Thirdly, the country was confronted with a rapidly growing population. Fourthly, more than before the war the Dutch economy was perceived to be dependent on international markets. It was generally felt that these four problems only could be solved by a major economic effort. All social groups agreed that economic growth was a national aim of the highest priority to which all group interests had to be subordinated[10]. The Dutch unions took over this culture of social harmony and accepted for years a so-called guided wage policy[11], i.e. the control of wages by national government aimed at the restriction of the wage level. The newly established trade-union federation EVC that refused to subordinate the workers' interests to this national priority, turned out to be a failure.

The economic policy of the 1950's that was made possible by the atmosphere of social harmony, succeeded in effecting a decrease of the labour share in income[12], a relative reduction of the tax burden and a relative increase of profits. This laid the foundation of the economic expansion of the 1960's and, as the labour market tightened, favoured the position of labour.

Guided wage policy became untenable in view of rising profits, economic growth and a tight labour market. Under the circumstances of the late 1950's it could no longer be regarded as a national interest. In 1959 guided wage policy was abolished and freedom of negotiation was restored. The 1960's were characterised by increasing claims on prosperity achieved by the efforts of the 1950's. These claims resulted, among other things, in public sector growth and an increase of the labour share in income.

In the second half of the 1960's two other tendencies appear. Firstly, the democratization of society touched the unions too. The authority of the traditional union elites was attacked by the rank and file. Of course, the history of guided wage policy sharpened this attack as the consent of the unions' elites to this policy elicited the suspicion that they primarily identified themselves with the ruling class. The tension between leadership on the one hand and rank and file on the other had always been latently present because of the consultative character of pillarized industrial relations. The combined effect of democratization and rising prosperity made it manifest. In due course new union elites came to power, which identified themselves primarily with the workers from which they were born. This implied that within the trade-union federations the different interests of each single union came more to the fore.

Secondly, the democratization of society led to an increasing emphasis on the company level. Democratization of the company was a major issue for the labour unions in the 1960's. This trend was strengthened by growing unease about the politics of multinational enterprises. It was believed that in this respect countervailing power could only be developed by international union cooperation and an increase of union power at the company level.

In the late sixties the economic climate started to deteriorate. The public sector expanded rapidly and continued to do so till the 1980's. In this respect the Netherlands ran ahead compared with other industrial nations (see table 6.3.). The large volume of the Dutch public sector made the administration to active participant in industrial relations. Especially when economic growth slowed down in the 1970's, the government had a strong interest in the outcome of negotiations between employers and employees. In fact it became the hidden ally of the employers.

While the unions were faced with a new and powerful adversary, their bargaining power was further weakened by the spectacular increase of unemployment of the late seventies. Government and employers presented to the unions the necessity of a restrictive wage policy. An increase of real wages would imply a rise of unemployment. Additional to this policy concept, it was stipulated by the government that the country could only recover from the economic downfall of the 1970's if free entrepreneurship and the functioning of market mechanisms - especially that of the labour market - would be restored.

Table 6.3. Total Government Expenditure as a Percent of GDP at Current Prices, France, Germany, Japan, Netherlands, UK, USA, 1880-1981

	1880	1913	1929	1938	1950	1960	1973	1981
France	11.2	9.9	12.0	21.8	27.6	33.9	38.8	48.7
Germany	10.0	17.7	30.6	42.4	30.4	33.4	41.2	47.7
Japan	9.0	14.2	18.8	30.3	19.8	20.9	22.9	34.1
Netherlands	n.a.	8.2	11.2	21.7	26.8	36.1	49.1	59.0
U.K.	9.9	13.3	23.8	28.8	34.2	32.9	41.5	46.4
U.S.A.	n.a.	8.0	10.1	18.5	22.5	27.9	32.0	34.4
Average	-	11.9	17.8	27.3	26.9	30.9	37.6	45.1

Source : A. Maddison, 1984, p. 57.

Faced with the threat of growing unemployment, the unions once again accepted a policy of wage restriction. Even a decline in the wages of civil servants in the 1980's did not provoke substantial revolt. Instead of promoting class struggle the unions tried to influence government policy into a more Keynesian direction. Their efforts to realize this aim were concentrated on the traditional mechanisms of consultation - like the SER and the political parties - that flowered during the 1950's. Their attempts failed, since they were confronted with right wing governments that time and again neglected the outcome of the consultations[13] when these did not fit into their policy concepts. One could say that the strategy used by the unions in the late seventies and the first half of the eighties was outdated because it was based on the rules of pillarized society which no longer existed at the time. The accomodative style of pillarized politics had been replaced by a more conflict oriented dualistic perspective, making the traditional mechanisms of conflict control on which the unions relied obsolete.

It would certainly be an exaggeration to argue that after 1967 Dutch industrial relations changed from a harmonious into a class struggle oriented phenomenon. From the modus operandi of the unions in the late seventies and the early eighties, it is clear that the institutional heritage of the pillarized past remained powerful. The fact, however, that the unions to some extent put trust into the old institutions of consultation, can be seen as an aftermath effect. The rules of pillarization on which the working of these institutions was based, no longer functioned properly. This can be concluded from the profound changes that took place in this domain: in the 1970's the deliberations between the parties involved had become tougher and more conflictual. Urged by their rank and file, the inclination to conflict of the union elites had increased. Negotiations had substituted consultation. Conflict solving mechanisms had become practically lacking. National government had become a third party instead of a more or less neutral conflict solving agency. Parties negotiated in public instead of secretly[14].

As far as the unions organized strikes these were aimed at influencing both their adversaries in the negotiations as well as the political parties as centers of the traditional consultative system. If, however, the ultimate outcome of political discussions turned out to be negative for labour, strikes were aborted. Thus, in the last instance, the unions recognized the existing political and social order.

Although de-pillarization weakened the institutional framework of Dutch industrial relations it did not result in a significant expansion of strike activity. The arithmetic mean and standard deviation of the period 1965-1980 with regard to the working days lost by industrial disputes are respectively 0.010 and 0.014, only slightly higher than that of the years 1950-1965, respectively 0.009 and 0.011[15]. Compared with other European countries the strike activity in the Netherlands in the sixties and the seventies has been rather low too (see table 5.10. and figure 5.4.).

In 1976 the Catholic trade-union federation NKV and the Social-Democratic trade-union federation NVV merged into one, the FNV (Federatie Nederlandse Vakbeweging). At the time of the merger the FNV embraced 61 per cent of all organized workers and became the most important Dutch trade-union federation[16]. This merger implied another important break with the pillarized tradition as it meant that the Catholic and the Social-Democratic unions from now on had a stronger identification with their common class position than with their respective pillars. Despite the fact that the Protestant CNV was intensively involved in the negotiations on the merger, it could not abandon its Protestant ideology and therefore preferred to continue its separate position. In its own right, however, the CNV was also subject to the trend of de-pillarization. In the early eighties it frequently chose the side of the FNV against the government that was dominated by the Christian-Democratic party CDA. Further, some Roman-Catholic unions which could not agree to a merger with the Social-Democratic NVV left the Roman-Catholic NKV and joined the Protestant CNV.

We can now put together the main conclusions concerning the changes in Dutch industrial relations since the late 1960's. Dutch industrial relations lost their pillarized character to a substantial degree, which manifested itself especially in changes of the ways conflicts were solved as well as in the merger of the Catholic and Social-Democratic trade-union federations into the FNV. De-pillarization of Dutch industrial relations, however, did not lead to a significant increase of strike activity in the 1970's, although in these years some rise of instability can be observed. The relatively low degree of strike activity in the 1970's may be explained by the deterioration of the economic situation which affected the unions' power. More important, however, seems to be the fact that in those years the union elites continued to trust the traditional institutions of consultation without challenging the existing social and political order.

Individual Behaviour: Social Distance

As explained in chapter V, ideological cleavage implied social distance between the members of the social categories involved. The indicator we used to measure social distance was the connubium. The more people tend to marry within their own ideological group, the greater the social distance between the members of the different groups will be. Since a dimension of de-pillarization is the lessening of ideological cleavage at the basis of society, it must ultimately result in a lessening of social distance between the ideological groups involved. In other words, the number of marriages involving partners of different ideological backgrounds must rise.

In figure 5.6. we depicted the percentages of the in the current year homogeneously married by denomination. The percentages for Roman-Catholics, Protestants and the total population start to decrease in the second half of the 1960's. The category "No Denomination" remains more or less constant, whereas the percentage of those with other denominations rises in the 1970's.

A more refined method developed in chapter V is the comparison between the actual number of people marrying a partner of the same denomination with the statistical probability of marrying a partner of the same denomination given the distribution of available marriage partners of the same and other denominations. The difference-rate expresses the differences between actual behaviour and its probability. These difference-rates were previously presented in the figures 5.7. till and including 5.10.

It can be seen that the difference-rate for Roman-Catholics started to decline as early as the second half of the 1950's. The same development can be seen for the category "No Denomination". The Protestant difference-rate goes down after 1965 which is about a decade later than that of the Roman-Catholics.

The subcategory "Others" has quite another story to tell as its difference-rate rises from the beginning of the 1970's onwards. Our analysis is too limited to give a definite interpretation of this phenomenon. However, I would like to put forward some tentative ideas.

If we take a closer look at the absolute figures of marriages contracted in the category "Others" (appendix B, table B.7.I.), we can see that these are rather low compared with those of other denominations. The category "Others" consists of small and orthodox religious groups that may have been less affected by a general trend of religious tolerance than such big denominations as Roman-Catholicism and Protestantism. The same can be seen in politics. The two small parties GPV and SGP representing the orthodox Calvinist tradition maintain over the years an almost constant electoral support. Further, it may well be that the exclusive character of the "religions" of the category "Others" was strengthened as a reaction to the loss of orthodoxy of official Catholicism and Protestantism. This may explain the upward trend of the difference-rate of the category "Others".

The general trend of tolerance as to marrying a partner of a different denomination is also reflected in public opinion data. In surveys of nationally representative samples church members were asked how they would react if their daughter would like to marry someone belonging to a different denomination. The posible answers were: *no objection whatsoever, it depends, would not like it*, and, *would resist it*.

The data of table 6.4. refer to possible consent or resistance of parents and not to attitudes of potential marriage candidates. In that sense table 6.4. is not completely comparable with the data on actual marriage choices. Nevertheless, table 6.4. makes clear that in twenty years the attitude to mixed marriage changed from predominantly negative to positive. It is remarkable, that this shift took place between 1965 and 1970, a period of only five years!

Table 6.4. Public Opinion of Church Members on the Marriage of a Daughter to a Partner of a Different Denomination (percentages), 1966-1986

	1965	1970	1975	1980	1986
no objection whatsoever	25	59	52	55	54
it depends	5	12	19	18	23
would not like it	53	21	25	24	21
would resist it	17	8	4	3	3

Source : *Sociaal en Cultureel Rapport 1986*, 1986, p. 377, annex table 11.32

Dissimilarities in the De-pillarization of Dutch Society

If we review the process of Dutch de-pillarization as monitored in the preceding paragraphs, we can conclude that in general de-pillarization took place after the second half of the 1960's. However, at the same time we are confronted with some dissimilarities within this process. Thus, at the political level the rules of the game changed and the phenomenon of the action group entered the scene. At the level of individual behaviour all ideological groups investigated became less inclusive, but they differ with regard to the moment in time on which the process started. At the level of organizational de-pillarization the trend is less clear. For some types of organization (broadcasting, Roman-Catholic and Social-Democratic unionism) de-pillarization can be observed, for others (health services, Protestant unionism) it can not. Education takes a special place, because from national surveys it follows that public attitudes became much more indifferent with regard to the relation between religion and elementary education, while, at the same time, the Dutch school system in its structure is still pillarized to a high degree.

Public discussion often asks whether or not Dutch society has de-pillarized. The above analysis leads to the conclusion that such a question is too simple. Pillarization and de-pillarization are complicated processes that develop on the attitudinal as well as on the structural level. They comprise the level of individual behaviour, the organizational and the political level as well as the interrelations between those three. This is exactly what I wanted to emphasize by the presentation of the ideal typical model of pillarization in chapter V. Therefore, questions regarding the degree of pillarization and de-pillarization should only be put forward with reference to the level of analysis and the groups involved.

One issue is still open for analysis, i.e. the degree of de-pillarization at the organizational level and how it is related to individual and political behaviour. If we look at the organizational level from the angle of the pattern of organizations, the triple structure of Protestant, Roman-Catholic and non-denominational organization is still present[17].

The general picture emerging is one of anachronism. While politics and individual behaviour have de-pillarized, the traditional pillarized organizational structure still stands[18]. Of course, it has eroded to some extent, but basically it is there. It points at the law of social inertia: it takes time for vested interests to adapt to changed conditions. This holds especially for those organizations that have monopolist characteristics and that have only a weak relation to client markets.

For example, the right of freedom of education is stated in the constitution. It has a strong tradition in Dutch politics. As long as the Christian-Democrats are in the centre of power, de-pillarization of the Dutch school system is taboo. The other option for educational reform would be the establishment of new schools by groups of parents. The law provides this option. However, it also stipulates that a new school will only be subsidized if it has a minimum number of enrollments. Given the availability of denominational and non-denominational schools and a decreasing birth rate, it has proved extremely difficult to establish new schools. The more so, as available educational facilities are generally of good standard and the differences between the curricula are in practice minimal.

In the field of economic organization we can see the same phenomenon. The established organizations are part of networks of consultations. Business is done within these institutional frames. For newcomers it will take time and considerable effort to fight their way in. Meanwhile their influence is practically zero and they are unable to deliver to their clients. Such is not the way of advertising for organizational growth.

At the organizational level, vested interests and, since the 1970's, a contracting economy made the development of new initiatives towards de-pillarization difficult. On the other hand, in some fields it was contraction that led to de-pillarization tendencies and increased efficiency, notably the mergers between home nursing services, hospitals and the Roman-Catholic and Social-Democratic trade-union federations.

As early as the 1950's Kruijt presented his model of pillarization as a set of concentric circles. Kruijt positioned the church in the centre. The more social activities were closely related to the church, the more they were placed to the centre of the set of circles. According to Kruijt the strain between social activity and legitimation increases as the distance to the core of the set of circles grows[19].

As I stated in chapter V, I cannot fully agree with Kruijt's theory because for him pillarization is exclusively related to religious organization. But if we adapt his theory by placing ideology instead of church in the centre, it may be useful to explain why some organizations de-pillarized more than others.

The adjusted hypothesis then runs as follows. The greater the distance between ideology and social activity, the more strain there will be on the legitimation of the social activity involved. When individual behaviour de-pillarizes, pillarized social activities will come under pressure. The nearer social activities are placed to the core of the set of circles, the more they will resist tendencies to de-pillarization. Therefore, organizational de-pillarization will develop from the outer circles towards the centre.

For reasons we explained before, our analysis did not cover all pillarized social activities. More research is needed before we can really accept the theory of the concentric circles. In the meantime I have indicated that such research may prove worthwhile. In the Netherlands education was traditionally much more ideologically oriented than health care. So were industrial relations, but less than education. Organizations like the Dutch Tourist Club (ANWB) and consumer organizations were never identified with ideology and have never been pillarized.

If for the moment we accept the theory of the concentric circles, it is clear that beside attitudinal changes scarcity may be an important factor too. It puts pressure on pillarized organizational structures to economize which may imply de-pillarization by mergers. But the way it does and whether or not it is successful again depends on the distance between social activity and the centre of the set of circles. Education is a good example. During the last decade the government has had to cut the education budget. Several studies indicated that de-pillarization of the educational system would largely contribute to an increase of its efficiency and lower costs. Nevertheless, due to pressure of the CDA the government evaded the discussion and distributed the cuts proportionally over all schools.

In fact, the stagnating de-pillarization at the organizational level leads to incompatibilities. As we have seen, in some cases it blocks social reform. In other cases it causes ineffective interest articulation. Thus the unions still bang the drum of the old pillarized consultations while in the meantime the politicians have changed the rules of the game. The rise of action groups is perhaps the best proof of the existence of incompatibilities between individual attitudes and traditional interest articulation. These groups concentrate on issues instead of ideology. They are outside the consultative system. They are mechanisms that attempt to influence politics directly or indirectly by putting pressure on the established organizations.

De-pillarization Explained

How can we explain that Dutch society de-pillarized after the 1950's? It is my thesis that it did because of the nature of accelerated modernization in the decades after the Second World War.

In his study on cultural change and de-pillarization Middendorp gives ample attention to the developments in the political and non-political culture[20]. These are explained by three structural "basic processes": an increase in the level of technological development, processes of enlargement of scale and increasing possibilities of individual behaviour[21]. The latter being defined as *"the increase of possibilities (often also the necessity) for many to determine behaviour according to individual wishes and desires"*[22]. This is regarded as the cumulative effect of technological development and enlargement of scale.

Middendorp's basic processes are part of the modernization process as we defined it in chapter II. Pillarization is the polarization between religious and non-religious groups in society that came about as part of nineteenth century modernization. The problem with Middendorp's explanation of cultural change in the direction of de-pillarization is that he overlooks this fact. Our analysis in chapter III makes clear that after World War II the Netherlands again underwent large processes of enlargement of scale and technological development. It thus seems logical to expect higher degrees of pillarization. However, instead we observed strong tendencies of de-pillarization. The problem then is how to explain that both pillarization and de-pillarization are apparently part of the broader process of progressive modernization.

The same argument can be put forward to Ellemers' explanation of Dutch de-pillarization. He mentions four factors: a) the old methods for solving problems fail to solve new problems; b) the passivity of the masses has disappeared because of the emancipation process that was the result of pillarization itself; c) a new democratic ideology was diffused by the mass media of modern society; d) the increase of the emphasis on means to end rationality and professionalization[23]. The factors c and d are part of the modernization process. Factor b refers to the emancipation hypothesis. This hypothesis has proved to have its weaknesses. Therewith it also fails as an explanation of de-pillarization in the way Ellemers implies. Factor a is not an explanation but only raises further questions. Furthermore, it is not clear whether Ellemers' factors are mutually exclusive.

The thesis that both pillarization and de-pillarization are integral parts of progressive modernization should now be enlightened. Twentieth century post-war modernization eroded pillarization because it produced social forces that weakened the basis of normative social control on which it was built. These forces were: 1) progressive enlargement of scale, 2) a continuing secularization and 3) the rise of the welfare state. The first two factors contributed also to nineteenth century pillarization. However, after World War II enlargement of scale changed qualitatively in such a way as to significantly affect role expectations and to create options to escape from the effects of informal sanctions. It was supported by a general trend of secularization, which had already started in the beginning of the century but accelerated in the decades after the 1950's.

De-pillarization and the Rise of the Welfare State

Recently Hellemans compared Dutch and Belgian pillarization and de-pillarization from the perspective of the Roman-Catholic pillar[24]. As I stated before, pillarization refers to the interdependencies between pillars. As Belgium is characterized by a large and dominant Roman-Catholic majority it is doubtful whether it can be regarded as a pillarized country. Nevertheless, the Belgian Roman-Catholics have built a firm network in which social and political activities and religious ideology are intertwined. Hellemans writes that since the 1960s the Dutch Roman-Catholics have become far more de-pillarized than the Belgian Catholics. To him major processes behind Belgian and Dutch Roman-Catholic de-pillarization are modernization and the crisis of the Roman-Catholic church. He mentions additional factors that explain the differences in the degree of Roman-Catholic de-pillarization between the two countries. Firstly, Belgium is much more homogeneously Roman-Catholic. This means that contrary to the Netherlands, Belgian Roman-Catholics have a lack of socio-cultural partners for mergers. By their traditionally strong majority position, Belgian Roman-Catholics have a stronger inclination to a authoritarian attitude and thus still elicit anti-clerical reactions.

Secondly, after World War II the Dutch showed more consensus, whereas Belgian society polarized around the issue of the presumed German sympathies of King Leopold III.

Thirdly, the Belgian Roman-Catholic pillar has traditionally been more coherent than the Dutch. In the Netherlands Roman-Catholic socio-economic and political activities were more dependent on the clerical hierarchy because of the historical minority position. As a result the crisis in the Roman-Catholic church affected the Dutch Roman-Catholics more than the Belgian.

Fourthly, the rise of the welfare state. In Belgium the rise of the welfare state developed by pillarized organization. While the state provided the legal framework and the necessary subsidies, social policies were executed via private pillarized organizations. Thus in the Belgian welfare state the Roman-Catholic organization acts as the distributor of social provisions. Besides, the Belgian political system functions as a political spoil system, an *amigocracy*.

"Promotion in the administration, postponement of military enlistment, the acquisition of building licenses ... it runs more smoothly if you have the right party membership card"[25].

In the Netherlands there has been a much sharper borderline between the public sector and private pillarized organization. There is no such thing as a political amigocracy. Welfare provisions are distributed by the state, by state subsidized foundations or by branch associations of employers and employees. Dutch bureaucracy is standing above the parties instead of being usurped by the parties.

Hellemans ends his article with the question why the Dutch pillars did not usurp the state apparatus the way the Belgian Roman-Catholics did. Although I agree with his provisional answer that more research is needed, my foregoing analysis of Dutch pillarization yields at least a plausible hypothesis.

A characteristic of Dutch society is what I called the culture of "living-apart-together". Since the seventeenth century this culture has prevented a breakdown of Dutch plural society. Although in those days the Dutch state showed strong tendencies towards patrimonialism[26] at a regional and local level, at the level of national government no province could usurp the state. Major decisions had to be taken by consensus.

In the nineteenth century the content of Dutch pluralism changed, but its structure and problems essentially remained the same. Instead of seven autonomous provinces, now different almost sovereign ideological groups had to live within the boundaries of one nation state. Contrary to Belgium there was no majority of one group over the others that would be able to usurp the state. Instead the national state acted in accordance with tradition as a guardian of plural equilibrium. It prevented favouritism of one pillar above others and allocated welfare provisions to the citizens according to universal criteria or it subsidized organizations according to the principle of proportionality.

The neutrality of the Dutch state bureaucracy implied that by the rise of the public budget and by a growth of welfare provisions its power increased at the cost of pillarized organizations. Thus the more the state executed universal policies on sickness, disability, unemployment benefits and the likes, the more obsolete became the activities of private organizations in this field. Of special importance was the introduction of the Social Assistance Act that is executed by the state. It provides a financial minimum to all citizens irrespective of the causes of their need or ideological background.

So I agree with Hellemans that in the Netherlands the rise of the welfare state has been one of the factors that promoted de-pillarization. State-directed professional supply gradually substituted the ideologically coloured pillar supply. It freed the Dutch from the necessity to rely on pillarized associations as a last resort in case of material need. It eroded the possibilities of social control by pillars.

But also in other respects the expansion of the state made citizens less dependent on pillarized organization. The increased regulations on the relation between employer and employee, and notably the protection of the latter, made employees less dependent on the unions. In the 1970s, the function of labour unions eroded further by the increasing control of the wage level by the state. In general, in the last decades Dutch society saw a growth of legal regulations according to universalistic criteria in all domains. At the same time the possibilities for appeal increased as did financial provisions to guarantee every citizen an equal admittance to procedures of appeal.

These developments of Dutch society had two effects. Firstly, the necessity of social bargaining and social conflict decreased as possible conflict items were removed from the agenda by increased regulation. Secondly, in case of conflict individual citizens could now find satisfaction independent of their relations to one of the pillars.

The rise of the Dutch welfare state itself remains problematic. As such it was part of an international development in the advanced countries. The experience of the devastations of the war that followed the unresolved economic crisis of the 1930's induced everywhere a strong will to strive for a better society. The welfare state was considered a line of defense against the recurrence of Fascism and the advancement of Communism. It was inextricably associated in Western Europe with the need for economic, political and social security[27].

Within this trend countries had different patterns. Compared to other countries, total Dutch government expenditure increased sharply in the years between 1950 and 1960 (see table 6.3.). The reasons why are not clear[28]. Possibly it had to do with the fact that in this period for the first time the Social-Democrats, who were in favour of state regulation, participated in the government[29].

De-pillarization and Enlargement of Scale

Pillarization as a mode of social organization can only function if certain prerequisites are fulfilled. These prerequisites are related to the institutionalization of power and the role of ideology in society. They are affected by processes of enlargement of scale.

Etzioni[30] has discerned three types of power by which social control can be exerted: coercive, renumerative and normative power. These correspond with three kinds of involvement: alienation, calculation and moral involvement. The latter

"designates a positive orientation of high intensity. The involvement of the parishioner in his church, the devoted member in his party and the loyal follower in his leader are all 'moral'"[31].

The positive involvement of moral power is highly based on internalized values and norms. The actor conforms because he feels morally obliged to do so. From this it follows that the success of normative power is dependent of the degree to which belief systems are internalized.

Normative power is often of an informal nature, i.e. moral involvement may be supported by formal sanctions, but it can not be enforced by those exclusively. So Roman-Catholic excommunication can only function as an ultimate sanction for those who strongly believe in the Roman-Catholic doctrine. It will have no effects whatsoever on adherents of other believe systems.

In theocracies the religious leadership can rely on law and state bureaucracy to control the individual formally. In Western democracies, however, freedom of religion is guaranteed by the constitution. In the Netherlands the law has never provided the possibilities for the application of formal coercive and renumerative power to comply with the ideologies of the pillars. Thus social control exerted by the Dutch pillars had to be of an informal normative nature. In some cases it had remunerative elements. The previously mentioned attempt by the Roman-Catholic bishops to keep Roman-Catholic construction activities in Roman-Catholic hands is an example. However, such cases were rare and have never been very successful.

As dimensions of modernization both secularization and enlargement of scale interdependently decrease the normative power base of pillars. Secularization has to do with a change of attitudes in a more worldly direction. It implies a direct change of the evaluation of social behaviour. It affects the success of the application of normative power by religious groups, because it weakens the impact of religious ideologies. In the next paragraph I will elaborate on secularization. Here I am concerned with the relevant effects of enlargement of scale.

By enlargement of scale I mean that social interaction develops on a larger scale than before. It refers to increase of the potential of social interaction. Enlargement of scale may be related to the social system as a whole or to parts of it. Further, the causes of enlargement of scale may be manifold but do not bother us here in particular as they fall outside the scope of study.

Enlargement of scale is indirectly related to social control for two reasons. Firstly, it affects the criteria by which social behaviour is evaluated. Secondly, it breaks down the ideological isolation of groups and provides the individuals with alternative meaning systems and networks of interaction. The increase of these options for substitution mitigate the effects of informal sanctions on those who deviate.

As a modernizing force enlargement of scale affects the evaluation of social behaviour. It refers to changes in structure that are associated with an increase in instrumental rationality. Since Max Weber we know that processes of economic rationalization are part of a broader long term cultural change, which he paraphrased as a "*disenchantement of the world*". Instrumental rationality demands a calculative mind that is able to consider the balance between costs and returns. Such a calculative involvement is highly different from a moral one.

Because in pillarized society the pillar is the main locus of social identification in the engagement of social interaction, individuals belonging to the same pillar have a strong preference towards each other. Norms and values of the pillar's ideology become, therefore, an integrated part of role expectations. A shopkeeper is not expected to behave just like a shopkeeper but like a Protestant shopkeeper, a Catholic shopkeeper, a Liberal or a Social-Democratic shopkeeper. This holds true not only for professional roles but for other social roles as well. Since most people are born into pillars, that segment of the social role in which the values and norms of the pillar are expressed can be regarded as ascriptive. So, in pillarized society most social roles contain expectations of an ascriptive nature as well as achievement oriented elements.

To make this clear, let us take as an example the role of a Protestant carpenter working for a Protestant employer. The role of carpenter as such is based on achievement. The individual involved chooses to be a carpenter. His employer expects certain skills from him, like how to handle hammer and nails. These are the achievement oriented elements of the role. But above these, the Protestant employer expects his Protestant carpenter to attend church, to send his children to a Protestant school, to read a Protestant newspaper, to vote for a Protestant political party and, if unionized at all, to be a member of a Protestant trade union.

Rationalization by enlargement of scale leads to a decreasing importance of ascriptive criteria in the evaluation of social roles. It means that social relations become dominated rather by universalism than by particularism, that the number of formal relationships tends to exceed those with an emphasis on affective meaning. In this respect it undermines one of the basic social dimensions of pillarization: the moral discrimination between members of different pillars.

The very idea of bureaucracy is based on achievement and formalised procedures which are meant to counteract moral particularism. Only in small firms can morally coloured relations vested on a common ideological background of employer and employee survive. Only the small employer can apply normative social control in this respect. In large bureaucratic settings hiring, firing and promotion are determined by achievement and by the application of formalised universalistic norms. More than in small firms there is a borderline between company and individual morality which is considered to be private. It is productivity that counts as well as commitment to formal and rational rules.

In chapter III we noted a drop in the number of small firms and a rise of large companies since the late 1950's. The latter often have a bureaucratic type of organization. But the wider Dutch society also underwent strong tendencies of bureaucratization. The increasing role of government in economic and social life and the expansion of the welfare state implied a growth of standardized universalistic and bureaucratic norms.

The fact that there is a large social distance between pillars means that social life is to a large degree restricted to the circle of adherents to the same pillar. This has two important effects. Firstly, because the individual is consistently confronted with the same values and norms, the pillar's belief system is reinforced. Secondly, because of the exclusiveness of social interaction, shunning by the pillar involves the loss of identity. For that reason expulsion from a pillar associated network of interaction is very effective as the ultimate sanction of normative power.

Enlargement of scale affects both the reinforcement of ideology and the effectiveness of normative social control. As the scale of social life enlarges the individual is confronted with other life styles and value patterns. His ideological horizon broadens. At the same time his dependency of the pillar oriented interaction network decreases because the number of alternatives available increases. In other words, the individual has more options for substitution at his disposal. As a result he can evade the negative effects of social control exerted by the pillar.

Let us first investigate how post-war modernization eroded normative power by decreasing the dependency of the individual on the pillar. Four factors can be identified: the rise of the welfare state, the growth of per capita income, mobility and the changes in the spatial environment.

We stated before that the rise of the welfare state involved the creation of provisions that made people no longer dependent of private pillarized welfare. This holds not only for income maintenance but also for community work, family care, the care for the elderly and the like. In these domains ideological motivations were gradually replaced by professional ethics. Voluntary welfare was substituted by state subsidized professional workers who applied universalistic criteria. To be eligible the potential client only had to state his needs.

The rise of per capita income has made people financially less dependent on primary relations like family and neighbourhood. It created an economic surplus that enabled the expansion of welfare provisions in the 1960's. Especially as in the 1960's and 1970's the Dutch benefit-wage ratio was extremely high[32], a loss of earnings did not seriously affect household income, at least not for the first two years. As a result a loss of earnings did no longer urge households to rely on charity, family or neighbours. Therewith the pressure to conform to other people's norms was lessened.

The increase of geographic mobility implied that many had to change their local interaction networks from time to time. If an individual is tied for life to the same community, his potential for substituting contacts is small. The extreme example is the small country side community that is homogeneous as to ideological background. Expulsion because of norm violation seems almost irreversible as the deviant has no alternatives for compensation. The increase of transportation and a higher degree of labour market mobility makes people less dependent on the social setting which is connected to his place of living.

Both inter- and intragenerational upward social mobility has as effects that the individual involved changes lifestyle, enters in new social situations and gets new friends and acquaintances. In many cases there rises a gap in lifestyle and ideological orientation between parents and social mobile adult children. Changes of social environment weaken the impact of normative power. Modern man has to adapt constantly to new situations and circumstances which requires a flexible response as to norms and values.

Of special importance for the decline of normative social control have been the changes in the spatial environment. Chapter III pointed to the rise of dormitory outskirts of urban centers, suburbanization and the decline of the small rural municipality. These developments can be summarized as a change of community life in the direction of increasing anonymity and heterogeneity. The low intensity of social contacts and the ample opportunities for social interaction weaken the strength of social control and the effectiveness of sanctions as the deviant has numerous alternatives for social interaction at his disposal.

Since the 1960's the reinforcement of pillar ideology was also weakened substantially by the enlargement of scale in the domain of information. Institutions of lower education are more suitable for the transmission of pillar's norms and values as socializing agents than higher education because the former are more engaged in the teaching of technical skills and abilities like reading, writing, masonry and so on. Professional and academic training on the other hand, involve subjects that have to do with philosophy and critical thinking. They confront the student with new views on the world and elicit a critical attitude and the ability to see ones own convictions as relative. The rise of the level of education that took place since the 1960's therefore provided for large parts of the Dutch youth new ideas and ideologies. Besides, for many the Vietnam War came as a shock that proved that the promise of a better world was far away. As in other countries, in the Netherlands this triggered off a sense of alienation and suspicion of the established order and authority.

Besides the rise of the level of education, the whole Dutch nation experienced a presentation of new ideas and ideologies. In the 1960's a second TV channel was opened and alsmost all households were in the possession of TV sets. People could see programmes made by all pillars. Further, as the number of broadcasting hours increased the number of foreign programmes broadcasted grew. The diffusion of competing ideas promoted the break-down of ideological barriers and moved Dutch culture in a more cosmopolitan direction.

De-pillarization and Secularization

Enlargement of scale and the rise of the welfare state are structural dimensions of modernization that exerted influence on pillarized structures. Having in mind Kruijt's theory on the relation between ideology and the legitimation of social activities, secularization is a cultural component of the modernization process that affected pillarization.

Secularization is the process by which religious ideologies lose their influence on the evaluation of social behaviour. As such secularization is a prerequisite necessary for the process of de-pillarization to develop because the latter implies a decreasing legitimization of social and political activities by religious ideologies. Although secularization and the decline of church membership often go together, analytically this need not be so. Hypothetically one could imagine that church membership declines while at the same time people still base their patterns of behaviour on religiously inspired norms and values. Logically it follows that the decline of church membership is not a wholly valid indicator for secularization.

A second reason to take a suspicious view towards statistical data on church membership is the way these data are gathered. In the Netherlands, data on church membership come from the census in which citizens are questioned whether or not they reckon themselves to a denomination. To some extent the resulting data are unreliable as indicators of secularization because they do not discriminate between, on the one hand, church members, who are attending church and are living according to the churches' norms, and, on the other hand, those who never attend church, do not live according to religiously inspired ideas but have some vague notion of belonging because they were brought up in a religiously oriented family.

If we would forget the above caveats and, nevertheless, have at least some confidence in the census data on church membership we can see that during the whole of the twentieth century the percentage of those who do not consider themselves to be adherents of any denomination increased steadily (table 6.5.). In the post-war period the decline was at its sharpest in the 1960's: in 1960, 18.4 per cent of the total population did not reckon itself to any denomination, in 1971 this percentage had risen to 23.6 per cent. Broken down by denomination, the percentage of Roman-Catholics increased between 1899 and 1971 from 35.1 to 40.4 per cent. The percentage of the Dutch Reformed and the category "other denominations" decreased substantially, while from 1909 onwards the Re-Reformed stabilized at about 9.5 per cent.

As the total Dutch population increased from 5.1 million in 1899 to 13.3 million 1971, we should interpret these percentages with care. Regarding the absolute figures, the Dutch Reformed grew during the whole period. The growth of the percentage of Roman-Catholics has to be explained by their high fertility rates which were inspired deliberately by their leaders as a mechanism of Roman-Catholic emancipation in a Dutch democracy based on proportional representation[33]. Thus the growth of the Roman-Catholic part of the nation was a logical effect of their significantly larger families over the years, compared with those of other denominations, and is not due to a rise in the number of converts to the Roman-Catholic religion.

As mentioned before, the bias of church membership as an indicator of secularization becomes evident if we switch to other variables which are more valid indicators of actual behaviour. Such a variable is the number of marriages consecrated as a percentage of all marriages contracted[34]. The consecration of marriage is a sacrament. It is one of the core rituals of every denomination, a pattern of behaviour by which the persons involved express compliance to the norms of the church. In accordance with our definition, a decline of those rituals therefore points to a trend towards secularization.

In figure 6.2. all marriages consecrated (homogeneous as well as mixed) are expressed as a percentage of the total number of all marriages contracted. The latter is the sum of all church marriages and all non-church marriages. From 1951 to 1965 we see a gradual rise in the percentage of marriages consecrated from 57.2 to 63.2 per cent. After 1965 the percentage begins to decline and in 1974 it is about equal to that of 1951: 57.3 per cent. After 1974 it decreases sharply to 38.8 percent in 1986.

Figure 6.2. Marriages Consecrated as a Percentage of All Marriages Contracted, 1951-1986

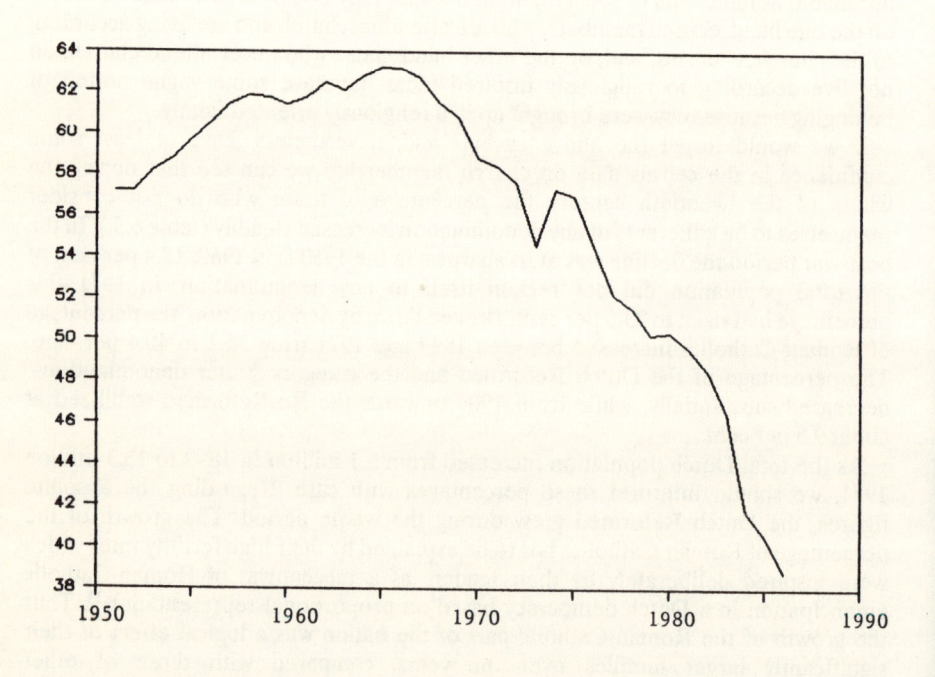

Source : Appendix D, table D.1.III.

Table 6.5. Church Membership by Denomination as a Percentage of Total Population, 1899-1981

Year	Catholic	Dutch Reformed	Re-Reformed	Others	No
1899	35.1	48.4	8.1	6.1	2.3
1909	35.0	44.2	9.7	6.1	5.0
1920	35.6	41.2	9.5	5.9	7.8
1930	36.4	34.5	9.4	5.3	14.4
1947	38.5	31.1	9.7	3.7	17.1
1960	40.4	28.3	9.3	3.6	18.4
1971	40.4	23.5	9.4	3.1	23.6
1981	37.5	21.8	8.8	5.1	26.8

Source : CBS, 1984, p. 48.

Figure 6.3. presents the number of homogeneous marriages consecrated by denomination as a percentage of the number of homogeneous marriages contracted by denomination. The Protestants are divided up into their major categories: the Dutch Reformed and the Re-Reformed. This figure differs from figure 6.2. because it refers only to homogeneous marriages. Further, in figure 6.2. marriages contracted

Figure 6.3. Roman-Catholic (a), Re-Reformed (b) and Dutch Reformed (c) Homogeneous Marriages Consecrated by Denomination as a Percentage of the Number of Homogeneous Marriages Contracted by Denomination, 1951-1986

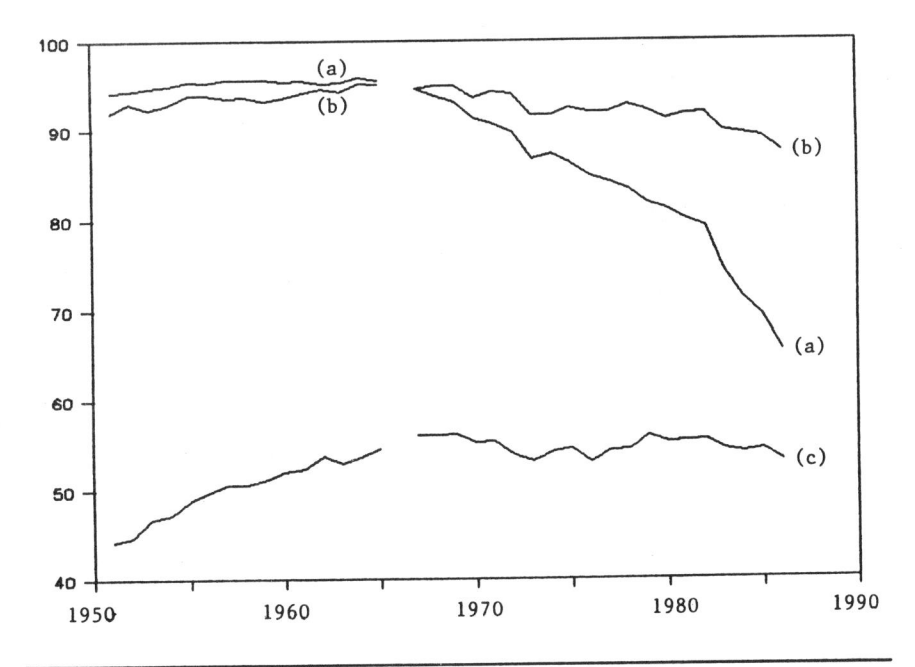

Source : Appendix D, table D.1.III.

by partners of the categories "other" and "no-denomination" are included in the total number of marriages contracted, while in figure 6.3. marriages consecrated are exclusively related to the total number of homogeneous marriages of the denomination involved. In other words, figure 6.3. relates to the issue of how many marriage partners belonging to the same denomination contracted their marriage before the church. It therefore points directly to the question how seriously the rules of the denomination involved are abided.

In the 1960's the secularization of Dutch society accelerated. However, there are striking differences between the three groups depicted in figure 6.3. We can infer that the level of Roman-Catholic compliance was high till the mid 1960's, but then eroded substantially. Over the whole period the Re-Reformed kept a high level of compliance despite a minor decrease in the 1970's and 1980's. The Dutch Reformed show a slight increase in the 1950's but have an overall pattern of low compliance compared to the Re-Reformed and the Roman-Catholics. Thus church members belonging to the two churches which together form the Protestant pillar differ considerably as to their compliance patterns.

In addition to the hypothesis we derived from Kruijt's theory, i.e. that resistance to the de-pillarization of social activities increases the closer these are positioned to the core of the ideology, we may likewise expect that the more a group's compliance pattern erodes, the more it will tend to de-pillarization and to mergers with other groups. As said before, systematic and conclusive research that covers all developments of pillarized social activities since World war II is lacking. Nevertheless there are some clues in this respect.

With regard to the merger between the Roman-Catholic party KVP and the two Protestant parties ARP and CHU, it was the KVP which pressed for the merger. Only after long internal discussions on the ideological bases of the new CDA to form, the ARP consented.

In the field of industrial relations, the Protestant CNV participated in the talks on a possible merger with the Roman-Catholic NKV and the Social-Democratic NVV. In the end it could not agree to a perceived loss of Christian identity which would result from the merger. So the NKV and the NVV merged to form a new general trade-union federation FNV. The CNV remained the independent Protestant federation it always was.

For many people religion came to play a less important part in their lives. Public opinion data point in the same direction. From 1966 onwards, representative samples of the Dutch population have been asked what kind of issues they consider most important in life. In 1966, 15 per cent of the sample answered: *a strong religious belief*. For the years 1975, 1980, 1983 and 1986, the percentages of the samples that considered a strong religious belief most important, were respectively 9, 5, 5 and 6 per cent[35]. This means a decrease of 9 per cent in twenty years. Furthermore, the most substantial part of this drop lies between 1966 and 1975: 6 per cent.

Secularization usually refers to the impact of religion on society. The concept of pillarization is wider as it encloses ideology in general. Can we trace developments within the Social-Democratic pillar that point to a loosening of the ideological grip on those who belong to that pillar? To my knowledge there are no studies available that investigate this issue. However, there are some phenomena from which a weakening of social-democratic ideology may be inferred.

The first is post-war Social-Democratic thinking on the so-called "break-through". It was inspired by experience under German occupation. Then, irrespective of ideology, many groups resisted Nazi oppression. After the war the Social-Democrats proclaimed their party open to all Dutchmen regardless of their religious background. Both Roman-Catholic and Protestant work communities were established within the Social-Democratic PvdA. Although the "break-through" failed, the attempt could only be made by the sacrifice of Social-Democratic ideological rigidity.

The second is the weakening of Social-Democratic organization. Till the 1960's the Social-Democrats considered themselves as a "red family". Their youth movement, the AJC, was a coherent organization with banners, rituals and the like. It's days of glory were in the 1930's. One of the countries leading newspapers, Het Vrije Volk (before the war: Het Volk) was advertised as a party paper. So was the broadcasting association VARA.

After the 1950's the AJC left the scene. Het Vrije Volk started to lose subscribers to the extent that in the 1980's it was reduced to a local Rotterdam newspaper that was officially no longer party aligned. In the 1970's the VARA was struck by endless internal conflicts on the interpretation of Social-Democratic doctrine and how to apply it in broadcasting programming. In the 1980's the VARA apparently takes the view that as a broadcasting association it now only has to compete with others for the favours of the audience. It is believed that the best way to do it is by creating attractive and competitive entertainment. Like the other organizations that were once part of the "red family" the VARA abstains from an identity too closely related to the PvdA. This is even true to the extent, that in the first half of the 1980's a theologian, once a leading figure in the Dutch Reformed church organization, was chairman of the VARA[36].

De-pillarization and the Individualization of Dutch Society

If individuals are inclined or forced to identify and to interact exclusively with one social collectivity, that social collectivity will be able to subdue the individuals involved to the highest degree of social control. Because of the exclusive nature of identification and interaction, members of the collectivity will be socially and psychologically almost totally dependent on it. Chances for deviation will be minimal. Under the condition that the collectivity is normatively coherent and homogeneous, i.e. that it is not divided by factions each claiming the true interpretation of the norms and values, its members will be confronted with a consistent set of norms. According to our model of Dutch pillarized society, these traits are logically connected with the phenomenon of pillarization.

De-pillarization implies that the exclusive tie between the pillar as a social collectivity and the individual is broken. For the individual the pillar is no longer the main locus of identification. If the relaxation of the tie between individual and pillar is not substituted by the former's total subjugation to another normative agent, the pillar becomes just one reference group among others. Social control exerted by the pillar will weaken and will be less successful. Interaction between individuals with different pillar backgrounds will increase and people will become more sensible to norms and values other than that of their pillars of origin.

De-pillarization thus affects the normative configuration of society. It does not mean that behaviour becomes normless. Contrary to the pillarized social setting, the total of norms to which an individual adheres in de-pillarized society is derived from a wide array of different sources. In different situations norms of different origins are applied. They are subject to constant changes, are diffuse and may be mutually inconsistent to a greater or lesser degree. Thus, de-pillarization means the individualization of society: the decrease of the exclusive normative orientation of individual behaviour to the norms and values of one well defined social collectivity.

If one wants to measure the changes over the years in the linkages between attitudes and pillarization and one has to rely on secondary data, the researcher has to live with the fact that perfectly fitting indicators are not present.

A second problem involves the time span. To picture the trend from pillarization to individualization, we would like to investigate the shifts in public opinion from 1945 to 1986. The surveys most suitable for our purpose are those held by the Sociaal en Cultureel Planbureau (SCP). These surveys have the advantage that over the years the same questions have been asked. However, in most cases questions were asked for the first time in the second half of the 1960's, some in the late fifties.

Regarding the attitudes to social control by religious groups, some remarkable changes in public opinion are indeed observable. In the SCP-surveys a question was asked on the obligation to conform to prescriptions of one's church or religious group. In table 6.6. the answers of church members only are presented as percentages of the total number of church members in the samples. The table demonstrates a significant change of normative commitment of church members in twenty years. It is remarkable that the major change took place in the four years between 1966 and 1970. Further, the figures of 1970 are more in favour of obeying religious rules than those of 1975, although the latter are still substantially less in favour to religious norms than those of 1966. To this phenomenon of "restoration" we will return later in this section. As the respondents were all church members, the observed change cannot be explained by a decline of church membership. It refers exclusively to the control of religious groups over their adherents.

Table 6.6. Attitudes of Church Members on the Obligation to Conform to the Prescriptions of Their Church or Religious Group (Percentages of Total Number of Church Members in the Samples), 1966-1986

	1966	1970	1975	1980	1986
feels obliged	51	27	28	18	20
thinks it depends	20	26	34	33	33
feels not obliged	29	47	38	49	47

Source : Derived from *Sociaal Cultureel Rapport 1986*, 1986, p. 377, annex table 11.32.

The wave of permissiveness from the late sixties onwards has not been restricted to the domain of religious rules only, but pervaded the whole of Dutch society. The SCP-surveys also give data on the attitudes towards some civil "liberties" (table 6.7.). Fast growing areas of permissiveness are the items "demonstrating" and "refusing military service". Both point at the erosion of authority.

The data presented indicate major trends. They are however, of a descriptive nature. To unravel the backgrounds of the shift from pillarization to individualization properly, we need tools that are more analytic. So we will refer to two studies on social change in the Netherlands. The first one by Middendorp is on de-pillarization, politization and restauration in Dutch politics in the 1960's and the 1970's[37]. The second one by Gadourek treats social change in the Netherlands as a redefinition of social roles[38]. Both studies are based on large surveys representative for the Dutch population.

Table 6.7. *Public Opinion on Some Civil Liberties. Respondents in Favour as a Percentage of All Responding on the Questions Involved, 1966-1986*

	1966	1970	1975	1980	1986
demonstrating	58	78	74	82	82
criticizing royalty	48	56	51	61	60
striking for wage rise	55	65	58	68	64
refusing military service	40	52	49	60	58
occupation of buildings	n.a.	30	24	31	35
freedom of writing	58	73	71	74	72
freedom of speech	n.a.	79	76	82	81

Source : Derived from *Sociaal Cultureel Rapport 1986*, 1986, p. 371, annex table 11.26.

The shifts Middendorp analyses can be considered as dimensions of the individualization process. He concludes for the second half of the 1960's a weakening of traditional attitudes concerning marriage, family, sexuality and religion[39]. At the same time contradictions increase between men and women, the youth and the elderly, the rich and the poor, between different religious groups and between church members and others[40]. The recession of the 1970's necessitated an adjustment of expectations. Therefore, the years from 1970 to 1975 are of a restorative nature as far as the political domain is concerned. In these years there is an increased desire for order and a distaste for radical reform. In the non-political domains the individualization of behaviour proceeds, be it at a more moderate tempo[41]. Middendorp labels the latter trend as "libertine".

In his study on social change Gadourek uses highly sophisticated analytical tools. Although his results are more or less similar to those of Middendorp, his interpretations are different. Dutch society of the late sixties shows strong hedonistic traits: the spread of drug abuse, the weakening of sexual norms and the rising of alternatives to the traditional nuclear family[42]:

> "Many of these were presented as tolerance. Instead of tolerance of the deviant, however, frequently the norm itself was deleted, abrogated.....To sum up, the many results of research in the normative domain suggest to me that at the end of the 1960's and the beginning of the 1970's in the Netherlands we are confronted with a wave of general norm weakening of which the Netherlands to some extent recovers after 1975".

Basing himself on Durkheim, Gadourek takes feelings of aimlessness as an indicator of anomia. For the adult population Gadourek found that in 1958 9 per cent felt aimless and in 1975 14 per cent; for the no-church affiliated the percentages were respectively 24 and 45[43].

Gadourek's conclusions on the basis of a multivariate analysis of his anomia indicator are enlightening: the elderly, the no-church affiliated and the singles feel more often aimless than others, whereas the factor "single" only has a strong influence in the category of the no-church affiliated. So the highest degree of anomic symptoms was found among single no-church affiliated people; the lowest degree among married church members[44].

To conclude, as far as empirical data are available, they all point in the direction of a trend towards individualization of behaviour. Since the late sixties the importance of pillars as the main locus of identification diminished, social distance between members of different pillars decreased as did normative social control.

With regard to the consistancy of the normset, no longitudinal data are available. From a theoretical point of view, however, it seems logical to assume that, when permissiveness increases and the locus of identification becomes diffuse, the consistancy of norms that guide the behaviour of an individual will decrease. When Gadourek uses the concept of anomia to describe the fading away of norms, he inspires us to this line of thought.

Goudsblom also supports the individualization thesis. He remarked that, parallel to the trend towards secularization of the last hundred years, an increase of ideological alternatives can be observed in the Netherlands. Both developments accelerated since the 1960's[45]. An increase of the supply of ideologies in itself does not need to imply a process of individualization. Goudsblom, however, also observes in our days a trend towards "*ideological abstinence*": more and more people state that they have an open mind regarding the different ideologies supplied. This means that the norms guiding the individual in different situations of social interaction do not need to be derived from the same ideological source. Such is the true meaning of individualization: the ability of the individual to refer in a situation of social interaction to those norms which he feels most suited. In contrast, in pillarized society all norms that guide the individual are derived from one single ideological source, leaving the individual no option. Therewith the chances of contradictions in the total normset of one individual are minimalised. Under ideological abstinence the chances for contradictions to occur are enhanced as the different ideologies from which the normset is derived may be overlapping or even contradictory.

Modernization and Pillarized Social Control: The Inverted U-Curve

Regarding the de-pillarization of Dutch society the main conclusions can be summarized as follows:
1. After the 1950s pillarization lost its effectiveness as a political strategy to guarantee stable constituencies.
2. Since the 1960s, in national politics the policy of accommodation has been gradually replaced by polarization between parties in power and opposition. This also holds for the relation between government and interest groups outside politics.
3. At the end of the 1960s, the phenomenon of the action group appeared on the scene. It meant a trend from interest articulation organized around ideology to interest articulation focussed on issues.
4. The data analysed point at dissimilarities in the process of Dutch de-pillarization. The conclusions 1 and 2 imply a de-pillarization of Dutch politics after the 1950s. At the organizational level developments are diverse. Some social activities de-pillarized, others did not. Within the latter category a distinction could be made between organizations that fully de-pillarized and those that kept pillarized structures while losing ideological attractiveness (e.g. education). I tried to explain the diverse patterns of organizational de-pillarization by referring to Kruyt's theory on the relation between legitimization of social activities and their distances to the core of the ideology. The smaller the distance, the greater the resistance to de-pillarization.

Additional factors that suggested an explanation of the relative lack of de-pillarization of some social activities are social inertion and scarcity.

Since the 1960s, at the level of individual interaction Dutch society can be clearly considered to be de-pillarized. However, like on the organizational level here too one is confronted with dissimilarities. Roman-Catholic de-pillarization of individual interaction set in relatively early compared to Protestant de-pillarization. Further, the degree of Roman-Catholic de-pillarization was higher.

The analysis suggests that one cannot properly speak of the de-pillarization of Dutch society. As such the statement makes little sense if it is not made explicit to what social domain and to what social group it refers.

5. The rise of the Dutch welfare state was an important factor that promoted tendencies of de-pillarization. It created a professional supply of welfare provisions, state-directed schemes of income maintenance and a wide array of bureaucratic regulations aimed at the management of social conflict. All these contributed to the erosion of social control by pillars.
6. A second feature associated with modernization and promoting de-pillarization is enlargement of scale. It induces in society a culture of calculation which is incompatible with moral involvement. It stresses achievement oriented aspects of social roles rather than the ascriptive ones which are basic to pillarized structures. It tends to change informal normative control into formalized bureaucratic rules. It creates possibilities for cultural substitution and diversification of interaction patterns. These factors also weakened the control of pillars' elites over the rank and file.
7. Economic growth created the economic surplus that enabled the foundation of the welfare state. It made individual households less dependent on churches, unions, family and all kinds of significant collectivities. To this also contributed the rise of discretionary income.
8. Secularization was an overall trend in twentieth century Dutch society. In general, secularization accelerated after the 1960s. However, it affected Roman-Catholics earlier and more profoundly than it did Protestants. I am not sure why this was so, but it may be an effect of the crisis in the Roman-Catholic world church. Secularization promotes de-pillarization because it weakens the relation between religious norms and behaviour. Although not religious, a weakening of normative coherence could also be observed within the Social-Democratic pillar.
9. De-pillarization promoted the individualization of Dutch society.

In the last decades of the nineteenth century Dutch society pillarized. Pillarization was part of the larger process of modernization that accelerated at the time. For two major reasons social structural and political modernization took the shape of pillarized structures. Firstly, pluralism had always been an essential characteristic of Dutch society and, secondly, a culture of "living-apart-together" was an almost inborn part of the nation's social codes.

At the end of the seventeenth century Republican geographic pluralism became incompatible with developments in the structure of the world economy. It prevented the Netherlands from adapting by infrastructural change. The French occupation at the end of the eighteenth and the beginning of the nineteenth century contributed to the levelling of this major social incompatibility. However, important vestiges of the old structure were still present when the Netherlands became a kingdom in 1813. It took till about the middle of the nineteenth century for them to disappear.

Although economic growth started as early as the first half of the nineteenth century, social, cultural and political modernization followed only in the second half of that century. The acceleration of modernization at that time had a profound influence on Dutch society. Processes like industrialization, enlargement of scale, the rise of income and social mobility shook the social order. In combination with liberalism as a force of enlightenment it induced secularization. The contradiction between religion and more worldly orientations did not restrict itself to the society at large, but could also be found within the religious groups where modernist and orthodox theologians had strong debates.

Pillarization arose as an attempt to control these developments, in particular to prevent the erosion of religious influence in society. In first instance it was directed against liberalism, later it attacked socialism. Gradually it became an effective strategy of social control.

Ellemers[46] considers pillarization a specific form of social differentiation brought about by progressive modernization. However, a close view on the nineteenth century developments suggests that pillarization meant a polarization of Dutch society. It can be understood as a reshuffling of social configurations along an orthodox-secular axis. From this cultural point of view it divided Dutch society in cohesive groups of either a traditional religious or a modern enlightenment orientation (Liberals, Social-Democrats).

It seems confusing that the acceleration of modernization after World War II and de-pillarization coalesce, because one would expect that progressive modernization would tighten up pillarized groups. This did not happen. Instead I made the point that post-war modernization effected an individualization of Dutch society.

We have seen that progressive modernization meant progressive secularization and progressive enlargement of scale. Although both factors are the same as in the late nineteenth century, in this chapter I demonstrated that post-war modernization changed enlargement of scale qualitatively in such a way that it affected pillarized social control. This development was strengthened by the rise of the welfare state which had the same impact. It is the qualitative change of enlargement of scale and the rise of the welfare state that explain why pillarized groups did not tighten up, but, on the contrary lost control over their followers. Instead of further polarization along the orthodox-secular axis, it led to the partial decomposition of the groups involved and thus contributed to the individualization of society.

Secularization is a general trend. It is connected to enlargement of scale, economic growth and technological development and thus part of the overall modernization process. As such we may expect that if overall modernization accelerates, so will secularization. Furthermore it seems plausible that after a certain point in the secularization process is passed, it will affect pillarization because it undermines religious ideologies. What we observed, however, was that the influence of the effects of secularization are dissimilar for the different religious groups. Notably the Roman-Catholics were heavily affected. Thus within the general trend of secularization subtrends can be discerned. I suppose that the Roman-Catholic deviation is due to an exogeneous factor, i.e. the developments in the Roman-Catholic world church. Protestant churches have a different type of organization: less hierarchical and more nationally, or even locally oriented. They are therefore more resistant to changes of international cultural trends.

Figure 6.4. The Inverted U-Curve Relation between Modernization and Pillarization

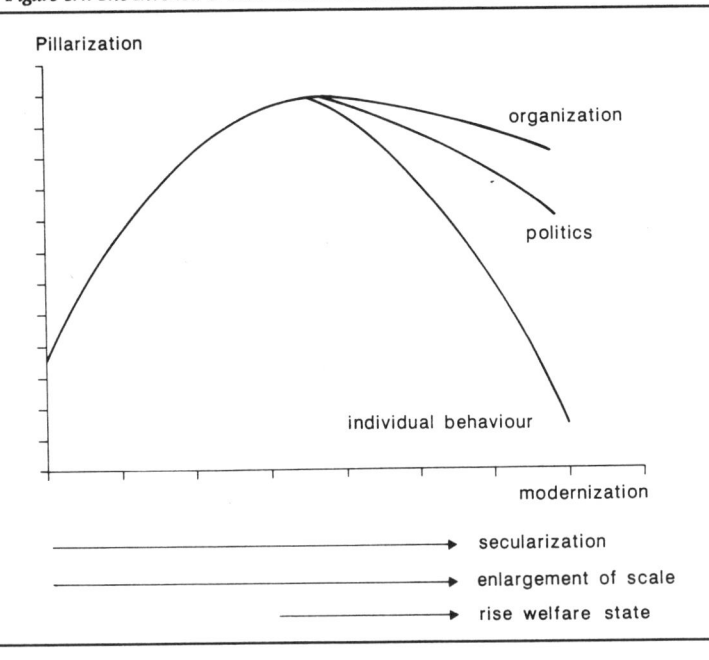

The implication of the above analysis is that late nineteenth century pillarization was an effective way of social control within the conditions of then prevailing modernization. For society at large it levelled the incompatibilities between the concomitants of enlargement of scale and industrialization on the one hand and the existing social structure and culture on the other hand in such a way as to prevent the outbreak of major social conflicts which could have setbacks on the process of modernization at large. Pillarization created new frames of social integration that were adjusted to the problems of the era.

After World War II incompatibilities arose as modernization accelerated. The exigencies of progressive enlargement of scale and the rise of the welfare state made social control by pillars obsolete.

The historical relation between modernization and pillarization then takes the shape of an inverted U-curve (figure 6.4). Pillarization increases as modernization progresses, but beyond a certain point it becomes counterproductive as a strategy of social control.

According to the data presented the turning point of the inverted U-curve should be located somewhere in the first half of the 1960s. Then the curve falls apart in three lines. The upper line represents the organizational level, the middle line the level of national politics and the lower line the level of individual interaction. The diversification of the inverted U-curve in the last decades suggests new incompatibilities in the process of Dutch modernization of the present. It points to the fact that individual behaviour, organizational interest articulation and politics do not make a perfect fit.

NOTES

Notes: chapter I

1. H. Daalder, 1985.
2. A. Lijphart, 1968-A and 1968-B.
3. J.E. Ellemers, 1984.
4. E.g. see C.P. Middendorp, 1979. The shift in attitudes and public opinion since the 1950's has been very thoroughly investigated by Gadourek. See I. Gadourek, 1982.

Notes: chapter II

1. Middendorp (1979) refers to an increase of technological development, enlargement of scale and an increase of the possibilities for individual behaviour. Ellemers (1980) mentions three factors: changes in the nature of Dutch Catholicism, the sudden and swift spread of television and the rise of new higher educated elites. The second factor has to do with technological development and is supposed to lead to a diffusion of cultural elements from one pillar to another, thus levelling the barriers between pillars. The third factor is the effect of the increase of the accessibility of the educational system.
2. See Eva Etzioni-Halevy, 1981, pp. 26-27.
3. Recently one can observe a revival of ideas as to the common origins of animal and human behaviour. The publication of E.O. Wilson's *Sociobiology: The New Synthesis* (1975) has been of major importance as it elicited firm discussions on the subject. See also Michael Ruse, 1979.
4. In his now classical volume on economic growth Lewis stated that the writing of his book was partly inspired by "*...the practical needs of contemporary policy-makers...*". See W. Arthur Lewis, 1955, p.5.
5. See Peter Flora, 1974, p. 13.
6. Daniel Lerner, 1968, p. 387.
7. Tomoyasa Fusé, 'Introduction', in: Tomoyasa Fusé (ed), 1975, p. 1.
8. See W.W. Rostow, 1966.
9. D. Riesman, 1961.
10. David McClelland, 1961.
11. Marion Levy, 1972, p. 3.
12. W. Arthur Lewis, 1955, p. 9.
13. According to Lerner this meant a major step ahead in theory formulation and research. See Lerner, 1968, p. 387.
14. E.g., Levy defines modernization as the growing ratio between inanimate and animate sources of power (see note 11 above). It is quite clear that the introduction of micro-electronics - which is sometimes loosely called the Second Industrial Revolution - has significant effects on the division of labour, the structures and procedures of organizations, the educational system and so on. It is not self-evident, however, that micro-electronics increases the ratio between inanimate and animate sources of power. Thus, using Levy's definition we might conclude to social stability whereas in fact important sectors of society vehemently change in such a way that we would like to call it modernization. Contrary to the purely economic character of his definition Lewis stresses throughout his book the complex intertwining of political, social and economic processes. See Lewis, 1955.
15. See Peter L. Berger, Brigitte Berger and Hansfried Kellner, 1973, p. 6.
16. On may argue that some of the old civilizations should be considered as modern because they used complex knowledge to build their empires. In these cases the technological development was either restricted to certain areas and did not spread throughout society as a whole, or it was aimed at improving the mechanisms of plundering and exploitation. An example of the first type is ancient China. The second type is represented by ancient Rome where abstract knowledge was used to improve the tools of administration and the art of war. One could even say that in the days of glory of ancient Rome substitution of labour by capital was low because of the accessibility to slave labour.
17. The criterion of self-sustained economic growth which is often considered as the crucial

distinction between non-modern and modern societies, can be related to the fact that in modern society technology is planned to raise productivity. See Lerner, 1968, p. 387.

18. This idea has been inspired by the definition of modernization given by Peter Berger, Brigitte Berger and Hansfried Kellner: *"the growth and diffusion of a set of institutions rooted in the transformation of the economy by means of technology"*. In my opinion this definition lacks the idea of interdependency of what the authors call *"the institutional concomitants of economic growth"* as well as the notion of purposefulness. See Berger, Berger and Kellner, 1973, p. 9.
19. See for a review of the theories involved L. Laeyendecker, 1984.
20. See for a critique on the technological theory of social change Percy S. Cohen, 1968, chapter 7, pp. 174-208.
21. In this respect there seems to be a deeply rooted misunderstanding as to Weber's exclusively idealistic explanation of the rise of capitalist society as opposed to that of Marx. We quote Etziony-Halevy: "Weber did not belittle the importance of economic factors as sources of social change as it is sometimes claimed. Indeed, interpreters who make this claim thereby show a lamentable ignorance of some of Weber's writings. For in his most famous work *The Protestant Ethic and the Spirit of Capitalism*, Weber indeed emphasized the importance of religious doctrines and ethical precepts, yet in his less famous *General Economic History* he emphasized economic factors no less than values and ideas. In this latter book Weber explained the role of economic factors as sources for the development of capitalism much in the vein of Marx himself. His analysis of the expansion of the markets, the transformation of arable land into pastures and the expulsion of peasants from the land differed little from that of Marx. In addition, he stressed the importance of the new methods for coal extraction and for the smelting of iron which gave such an enormous thrust to the development of industry. This analysis, however, brought Weber little fame, because it only adds marginally to what Marx had already said so well. In addition, Weber pointed to the role of the modern national state, in making rational, capitalist activity possible, but this analysis too did not earn him great recognition". Etzioni-Halevy, 1981, pp. 22-23.
22. Lewis, 1955, pp. 146-147.
23. Quoted in Berger, Berger & Kellner, 1973, p. 6.
24. E.g.: P.A. Baran, 1957; A.G. Frank, 1970, pp. 4-17; S. Amin, 1974; S. Amin, 1976; I. Wallerstein, 1974-A; I. Wallerstein, 1974-B, 16, pp. 387-415.
25. See Etzioni-Halevy, 1981, pp. 74-75.
26. See Flora, 1974, p. 24.
27. This is not surprising as modernization and industrialization coincided at the time.
28. See Flora, 1974, pp. 24-25.
29. T. Parsons and E.A. Shills (eds), 1951; See for an assessment of Parsons' scheme M.J. Mulkay, 1971, chapter 3, pp. 36-66.
30. See Flora, 1974, p. 25.
31. See Etzioni-Halevy, 1981, p. 37.
32. D. Bell, 1973.
33. A. Touraine, 1971.
34. Jan Tinbergen is the most outstanding representative of the convergence theorists. His article *The theory of the optimum regime* is generally regarded as a key publication. See Jan Tinbergen, 1959.
35. See D. Bell, 1973, and A. Touraine, 1971.
36. E.g.: Alvin Toffler, 1970.
37. B.F. Hoselitz, 1960, p. 194.
38. Rostow discerns five stages of economic development and modernization. The first relates to traditional agrarian society and is characterized by a high degree of stability. Substantial economic growth does not occur. In the second stage the conditions for economic growth are created: a change of values and attitudes and the rise of an enterpreneurial elite. The third and most important stage is that of economic take-off. In this relatively short stage, investments and per capita income rise spectacularly. There is a dramatic change in production technology. The take-off is decisive for the whole process: it changes the traditional economy in one that generates economic growth. In the fourth stage this development matures. In the fifth stage economic growth is continued but accompanied by mass consumption. See W.W. Rostow, 1966.
39. Etzioni-Halevy, 1981, p. 37.
40. E.g.: B.F. Hoselitz, 1964, 16, pp. 237-251.

41. See Mulkay, 1971, pp. 76-83.
42. Rostow, 1966, p. 12.
43. W. W. Rostow (ed), 1969, p. XXIII.
44. S. Kuznets, 1969.
45. Flora, 1974, p. 50.
46. D.C. North, 1969.
47. H.J. Habakkuk and P. Deane, 1969.
48. W.G. Hoffmann, 1969.
49. J. Marczewski, 1969.
50. S. Tsuru, 1969.
51. A. Gerschenkron, 1969.
52. W.W. Rostow, 1969, p. XVI.
53. Flora, 1974, p. 51.
54. W.W. Rostow, 1971.
55. C. Kerr, John T. Dunlop, Frederick H. Harbison and Charles A. Myers, 1960; Clark Kerr, 1983.
56. See Chie Nakane, 1973.
57. Fusé, 1975, p. 3.
58. See e.g. H. Braverman, 1975 and the works of Herbert Marcuse, especially his *One Dimensional Man: Studies in the Ideology of Advanced Industrial Society*, 1964.
59. Emile Durkheim, 1964 (1933).
60. Talcott Parsons, 1966.
61. N.J. Smelser, 1973.
62. See Lewis, 1955.
63. In view of recent research this thesis has become highly questionable. Historical demographers have doubted if the extended family type has ever been as general a phenomenon in pre-industrial society as supposed by the structural functionalist school. See Peter Laslett, 'Mean household size in England since the sixteenth century' in: Peter Laslett (ed), 1972.
64. McClelland, 1961.
65. R. Bendix, 1966.
66. P. Flora, 1974, p. 15.
67. P.A. Baran and P.M. Sweezy, 1966.
68. See Nakane, 1973.
69. Peter L. Berger, 1977, p. 75.
70. Peter L. Berger, 1977, p. 73.
71. See Clark Kerr, 1983.
72. Kerr, Dunlop, Harbison and Myers, 1960.
73. J. Tinbergen, 1959; J. Tinbergen, 1964.
74. See Raymond Aron, 1955.
75. Robert A. Dahl and Charles E. Lindblom, 1953.
76. Dahl and Lindblom, 1953, p. 3.
77. Dahl and Lindblom, 1953, p. 5.
78. J. van den Doel, 1971.
79. See Michael Ellman, 1980.
80. Seymour Martin Lipset, 1977.
81. Kerr, Dunlop, Harbison and Myers, 1960.
82. Kerr, 1983, p. 19.
83. Leon Gouré, Foy D. Kohler, Richard Soll & Annette Stiefbold, 1973.
84. Paul M. Sweezy and Charles Bettelheim, 1971.
85. Zbigniew K. Brzezinski, 1967, especially chapter 19.
86. Alexander Gerschenkron, 1962, pp. 191, 193; Alexander Gerschenkron, 1968, p. 495.
87. Wilbert E. Moore, 1979, pp. 50, 156-157.
88. Frederic L. Pryor, 1973.
89. Joseph A. Schumpeter, 1942, pp. 12-13, 182.
90. Bertram D. Wolfe, 1968, p. 70.
91. Bertram D. Wolfe, 1968.
92. S.N. Eisenstadt, 1977.
93. See Eisenstadt, 1977; S.N. Eisenstadt, 1973; S.N. Eisenstadt, 1978.
94. A. Etzioni, 1961.

95. This is not to say that violence or force is always absent in Western industrial relations, but rather that the emphasis is on calculative bargaining in which human freedom is in principle respected.
96. Kerr, 1983, pp. 33-76.
97. Kerr investigated also the domain of economic structure (ownership of property, management of enterprises, goals of management and competitors of managerial authority) and the domain of political structure (methods of selecting state leaders, mechanisms of developing public policy, means for administering the state apparatus, system of law). Regarding these two domains Kerr concludes to bipolarity between capitalism and communism.
98. In modern society people's life chances, status and power are highly dependent on educational achievement. Education is the first threshold to be passed before entering the world of work and profession. The strategic choice of the educational career one is going to follow, is usually made many years before labour market participation starts.

Notes: chapter III

1. Jan de Vries, 1984, p. 149. See also J. A. de Jonge, 1976.
2. Simon Kuznets, 1979, pp. 119-120.
3. Angus Maddison, 1982, table 1.4., p. 8.
4. Maddison, 1982, p. 29.
5. J.A. van Houtte, 1979, p. 73.
6. J.A. Faber, 1979, p. 154.
7. Maddison, 1982, p. 32.
8. Jan de Vries, 1984.
9. See Jan de Vries, 1984, note 6, p. 186-187.
10. J. Teijl, 1971, pp. 232-263.
11. J. H. van Stuijvenberg, 1981, p. 107.
12. Richard T. Griffiths, 1980, p. 10.
13. Jan de Vries, 1984, p. 153.
14. Gregory King, 1973.
15. Jan de Vries, 1984, pp. 153-157.
16. Maddison, 1982, pp. 31-32, 165-166.
17. According to Maddison, the core of this argument is that in the eighteenth century *"commerce was favoured at the expense of manufacture, and that the Dutch life-style was so satisfying and Dutch technology so close to the frontier of contemporary best-practice that there was little incentive to innovate"*. Maddison agrees with the latter, but does not see that this should lead to a loss of entrepreneurial dynamics. For him the primary cause of Dutch economic decline *"was British and French damage to Dutch foreign markets, together with an overdevelopment of the banking interest which kept the currency overvalued and further weakened export potential"*. See Maddison, 1982, pp. 33-34.
18. See J. de Vries, 1984, note 31, p. 188.
19. J. de Vries, 1984, p. 161.
20. De Vries refers to the works of Lindert and Williamson, Knick Harley and Crafts. See J. de Vries, 1984, notes 38, 39 and 41, p. 188.
21. J. de Vries, 1984, pp. 167-168.
22. J. de Vries, 1984, p. 169.
23. J. de Vries, 1981.
24. J. de Vries, 1984, p. 170.
25. J. de Vries, 1984, pp. 170-173.
26. J. de Vries, 1984, pp. 177-179.
27. J. de Vries, 1984, p. 181.
28. A. Maddison, 1987, pp.19-31.
29. A. Maddison, 1982, p. 165.
30. Joh. de Vries, 1983 (1977), pp. 13-30. This book has been translated in English as Joh. de Vries, 1978. I refer to the fourth printing of the original Dutch edition.
31. Joh. De Vries, 1983, p. 30.
32. Joh. de Vries, 1983, p. 51.
33. Maddison, 1982, p. 162, see also Maddison's Annex, table C-10, p.281.
34. See Maddison, 1982, table C-10, p. 281.

35. Average Annual Compound Growth Rates computed on the basis of Maddison, 1982, table C-10, p. 281
36. Joh. de Vries, 1983, pp. 38-41. De Vries analyses the phases of this process of catching up from the nineteenth century onwards.
37. J.A. de Jonge, 1976, p. 239 f(f).
38. I.J. Brugmans, 1961, p. 445, p. 473 and p. 546.
39. M. Weisglas, 1969, p. 1280.
40. Joh. de Vries, 1983, p.41.
41. R. Dekker, 1982, p. 20.
42. See J.C. Riley, 1980, pp. 16, 84.
43. I.J. Brugmans, 1977, p. 122.
44. H. van Dijk, 1976, pp. 132-133.
45. Van Dijk refers to W.A. Armstrong, 1974, p. 95 and A. Daumard, 1970, p. 43.
46. Van Dijk, 1976, p. 136.
47. Brugmans, 1977, pp. 123-124.
48. Th. van Tijn, 1977, p. 135.
49. W.M. Zappey, 1979, pp. 201,217.
50. D. Damsma and L. Noordegraaf, 1977.
51. H.P.H. Jansen, 1982, p. 157.
52. W.M. Zappey, 1979, pp. 216-218.
53. Th. van Tijn, 1979.
54. See Van Tijn, 1979 and also I.J. Brugmans, 1961 and De Jonge, 1976.
55. De Jonge, 1976, pp. 295-296.
56. De Jonge remarks that the increase of clerks has been greater from 1889 to 1909 than from 1849 to 1889, i.e. 4.5 per cent in twenty years versus 5.5 per cent in forty years. However, by comparing these periods using average annual compound growth rates and relating absolute numbers to the volume of the labour force, I reach a conclusion contradictory to that of De Jonge. See De Jonge, 1976, p. 297.
57. De Jonge, 1976, pp. 297-298.
58. Van Dijk, 1976.
59. *Ibidem*, p. 148.
60. *Ibidem*, pp. 151-152.
61. J. de Vries, 1984, p. 179.
62. Van Tijn, 1977, pp. 139-140.
63. *Enquête betreffende werking en uitbreiding der wet van 19 september 1874 en naar de toestand van fabrieken en werkplaatsen*, 1887.
64. Joh. de Vries, 1983, p. 40.
65. See for a review of theoretical and empirical problems of Dutch social mobility research: W.C. Ultee, 1984.
66. J.J.M. van Tulder, 1962.
67. F. van Heek and E.V.W. Vercruysse, 'De Nederlandse beroepsprestigestratificatie', in: F. van Heek (ed), 1958, pp. 11-48.
68. See also Ultee, 1984, p. 13.
69. See Van Tulder, 1962, p. 22.
70. Van Tulder, 1962, p. 215-217.
71. Van Tulder, 1962, p. 220.
72. Van Tulder, 1962, p. 220 and p. 294, table D.23.
73. Van Tulder, 1962, p. 294, table D.24.
74. H. Ganzeboom and P. de Graaf, 1983, p. 31.
75. Ganzeboom and De Graaf, 1983, p. 32.
76. Ganzeboom and De Graaf, 1983, p. 50.
77. H. Sixma and W.C. Ultee, 1983, pp. 360-382.
78. See for a summary of recent research: A. Szirmai, 1988, pp. 17-26.
79. J. Pen and J. Tinbergen, 1976, and also J. Pen, 1979.
80. Szirmai, 1988, p. 25.
81. See Szirmai, 1988, p. 25 and also his paragraph on wage and incomes policy 1945-1980, pp. 26-37.
82. G.A. van der Knaap, 1978.
83. In a recent study Deurloo and Hoekveld compare the average annual growth rates of the total

number of inhabitants of Dutch urbanized municipalities with the average annual growth rates of the individual urbanized municipalities as well as with the average annual growth of the Dutch population. The authors do not give many details as to their sources and methods of computation. Besides, their category of urbanized municipalities is of a scope too broad as it covers municipalities with numbers of inhabitants ranging from 2,000 to over 100,000. For these reasons I regard their data less suitable for the analysis of Dutch urbanization than Van der Knaap's. However, the average annual growth rates Deurloo and Hoekveld present, lead to similar conclusions as my computations on Van der Knaap's data: since 1860 the urbanized municipalities grew faster than the Dutch population. With an acceleration after 1880, the growth of urbanized municipalities continued till 1930. After 1940 the growth of urbanized municipalities as a whole lags behind population growth. See M.C. Deurloo and G.A. Hoekveld, 1981.
84. See J.J. Harts and L. Hingstman, 1986, pp. 18-21.
85. See L. Brunt, 1973.
86. A. van Braam, 1957, p. 33.
87. *Ibidem*, pp. 36-61.
88. T. Jaspers, 1980, p. 329.
89. *Ibidem*, p. 336.
90. On the history of the Dutch educational system I derived information from Ph.J. Idenburg, 1964 and from J.A. van Kemenade (ed), 1981, pp. 36-41.
91. Van Kemenade, 1981, p. 71.
92. Idenburg, 1964.
93. J.M.H.J. Hemels, 1969.
94. *Ibidem*, pp. 359-360.

Notes: chapter IV

1. C. Bagley, 1973, p. 2.
2. H. Daalder, 1966, p. 188.
3. Daalder, 1966, p. 192.
4. J.E. Ellemers, 1984, p. 130.
5. Ellemers, 1984, pp. 130-131.
6. H. Wansink, 1980. In English published as 'Holland and Six Allies: the Republic of the Seven United Provinces', in: J.S. Bromley and E.H. Kossmann, 1971, pp. 133-155. We refer to the Dutch text.
7. J.W. Smit, 1970.
8. Smit, 1970, p. 54.
9. D.J. Roorda, 1980. In English published as D.J. Roorda, 'The ruling classes in Holland in the Seventeenth century', in: J.S. Bromley and E.H. Kossmann, 1964, pp. 109-132. We refer to the Dutch text.
10. D.J. Roorda, 1961, p. 3.
11. R. Dekker, 1982, p. 20.
12. Wansink, 1980, pp. 208-217.
13. Roorda, 1980, p. 233.
14. Roorda, 1961, pp. 2-9.
15. Roorda, 1961, p. 3.
16. Roorda emphasizes that the distinction between parties and factions as respectively oriented towards idealistic and materialistic goals is an ideal typical one. In reality the two often overlapped as to goals and personnel. Roorda, 1961, pp. 5-7.
17. Roorda, 1961.
18. Dekker, 1982.
19. Regarding the issue of the relation between social structure and uprisings, Dekker refers to the discussion in France between Porchnev and Mousnier. Porchnev interprets rioting in seventeenth century France as a type of class struggle, whereas Mousnier regards it as regional reactions to centralizing tendencies of the French monarchy. As to Dutch rioting in the seventeenth and eighteenth century, Dekker tends to accept the Mousnier thesis. Dekker, 1982, pp. 138-139.
20. Dekker, 1982, pp. 41-50, 142-145.

21. See for a discussion on these estimates: *A.A.G. Bijdragen*, 1965, pp. 149-180.
22. Lijphart sees this informal tolerance of Catholics as an important factor that explains why there has never been a separatist movement in the Catholic South of the country. Instead the Catholic minority always fought for emancipation within the nation. See A. Lijphart, 1982, p. 88.
23. E.g., in the first quarter of the seventeenth century differences within the Dutch Reformed church between Remonstrants and Contra-Remonstrants led to the most violent riots of the era. The battle was fought between clergymen, regents and believers.
24. According to Maddison, around 1700 Dutch investment in infrastructure was high for the period. See A. Maddison, 1982, p. 31.
25. J.A. van Houtte, 1979.
26. It is precisely because of the regional character of water management that Wittfogel did not regard the Netherlands as a hydraulic society. See K.A. Wittfogel, 1978, p. 12.
27. J.C. Boogman, 1962.
28. Daalder, 1966, pp. 191-192.
29. Wansink, 1980, p. 218.
30. Roorda, 1980, p. 227.
31. Roorda, 1980, p. 237.
32. Wansink, 1980, p. 218.
33. Wansink, 1980.
34. Lijphart, 1982.
35. See also Daalder, 1966.
36. H.F.J.M. van de Eerenbeemdt and J. Hannes, 1981, pp. 141-142.
37. H. van Dijk, 1986.
38. Th. van Tijn, 1977, p. 135.
39. Ph. J. Idenburg, 1960.
40. L.F. van Loo, 1981-A, pp. 47-48.
41. L.F. van Loo, 1981-B, p. 418.
42. Van Tijn, 1977, pp. 136-138.
43. Th. van Tijn, 1980, pp. 103-104.
44. Van Tijn, 1980, p. 104.
45. Daalder, 1966, p. 204.
46. Van Tijn, 1980, pp. 108-109.
47. F.L. van Holthoon, 1985-B, p. 162.
48. Van Loo, 1981-B, pp. 420-421.
49. Van Holthoon, 1985-B, pp. 163-164.
50. Daalder, 1966, p. 200.
51. Van Holthoon, 1985-B, p. 166.
52. Van Holthoon, 1985-B, p. 166, transl. mine.
53. Daalder, 1966, p. 201, note 31.
54. D. Th. Kuiper, 1972, p. 493.
55. Van Tijn, 1980, p. 114.
56. Lijphart, 1982, pp. 105-106.
57. Lijphart, 1982, p. 11.
58. Lijphart, 1982, p. 106.
59. S. Stuurman, 1983, p. 132.
60. F. van Heek, 1954, p. 123.
61. W. Goddijn, 1957, p. 53.
62. Van Heek, 1954.
63. J. Hendriks, 1971, pp. 82-87.
64. With the exception of a Roman-Catholic press. Already in 1845 the Catholic newspaper *De Tijd* was founded.
65. Van Tijn, 1977, pp. 141-142.
66. G. Harmsen, 'De arbeiders en hun vakorganisaties', in: F.L. van Holthoon (ed), 1985-B, pp. 261-283.
67. Van Tijn, 1977, pp. 142-143.
68. Van Tijn, 1977, p. 142.
69. Harmsen, 1985, pp. 263-264.
70. Harmsen, 1985, p. 264.

71. Harmsen, 1985, p. 272; transl. mine.
72. J.P. Windmuller and C. de Galan, 1979, p. 28.
73. Harmsen, 1985, p. 273.
74. Windmuller and De Galan, 1979, p. 29.
75. Windmuller and De Galan, 1979, p. 33.
76. Stuurman, 1983, p. 319, transl. mine.

Notes: chapter V

1. J.C.H. Blom, 1981, p. 11.
2. J.A.A. van Doorn, 1956; J.A. Ponsioen, 1956.
3. J.P. Kruijt and W. Goddijn, 1962, table IV, p. 243.
4. See e.g. J.M.G. Thurlings, 1971.
5. J.P. Kruijt, 1959.
6. Righart has recently remarked that this thesis is value laden as in this case the perception of strain is a subjective matter. E.g. whether or not religion and unionism are related, depends on the value orientation of the observer. See H. Righart, 1986, p. 12.
7. After 1920 the Social-Democrats gradually dropped the idea of class truggle. See also F.L. van Holthoon, 1985-A, p. 159.
8. In the first years after World War II the Dutch Social-Democrats in fact proclaimed the idea of a *"Break-Through"* (in Dutch: Doorbraak). This was an attempt to integrate the denominational organized parts of he working class into the Social-Democratic party. The *"Break-Through"* had only a limited success and is generally considered to be a failure.
9. S. Stuurman, 1983, pp. 62-67.
10. Lijphart, 1982 (1968-A), p. 28.
11. Lijphart, 1982 (1968-A), p. 29.
12. This argument has been put forward by Hans Righart. It seems especially valid with regard to consociational democracy theory. See Righart, 1986, p. 15.
13. It is difficult to define the exact boundaries of the liberal subsystem. To do so an inventory of all Dutch social organizations as well as their ideological orientations is needed. If at all possible, the construction of such an inventory is a time consuming task. As our remarks on the boundaries of the liberal subsystem were of a theoretical nature and the availability of the inventory is not needed for the empirical analysis in this study, I consider its construction to fall outside the scope of my work.
14. Daalder, 1985, p. 57.
15. Righart, 1986.
16. L. Brunt, 1972; H. Verwey-Jonker, 1961; F. van Heek, 1954; W. Goddijn, 1957; M.A.J.M. Matthijssen, 1958; J. Hendriks, 1971; D. Th. Kuiper, 1972.
17. See R. Steininger 1975 and 1977.
18. Daalder, 1985, p. 54, transl. mine.
19. Righart, 1986, p. 29.
20. Thurlings, 1971 and Righart, 1986.
21. Righart, 1986, p.347.
22. Righart, 1986, p. 31.
23. Van Doorn, 1956; Stuurman, 1983.
24. M.P.C.M. van Schendelen, 1978.
25. J.A.A. van Doorn, 1971; I. Schöffer, 1973,; I. Scholten, 1980; R. Steininger, 1975 and 1977; C.G.A. Bryant, 1981.
26. M. Fennema, 1976; R. Kieve, 1981; S. Stuurman, 1983.
27. Stuurman, 1983, p. 319 et passim.
28. Kieve, 1981.
29. Lijphart, 1987, p. 204.
30. See for a review of criticisms on Lijphart: M.P.C.M. van Schendelen, 1984.
31. Lijphart, 1968-A, p. 1-2, 15.
32. Lijphart, 1968-A, chapter V.
33. Lijphart, 1968-A, p. 103.
34. Van Schendelen, 1984, p. 22.
35. Lijphart 1987, p. 182.

36. See also Van Schendelen, 1984, pp. 26-28.
37. R. Dahl, 1971.
38. Lijphart, 1977, p. 25.
39. J. Steiner, 1981.
40. Van Schendelen, 1984, p. 31.
41. S. Nilson, 1979.
42. Lijphart, 1987, p. 184-185.
43. The exact content of Lijphart's concept *democratic system* is not made unambiguously clear. See Van Schendelen, 1984, pp. 32-33.
44. Van Schendelen, 1984, p. 37.
45. H. Daudt, 1980, p. 186.
46. I. Scholten, 1980, pp. 332-335.
47. Daalder, 1985, p. 57, transl. mine.
48. S. Rokkan, 1970.
49. R. Dahl, 1966 and 1971.
50. G. Lehmbruch, 1967-A and 1967-B; V.R. Lorwin, 1971; J. Steiner, 1974; S. Rokkan 1968, 1970 and 1975; Robert A. Dahl integrated the theories of this group in chapter 7 of his book *Polyarchy: Participation and Opposition* (Dahl, 1971). See for the consociational democracy school also Daalder, 1984.
51. Daalder, 1984, p. 107.
52. Daalder, 1966, summarized in Daalder, 1984, p. 107-108.
53. It is not quite clear if Daalder means that the Dutch state bureaucracy has ever been a-political. In fact there are examples in recent history of ministries having a definite political colour, e.g. the ministries of welfare and agriculture.
54. Daalder, 1984, pp. 107-108.
55. Sources: figures for the elections of 1880, 1887 and 1890 are found in J.N. Romein and A.H.M. Romein-Verschoor, 1979, p. 482. Figures for 1900 in J.N. Romein and A.H.M. Romein-Verschoor, *Op.cit.*, p. 485 and H.P.H. Jansen, 1982, p. 164; figures for 1910 in A. Lijphart, 1982, p. 103. For 1890 Lijphart gives a figure of 26.8 per cent which is even lower than that given by Romein and Romein-Verschoor.
56. Including Dutch Reformed and Re-Reformed.
57. Source: Peter Flora, 1983, p.60. In the 1950's no census was taken.
58. J.P. Kruijt and W. Goddijn, 1962, table IV, p. 243.
59. Kruijt and Goddijn also present a more dynamic approach by which they measure the number of pillarized organizations for 1914, 1925, 1932, 1939 and 1956. However, the methodology they followed as well as the sources they used remain unclear. In a footnote they state that the figures are not exact. See Kruijt and Goddijn, 1962, pp. 228, 239, 244 and 245.
60. Kruijt and Goddijn also presented figures on voting behaviour. Because we treated those before, they are left out from this paragraph.
61. Beside public universities in the Netherlands there are two Catholic universities and one Protestant university.
62. It is strange that for Dutch denominational schools the interference of ideology with the transmission of knowledge has been accepted as a necessary requirement for educational quality, while public education is expected to be value neutral.
63. See I. Gadourek, 1956.
64. Ph.J. Idenburg, 1964, pp. 443-444.
65. Kruijt and Goddijn, 1962, p. 243.
66. See Lijphart, 1968-B; C. Bagley, 1973, J.P. Kruyt, 1959.
67. Quotation from Kruyt, 1959, pp. 41-42.
68. Abbreviated: EVC. The Dutch name would be in English something like Trade-Union Federation of Unity.
69. See for a historical analysis of Dutch labour relations since 1860: John P. Windmuller, 1969. For my analysis of the effects of pillarization for the industrial relations I used the chapters III and VII of this standard work as a source.
70. In 1946 the EVC attracted 20 percent of all organized workers. After 1946 the number of its members steadily decreased: 18 percent in 1947, 17 percent in 1948, 15 percent in 1949 and 14 percent in 1950. Source: J.P. Windmuller, 1970, p. 129, table 1. This is the in Dutch translated edition of Windmuller, 1969. In referring to this work, I cite the Dutch edition.

71. See D.A. Hibbs jr., 1978, p. 155.
72. See for the desaggregated data W.F. de Nijs, 1983, pp. 230-231.
73. See Windmuller, 1969.
74. See Albeda, 1975, p. 117.
75. See A.M. Ross and P.T. Hartman, 1960 and W.F. de Nijs, 1983, pp. 260-261.
76. According to Hibbs this is a major factor promoting stability in industrial relations. See D.A. Hibbs jr., 1978, p. 165.
77. For the early fifties it holds that the EVC was regarded as an unofficial union. Therefore, strikes organized by the EVC were considered as wildcat strikes, which explains the large number of wildcat strikes at the beginning of the 1950's. There were some big strikes in 1946 (273,000 man years lost) and in 1947 (127,900 man years lost). These strikes were organized by the EVC and thus according to our argument not representative for Dutch industrial relations.
78. CBS, 1951, p.23.
79. See G. Dekker, 1965, pp.105-106.
80. G. Dekker, 1965, pp.100-105.

Notes: chapter VI

1. See J.E. Ellemers, 1980 and C.P. Middendorp, 1979.
2. Here the word "periods" is not used in a connotation derived from evolutionary theory. It merely serves to point at periods in the modernization process which are dissimilar.
3. C. van der Eijk and B. Niemöller, 1983, no.2, p. 176.
4. A.A.G. Pijnenburg and J.J.M. Holsteyn, 1987, pp. 41-60.
5. See A. Lijphart, 1982 (1968-B), pp. 18-24.
6. See for the items which form the pillarization-index: *Sociaal en Cultureel Rapport 1986*, 1986, pp. 373-374, note 1.
7. *Sociaal en Cultureel Rapport 1986*, 1986, p. 337, transl. mine.
8. See for a graphical representation: *Sociaal en Cultureel Rapport 1986*, p. 336, graph 11.15.
9. On the one extreme we see marxist theories like G.K. Inghams's (1974), on the other extreme those of institutionalists like Ross and Hartman (1960).
10. See J.P. Windmuller and C. de Galan, 1979, part II, p. 140.
11. See for an elaborate exposé on the Dutch guided wage policy J.P. Windmuller, 1970.
12. I.e. labour share in net value income of the private sector at factor prices (including imputed income of self-employed).
13. Often these consultations did not result in a unanimous advise, which made it easier to neglect them.
14. See Windmuller and De Galan, 1979, pp. 182-184.
15. Computed on the data of appendix B, table B.6.I.
16. The Protestant CNV embraced 13.3 per cent of all organized workers; the remaining 25.3 were organized in several other small unions. See CBS, 1981, p. 21.
17. Although the denominational and the Social-Democratic broadcasting associations lost a substantial share of the market to newcomers, they still defend the system of plurality in broadcasting and publicly state their ideological backgrounds. While the pillarized school system is loosing public support, it still stands. Adherents of the denominational school fiercely defend its existence on the basis of believed significant differences between denominational and non-denominational schools. The powerful farmers' associations are still organized on the basis of religion. So are the employers' organizations. There is still a distinction between Roman-Catholic, Protestant and non-denominational hospitals.
18. In a recent article Hellemans observes the same phenomenon in Belgium. Although the figures of mass attendance decreased substantially between 1967 and 1978 the Belgian Roman-Catholic organizations have maintained their power. See S. Hellemans, 1988, p. 49.
19. J.P. Kruijt, 1959.
20. C.P. Middendorp, 1979, p. 9.
21. Middendorp, 1979, p. 28.
22. Middendorp, 1979, p. 33.
23. J.E. Ellemers, 1984, pp. 129-143.
24. Hellemans, 1988.
25. Hellemans, 1988, p. 53, trans. mine.

26. Especially with regard to the allocation of public offices.
27. See Pat Thane, 1982.
28. To my knowledge there is no systematic research regarding the contribution of pillarization to the creation of the Dutch welfare state. It would be interesting to know whether in the first half of the twentieth century the Dutch lagged behind because of antagonisms between the socialist and denominational parties. One could also think of possible detrimental effects of state intervention on the power of pillarized organizations. Under the post-war conditions the principle of proportionality associated with pillarization might have promoted an upswing of government expenditures. However, these intriguing questions fall outside the scope of this study.
29. See H. Spoormans, 1988.
30. A. Etzioni, 1961.
31. A. Etzioni, 1961, p. 10.
32. In case of unemployment, sickness and disability the before taxes replacement rate was 80 per cent. After taxes the difference between wage and benefit was for the bulk of the labour force much smaller than 20 per cent. The actual sickness benefit even equaled the wage level.
33. Therefore, in a way Roman-Catholic fertility rates were the very result of pillarization itself. See: F. van Heek, 1954.
34. For the Dutch law only civil marriage is valid. So for every couple that wants to get married a registered-office wedding is obligatory and sufficient. Furthermore, one can marry in a church which, however, is not officially recognized by law as a marriage contract. Church marriage, therefore, indicates an orientation to church norms.
35. SCP, 1986, Annex table 11.2 p. 349.
36. Before I stated that because of its ideology of individualism Dutch liberalism never took a pillarized structure the way the religious groups and the social-democrats did. Because of its ideology of individualism it does not make sense to speak of a lessening of compliance to the ideology of liberalism.
37. C.P. Middendorp, 1979.
38. I. Gadourek, 1982. A preview of this study is: I. Gadourek, 1980.
39. Middendorp, 1979, pp. 165-167.
40. Middendorp, 1979, p. 168.
41. Middendorp, 1979, p. 169.
42. Gadourek, 1980, p. 147.
43. Gadourek, 1980, pp. 144-145.
44. Gadourek, 1980, pp. 145-146.
45. See J. Goudsblom, 1985, pp. 4-5.
46. See J.E. Ellemers, 1984.

APPENDIX A: CHAPTER III

A.1 Per Capita National Income 1900-1986

Table A.1.I Dutch Per Capita National Income (guilders of 1900, net market prices), 1900-1986.

Year	Per Capita Income	Year	Per Capita Income
1900	349	1944	253
1901	336	1945	257
1902	348	1946	426
1903	351	1947	481
1904	344	1948	524
1905	350	1949	561
1906	361	1950	572
1907	358	1951	575
1908	353	1952	583
1909	365	1953	631
1910	364	1954	668
1911	370	1955	709
1912	385	1956	726
1913	391	1957	737
1914	373	1958	720
1915	379	1959	744
1916	382	1960	799
1917	350	1961	813
1918	323	1962	830
1919	398	1963	849
1920	407	1964	912
1921	428	1965	945
1922	445	1966	954
1923	447	1967	995
1924	473	1968	1,046
1925	485	1969	1,105
1926	518	1970	1,149
1927	533	1971	1,180
1928	553	1972	1,206
1929	549	1973	1,256
1930	538	1974	1,298
1931	494	1975	1,268
1932	478	1976	1,325
1933	470	1977	1,348
1934	455	1978	1,363
1935	468	1979	1,385
1936	494	1980	1,383
1937	518	1981	1,359
1938	499	1982	1,327
1939	529	1983	1,344
1940	461	1984	1,378
1941	432	1985	1,411
1942	391	1986	1,428
1943	380		

Source : C.A. van Bochove and Th.A. Huitker, *National Accounts. Occasional Paper, NA-017. Main National Accounting Series 1900-1986*, CBS, Voorburg, 1987. I derived the above series of Dutch per capita national income in guilders of 1900 (net market prices) by using the series national income in net market prices (table H2, column 3, p. 9), percentage volume changes 1901-1986 (table H2, column 8, pp. 9-11) and average population (table H1, column 9, pp. 6-8) presented in this publication.

A.2 Employment by Sector of Industry

Table A.2.I Employment by Sector of Industry. Absolute Numbers x 1000, 1899-1981

Year	Agriculture	Industry	Service	Other Professions and Enterprises	Total
1899	592.3	598.0	696.6	34.0	1,920.9
1909	640.8	723.6	871.5	23.2	2,259.1
1920	640.9	937.2	1,111.3	29.8	2,719.2
1930	655.4	1,125.6	1,365.0	33.2	3,179.2
1947	747.1	1,385.7	1,677.1	56.5	3,866.4
1960	446.8	1,717.6	1,908.0	96.2	4,168.6
1971	290.6	1,766.8	2,226.4	451.3	4,735.1
1975	295.6	1,718.3	2,656.7	15.4	4,686.2
1981	275.6	1,597.0	3,156.9	66.6	5,096.0

Note : Industry includes industry, mining and construction. Services include commerce, banking and insurance, transportation, communication, public utilities, government and domestic services. From 1971 onwards excluding the unemployed.
Professions and enterprises unknown are included in the category "other professions and enterprises"; for 1971 the number in this category is extremely high as a result of census non-response.
Source : CBS, *1899-1984. Vijfentachtig jaren statistiek in tijdreeksen*, Staatsuitgeverij, The Hague, 1984, p. 76.

Table A.2.II Employment by Sector of Industry as a Percentage of Total Employment, 1899-1981

Year	Agriculture	Industry	Service	Other Professions and Enterprises
1899	30.8	31.1	36.3	1.8
1909	28.4	32.0	38.6	1.0
1920	23.6	34.5	40.9	1.1
1930	20.6	35.4	42.9	1.0
1947	19.3	35.8	43.4	1.5
1960	10.7	41.2	45.8	2.3
1971	6.1	37.3	47.0	9.5
1975	6.3	36.7	56.7	0.3
1981	5.4	31.3	61.9	1.3

Source : Table A.2.I

A.3 Urbanization

To compute changes of the degree of urbanization I used data presented by Van der Knaap, *A Spatial Analysis of the Evolution of an Urban System: The Case of the Netherlands*, Utrecht, 1978. He selected 502 Dutch municipalities which in 1970 had more than 5000 inhabitants. From 1840 onwards, for every decade Van der Knaap counted the number of inhabitants of each municipality selected and classified the total population of all municipalities under investigation in classes ranging from municipalities with less than 2,500 inhabitants to cities with 250,000 to one million inhabitants.

Van der Knaap corrected for frontier modifications which occurred during the period 1840-1970. As an example: On January 1st 1920 the municipalities of Gestel, Stratum, Strijp, Tongelre and Woensel were added to the municipality of Eindhoven. As a result, in 1920 the number of inhabitants of Eindhoven increased substantially.

Van der Knaap corrected such a bias by adding to Eindhoven the number of inhabitants of Gestel, Stratum, Strijp, Tongelre and Woensel before 1920 as well. In this way the areas of the municipalities selected remain constants for the whole period from 1840 to 1970.

In table A.3.II. the total population of all municipalities selected is given by municipality class and by year. In table A.3.III. the number of municipalities in each municipality class is given for the successive years. On the basis of the figures of the tables A.3.II. and A.3.III. I computed for each year the coefficient of variation V (see formula below). The coefficients of variation computed are presented in table A.3.I. below

Table A.3.I Coefficient of Variation V, 1840-1970

Year	V	Year	V
1840	2.52	1910	3.36
1850	2.41	1920	3.45
1860	2.57	1930	3.44
1870	2.62	1940	3.33
1880	2.82	1950	3.11
1890	3.21	1960	2.78
1900	3.37	1970	2.37

Coefficient of variation V:

$$V = \frac{\sqrt{\frac{\sum_{1}^{s}(k_s - \bar{x})^2 \cdot f_s}{n}}}{\bar{x}}$$

where
V   = coefficient of variation
k_s = average number of inhabitants in municipality class s
f_s = number of municipalities in municipality class s
n   = total number of municipalities selected (= 502)
$\bar{x}$ = average number of inhabitants of all 502 municipalities

Table A.I Total Population of 502 Dutch Municipalities by Municipality Class (Number of Inhabitants) and Year, 1840 1970

Municipality Class	1840	1850	1860	1870	1880	1890	1900	1910	1920	1930	1940	1950	1960	1970
< 2,500	360,873	330,864	311,626	289,426	258,970	232,003	207,404	152,760	89,262	51,622	27,426	1,822	0	0
2,500 - 5,000	651,065	648,242	677,176	703,853	701,853	711,102	677,106	711,792	726,073	688,901	619,552	488,026	290,298	
5,000 - 10,000	451,700	524,153	551,025	626,793	691,908	729,412	849,337	965,931	1,095,093	1,169,647	1,346,480	1,324,519	1,484,146	1,550,857
10,000 - 30,000	370,544	459,289	493,577	542,426	733,887	721,436	855,109	1,030,584	1,258,888	1,497,342	1,644,441	2,256,498	2,778,255	3,445,049
30,000 - 50,000	78,100	77,466	173,877	217,417	250,615	274,008	384,565	449,723	383,880	488,821	704,729	762,816	676,141	1,159,872
50,000 - 100,000	217,413	126,046	137,636	156,345	74,481	308,506	256,017	455,473	818,672	903,418	1,102,262	1,011,796	1,365,530	1,809,203
100,000 - 250,000	223,381	349,469	131,663	151,024	314,922	407,811	319,096	124,385	147,466	396,633	581,005	1,125,589	1,318,327	1,448,922
250,000 - 1,000,000	0	0	257,078	281,721	343,178	447,045	868,801	1,313,025	1,582,292	1,808,446	1,980,801	2,074,889	2,465,735	2,347,628

Table A.II Total Number of 502 Dutch Municipalities Selected by Municipality class (Number of Inhabitants) and Year, 1840 1970

Municipality Class	1840	1850	1860	1870	1880	1890	1900	1910	1920	1930	1940	1950	1960	1970
< 2,500	227	205	188	170	148	130	113	80	46	25	12	1	0	0
2,500 - 5,000	184	185	190	195	193	196	189	196	196	181	159	121	68	
5,000 - 10,000	64	78	82	92	104	111	127	140	156	166	187	191	213	224
10,000 - 30,000	21	28	33	35	47	49	55	64	78	97	104	145	170	208
30,000 - 50,000	2	2	5	6	6	8	10	11	10	13	18	19	17	31
50,000 - 100,000	3	2	2	2	1	5	4	7	12	14	15	14	20	25
100,000 - 250,000	1	2	1	1	2	2	2	1	1	3	4	8	10	10
250,000 - 1,000,000	0	0	1	1	1	1	2	3	3	3	3	3	4	4

Sources table A.I. and A.II: G.A. van der Knaap, *A Spatial Analysis of the Evolution of an Urban System: The Case of the Netherlands*, Utrecht, 1978, table 2.4 A & B, p. 47.

A.4 Size of Enterprises by Employed Persons

Sources of table 3.9 :

Data for 1859, 1889 and 1909 derived from J.A. de Jonge, *De industrialisatie van Nederland tussen 1850 en 1914*, Sun, Nijmegen, 1976 (1968), p. 232. Figures for 1930 and 1950 were computed on the basis of CBS, *2e Algemene bedrijfstelling 16 oktober 1950. Deel 5: Statistiek der ondernemingen*, Zeist, 1957, p. 15. Data for 1963 derived from CBS, *Derde algemene bedrijfstelling 15 oktober 1963. Deel 4: ondernemingen*, The Hague, 1969, p. 15, p. 48. For 1978: CBS, *Vierde algemene bedrijfstelling 1978. Deel 2: algemene sectorale gegevens: A. delfstoffenwinning, industrie, openbare nutsbedrijven, bouwnijverheid en bouwinstallatie*, The Hague, 1985, pp. 14-15, p. 34, table 1a, rows 4 till and including 11.

Due to different classifications applied in the respective sources used, the classes of table 3.9. are for some years slightly dissimilar from eachother. As to large size enterprises, in the sources for 1859, 1889, 1909, 1930 and 1950 enterprises with more than 50 employed persons are defined as such, while for 1963 and 1978 enterprises with more than 49 employed persons are considered large scale. Regarding the border between small and medium size enterprises, for 1859, 1889, 1909, 1963 and 1978 the class "medium size enterprises" starts with 10 employed persons. For 1930 and 1950 with 11 employed persons.

Different from 1950, the statistical sources available for 1963 and 1978 count as employed persons only those who worked 15 hours or more per week.

Further, in the sources for 1963 and 1978 part of the concept of "enterprise" is the condition that it must be a legal body. This condition was not incorporated in the definition of "enterprise" for 1950.

Finally, the industrial sector of the economy is differently defined in the sources used. We took the classification of the source for 1963 as a starting point. In this source the industrial sector is defined as comprising mining, industry and crafts, construction and public utilities. Therefore, for 1978 I took the classes 2 till and including 5 (industry, public utilities, construction and installation) as comprising the industrial sector. Because it was not clear whether or not mining was included in the sources for 1859, 1889 and 1909, mining was excluded in all sources.

Table A.4.1 Enterprises by Size of Personnel as a Percentage of Total Number of Enterprises, 1930, 1950, 1963 and 1978

Year	0-5	5-10	10-50	>50	Total (abs)
1930	91.20	4.80	3.20	0.80	384,175
1950	86.20	7.10	5.50	1.20	384,926
1963	74.71	13.51	9.53	2.26	359,000
1978	72.84	13.92	10.93	2.31	338,572

Notes : In 1930 and 1950 personnel includes all who were economically active in the enterprise at the moment of counting and includes also employees on sickness leave and on holiday. In 1963 personnel equals all persons who worked on average 15 hours or more per week including enterprise owners and working family. In 1978 personnel covers all men and women including owners, who on September 29th 1978 worked 15 hours or more in the enterprise.

Sources: 1930 and 1950: CBS, *2e Algemene bedrijfstelling 16 oktober 1950. Deel 5: Statistiek der ondernemingen*, Zeist, 1957, p. 15.

1963 : CBS, *Derde algemene bedrijfstelling 15 oktober 1963. Deel 4: ondernemingen*, Staatsuitgeverij, The Hague, 1969, p. 15.
1968 : CBS, *Vierde algemene bedrijfstelling 1978. Deel 2: algemene sectorale gegevens*, Staatsuitgeverij, The Hague, 1985, p. 12-15.

A.5 Ministry Personnel as a Percentage of the Labour Force and of Total Population

I used Peter Flora, *State, Economy, and Society in Western Europe 1815-1975. A Data Handbook, Vol. I*, Frankfurt/London/Chicago, 1983, pp. 223-227 as to the volume of ministry personnel as a percentage of the labour force and of total population for the period 1877-1975. To estimate the volume of ministry personnel in 1849 I used figures from A. van Braam, *Ambtenaren en bureaucratie in Nederland*, The Hague, 1957.

Van Braam presents data of government office personnel excluding the military (OP) and of total government personnel (TGP). In table A.5.I. these data are given for 1849, 1899, 1909 and 1930 as well as Flora's data on ministry personnel for the corresponding years.

Table A.5.I Government Office Personnel (OP), Total Government Personnel (TGV) and Ministry Personnel (MP) 1849, 1899, 1909, 1930

	OP*	TGP*	MP**
1849	n.a.	20,000	n.a.
1899	20,000	69,000	20,151
1909	25,000	91,000	28,795
1930	60,000	212,000	60,652

Source : * A. van Braam (1957), p. 22, p. 34
** P. Flora (1983), pp. 225-226

I estimated OP (1849) by:

$$\text{OP (1849)} = \frac{\frac{OP(1899)}{TGP(1899)} + \frac{OP(1909)}{TGP(1909)} + \frac{OP(1930)}{TGP(1930)}}{3} * TGP(1849)$$

OP (1849) then equals 5,651. As Flora's figures for ministry personnel in 1899, 1909 and 1930 roughly equals Van Braam's volumes of government office personnel in the corresponding years, I use OP (1849) as an estimation of MP (1849).

Table A.5.II Ministry Personnel (MP), Ministry Personnel as a Percentage of the Population, Ministry Personnel as a Percentage of the Labour Force, Population (PO) and Labour (LF), 1849-1975

Year	MP	MP as a Percentage of Population	MP as a Percentage of Labour Force	Population (PO)	Labour Force (LF)
	(1)	(2)	(3)	(4)	(5)
1849	5,651	0.18	0.49	3,056,879	1,159,000
1877	12,746	0.33	0.86	3,910,000	1,487,000
1887	15,430	0.35	0.93	4,392,000	1,650,400
1897	18,567	0.37	0.99	4,954,000	1,869,500
1900	20,851	0.41	1.07	5,142,000	1,940,000
1913	33,308	0.54	1.41	6,164,000	2,365,000
1929	59,990	0.77	1.95	7,782,000	3,075,000
1938	65,483	0.75	1.86	8,685,000	3,517,000
1950	103,596	1.02	2.78	10,114,000	3,728,000
1955	102,133	0.95	2.59	10,751,000	3,940,000
1960	106,722	0.93	2.57	11,486,000	4,152,000
1965	119,114	0.97	2.66	12,292,000	4,475,000
1970	132,694	1.02	2.77	13,032,000	4,798,000
1975	146,940	1.08	2.92	13,654,000	5,027,000

Source : Ministry personnel was derived from P. Flora, 1983, pp. 224-226. The figure of 1849 was estimated as above.
Population:
From 1877 to and including 1975: A. Maddison, 1982, pp. 182-187, table B2-B4. 1849: census figure, Februari 19th derived from P. Flora, 1983, p.47.
Labour Force:
Maddison gives figures for the following years: 1870, 1880, 1890, 1900, 1913, 1929, 1938, 1950, 1960, 1970, 1973 and 1978. Figures for intermediate years were calculated by interpolation. See A. Maddison, 1982, p. 209, table C7.
The Labour Force of 1849 is estimated in the following way:

$$\frac{\frac{LF(1877)}{PO(1877)}+\frac{LF(1887)}{PO(1887)}+\frac{LF(1900)}{PO(1900)}+\frac{LF(1913)}{PO(1913)}}{4} * PO(1849)$$

Then LF(1849) equals to 1,159,000.

A.6 Civil Cases Brought to Courts of Justice

Table A.6.I Total Number of Civil Cases Brought to All Courts of Justice (1 + 2) and All Courts of Justice excluding District Courts (3) and Population (4), 1852-1985

Year	All Courts (1)	All Courts (2)	All Courts excl. District Courts (3)	Population x 1000 (4)
1852	9,852		7,022	3,168
1859	11,114		7,888	3,357
1873	13,428		9,325	3,716
1883	22,505		15,209	4,225
1890	23,539		17,421	4,565
1895	25,032		18,669	4,859
1900	25,984		18,848	5,179
1905	28,976		20,716	5,591
1910	37,291		26,881	5,946
1916	36,817		24,603	6,583
1920	53,720		33,427	6,865
1925	75,987		47,477	7,416
1930	75,846		49,508	7,936
1935	80,018		51,798	8,475
1938	69,682		45,047	8,729
1947	64,277		34,535	9,716
1948	77,193		45,189	9,884
1949	92,858		57,996	10,027
1950	91,448		56,834	10,200
1951	93,192		56,193	10,328
1952	93,170		56,595	10,436
1953	91,125		54,080	10,551
1954	90,620		54,234	10,680
1955	89,708		53,491	10,822
1956	86,010		52,040	10,957
1957	91,472		54,901	11,096
1958	98,983		59,270	11,278
1959	95,949		56,676	11,417
1960	97,585		56,434	11,556
1961	97,011		58,062	11,721
1962	98,348		58,636	11,890
1963	100,073		58,647	12,042
1964	101,271	82,623	58,237	12,212
1965	99,663	81,215	55,396	12,377
1966		85,306	57,492	12,535
1967		91,381	60,448	12,661
1968		95,329	62,485	12,798
1969		101,016	66,506	12,958
1970		108,243	71,820	13,119
1971		120,157	85,994	13,270
1972		128,349	89,066	13,388
1973		131,729	92,593	13,491
1974		134,989	93,260	13,599
1975		137,289	91,757	13,734
1976		140,323	93,000	13,814
1977		140,336	90,411	13,898
1978		140,638	93,985	13,986
1979		146,070	96,201	14,091
1980		162,251	104,543	14,209

Table A.6.1 (continued) Total Number of Civil Cases Brought to All Courts of Justice (1 + 2) and All Courts of Justice excluding District Courts (3) and Population (4), 1852-1985

Year	All Courts	All Courts	All Courts excl. District Courts	Population x 1000
	(1)	(2)	(3)	(4)
1981		183,260	119,782	14,286
1982		203,141	134,804	14,340
1983		210,676	144,123	14,395
1984		203,156	141,563	14,454
1985		204,099	145,373	14,529

note : From 1967 onwards civil cases brought to district courts were collected by questionnaires of the Ministry of Justice. For that reason the time series of the number of civil cases brought to all courts before and after 1964 are not comparable. Because of that we present the civil cases brought to all courts by two different columns (column 1 and column 2). Similarly in the relevant figure in chapter III two separate lines with regard to civil cases brought to all courts are depicted. To get also a time series which is comparable over the years for the whole period I computed a time series of civil cases brought to all courts excluding district courts (column 3).

Sources : Data of the number of civil cases brought to all courts and all courts excluding district courts were derived from:

1852	: *Geregtelijke statistiek van het Koninkrijk der Nederlanden, 1852*, The Hague, 1853, pp. XX-XXV.1859:*Ibidem 1859*, The Hague, 1860, pp. XXVI-XXXI.
1873	: *Ibidem 1873*, The Hague, 1875, pp. XXXIII-XXXVII.
1883	: *Ibidem 1883*, The Hague, 1884, pp. XXXVI-XLII.
1890	: *Ibidem 1890*, The Hague, 1891, pp. XXI-XXVII.
1895	: *Ibidem 1895*, The Hague, 1896, pp. XXI-XXVII.
1900	: CBS, *Justitieële statistiek van het Koninkrijk der Nederlanden over het jaar 1900*, Belinfante, The Hague, 1901, pp. XXII-XXX.
1905	: *Ibidem 1905*, The Hague, 1906, pp. XXIV-XXXII.
1910	: *Ibidem 1910*, The Hague, 1911, pp. XXVIII-XXXV.
1916	: *Ibidem 1916*, The Hague, 1918, pp. XXIX-XXXIII.
1920	: *Ibidem 1920*, The Hague, 1922, pp. XXVIII-XXXI.
1925	: CBS, *Justitieële statistiek 1926*, Algemeene Landsdrukkerij, The Hague, 1928, pp. 26-31.
1930	: *Ibidem 1930*, Drukkerij Trio, The Hague, 1931, pp. 64-67. On page 28 of this source the figure 26,267 for the number of civil cases brought to district courts.
1935	: CBS, *Justitieële statistiek en faillissements-statistiek over het jaar 1935*, Algemeene Landsdrukkerij, The Hague, 1936, pp. 54-57.
1938	: *Ibidem 1938*, Rijksuitgeverij, The Hague, 1939, pp. 64-67.
1947, 1948	: CBS, *Justitieële Statistiek over het jaar 1951*, De Haan, Utrecht, 1952, pp. 21-25.
1949-1952	: *Justitieële Statistiek 1953*, De Haan, Utrecht, 1954, pp. 21-25.
1953-1956	: *Ibidem 1957*, De Haan, Zeist, 1959, p.28-30.
1957-1960	: *Ibidem 1961*, De Haan, Zeist, 1963, p.31-34.
1961-1965	: Column (1): *Ibidem 1965*, Staatsuitgeverij, The Hague, 1967, p.42-45.
1964-1966	: Column (2): *Ibidem 1968*, Staatsuitgeverij, The Hague, 1970, pp. 39-43.
1967-1970	: *Ibidem 1971*, Staatsuitgeverij, The Hague, 1973, pp.32-35.
1971-1972	: *Ibidem 1973*, Staatsuitgeverij, The Hague, 1975, pp. 29-31.

1973	: *Ibidem 1974-1975*, Staatsuitgeverij, The Hague, 1977, pp. 31- 34.
1974-1975	: *Ibidem 1976*, Staatsuitgeverij, The Hague, 1980, pp. 27-29.
1976	: *Ibidem 1978*, Staatsuitgeverij, The Hague, pp. 27-29.
1977-1979	: *Ibidem 1979*, Staatsuitgeverij, The Hague, 1982, pp. 29-32.
1980-1981	: *Ibidem 1981*, Staatsuitgeverij, The Hague, 1983, pp. 29-32.
1982-1983	: CBS, *Burgerlijke en administratieve rechtspraak 1983*, Staatsuitgeverij, The Hague, 1985, pp. 11-15.
1984-1985	: *Ibidem 1985*, Staatsuitgeverij, The Hague, 1987, pp. 9, 11, 14 and 17.

Population data were derived from:

1852	: *Staatkundig en Staathuishoudkundig Jaarboekje voor 1854*, Amsterdam, 1854, p. 1.
1859	: *Ibidem 1861*, Amsterdam, 1861, p. 2.
1873:	*Ibidem 1875*, Amsterdam, 1875, p. 7.
1883	: *Ibidem 1884*, Amsterdam 1884, p. 11.
1890, 1895	: CBS, *Jaarcijfers voor het Koninkrijk der Nederlanden. Rijk in Europa 1898*, The Hague, 1899, p. 3.
1900-1938	: CBS, *1899-1969. Zeventig jaren statistiek in tijdreeksen. 1899- 1969*, Staatsuitgeverij, The Hague, 1970, p. 14.
1947-1983	: CBS, *1899-1984. Vijfentachtig jaren statistiek in tijdreeksen*, Staatsuitgeverij, The Hague, 1984, p. 20.
1984-1985	: CBS, *Statistisch Zakboek 1987*, Staatsuitgeverij, The Hague, 1987, p. 57.

Table A.6.II Number of Civil Cases Brought to All Courts of Justice (1+2) and All Courts of Justice excluding District Courts (3) per Thousand of the Population, 1852-1985

Year	All Courts	All Courts	All Courts excl. District Courts
	(1)	(2)	(3)
1852	3.1		2.2
1859	3.3		2.3
1873	3.6		2.5
1883	5.3		3.6
1890	5.2		3.8
1895	5.2		3.8
1900	5.0		3.6
1905	5.2		3.7
1910	6.3		4.5
1916	5.6		3.7
1920	7.8		4.9
1925	10.2		6.4
1930	9.6		6.2
1935	9.4		6.1
1938	8.0		5.2
1947	6.6		3.6
1948	7.8		4.6
1949	9.3		5.8
1950	9.0		5.6
1951	9.0		5.4
1952	8.9		5.4
1953	8.6		5.1
1954	8.5		5.1
1955	8.3		4.9
1956	7.8		4.7
1957	8.2		4.9
1958	8.8		5.3
1959	8.4		5.0
1960	8.4		4.9

Table A.6.II (continued) Number of Civil Cases Brought to All Courts of Justice (1+2) and All Courts of Justice excluding District Courts (3) per Thousand of the Population, 1852-1985

Year	All Courts (1)	All Courts (2)	All Courts excl. District Courts (3)
1961	8.3		5.0
1962	8.3		4.9
1963	8.3		4.9
1964	8.3	6.8	4.8
1965	8.1	6.6	4.5
1966		6.8	4.6
1967		7.2	4.8
1968		7.4	4.9
1969		7.8	5.1
1970		8.3	5.5
1971		9.1	6.5
1972		9.6	6.7
1973		9.8	6.9
1974		9.9	6.9
1975		10.0	6.7
1976		10.2	6.7
1977		10.1	6.5
1978		10.1	6.7
1979		10.4	6.8
1980		11.4	7.4
1981		12.8	8.4
1982		14.2	9.4
1983		14.6	10.0
1984		14.1	9.8
1985		14.0	10.0

Source : table A.6.I

A.7 Public Expenditures

Data on public expenditures for the years 1850 till and including 1890 were derived from Peter Flora, *State, Economy and Society in Western Europe, 1815-1975. A Data Handbook. Vol.I*, Campus, Frankfurt/London/New York, 1983, p. 413. Public expenditures of the years 1900-1910 and 1925-1980 were derived from CBS, *1899-1984. Vijfentachtig jaren statistiek in tijdreeksen*, Staatsuitgeverij, The Hague, 1984, p. 166, column 12. The year 1921 was taken from CBS, *1899-1979. Tachtig jaren statistiek in tijdreeksen*, Staatsuitgeverij, The Hague, 1979, p. 150. The figures from P. Flora include debt redemption, CBS figures do not.

GDP figures from 1921 onward were taken from C.A. van Bochove and Th.A. Huitker, *National Accounts. Occasional Paper NA-017. Main National Accounting Series 1900-1986*, CBS, Voorburg, 1987, pp. 6-8. Bochove and Huitker state that their figures for the period 1921-1939 are in harmony with current concepts (p. 3). However, there is a conceptual break in 1969. Therefore, I multiplied the figures of 1950, 1955, 1960 and 1965 with the quotient of the 1969 GDP's after and before revision.

GDP figures for 1850-1910 in real prices (US $ 1970) including a correction for purchasing power could be derived from Angus Maddison, "Recent Revisions to British and Dutch Growth, 1700-1870 and Their Implications for Comparative Levels of Performance", *Economic Growth in Northwestern Europe: the last 400 years*, Memorandum from Institute of Economic Research, Faculty of Economics, University of Groningen, nr. 214, Groningen, March 1987 and from Angus Maddison, *Phases of Capitalist Development*, Oxford University Press, Oxford, 1982.

Maddison (1987) presents GDP in real prices (US $ 1970) including a correction for purchasing power for the years 1820, 1870 and 1913. GDP in real prices (US $ 1970) for 1900 and 1910 were calculated by relating the indices (1913 = 100) in Maddison, *Phases of Capitalist Development*, table A.6., p. 173 to the above mentioned GDP-figure of 1913. GDP in real prices (US $ 1970) for 1850 and 1860 were calcuted by interpolating the GDP figures of 1820 and 1870 by the use of the average annual compound growth rates. In the same way GDP figures for 1880 and 1890 were calculated by using the GDP figures of 1870 and 1900 as basis for the average annual compound growth rates.

In the above way the time series (1850-1910) of GDP in real prices (US $ 1970) including a correction for purchasing power were established. The next step was to transfer the GDP data from US $ 1970 to guilders. I did so by applying the exchange rate of 3.6166 (1970) as given by I.B. Kravis, A. Heston and R. Summers, *International Comparisons of Real Product- and Purchasing Power*, Johns Hopkins, 1978, p. 8. Further I corrected for the ratio of purchasing power of currency to exchange rate (Maddison, 1982, p. 160).

Finally I transformed GDP in real prices to GDP in current prices. For 1850, 1860 and 1870 I used whole sale prices and for 1880 and 1890 index numbers of the cost of living as presented by J.H. van Stuyvenberg and J.E.J. de Vrijer, 'Prices, population and national income in the Netherlands 1620-1978', in: *Journal of European Economic History*, 1982, p. 708. For 1900-1910 I used price index numbers given by the Dutch Statistical Bureau CBS (CBS, *1899-1969. Zeventig jaren statistiek in tijdreeksen*, Staatsuitgeverij, The Hague, 1970, p. 151).

Table A.7.I Public Expenditures and Gross Domestic Product (current market prices (x 1,000,000 guilders)), 1850-1982

Year	Public expenditures	GDP
	(1)	(2)
1850	70	1,099
1860	85	1,642
1870	99	1,933
1880	113	1,957
1890	166	1,891
1900	151	2,272
1910	199	2,987
1921	1,032	5,679
1925	699	5,644
1930	738	6,248
1935	989	4,849
1938	1,045	5,446
1950	4,968	19,749
1955	8,064	31,291
1960	10,376	44,421
1965	19,505	71,985
1970	35,008	121,180
1975	76,457	219,960
1980	129,147	336,740
1982	149,315	368,860

A.8 Education

Table A.8.I Participants Between 3 and 25 Years of Age in Formal Education, Absolute Numbers (1) and Per Thousand (2) of the Population Between 3 and 25 Years of Age (3), 1850-1960

Year	Absolute Number Participants	Participants Per Thousand	Population of 3 - 25 years
	(1)	(2)	(3)
1850	437,083	307	1,424,925
1860	487,943	326	1,495,330
1870	581,833	370	1,572,337
1880	676,865	376	1,800,567
1890	792,394	379	2,091,592
1900	924,266	388	2,381,550
1910	1,182,104	436	2,713,218
1920	1,412,092	449	3,144,538
1930	1,696,905	483	3,510,865
1938	1,776,794	485	3,663,747
1950	2,205,226	542	4,070,486
1960	2,971,795	639	4,649,413

Source : Figures on formal education and population were derived from CBS, *De ontwikkeling van het onderwijs in Nederland. Edition 1966. Part I: Tables*, Staatsuitgeverij, The Hague, 1966, pp. 28-29. Formal education includes pre-school education, primary and secondary education, vocational education and higher education. Column 1 is the total of these types of education. The construction of more desaggregated time series was not possible due to differences of definition over the years. For more details the reader is referred to part 3 pages 9 and 10 of the publication mentioned above.

Table A.8.II Absolute Number of Enrollments of Universities and Higher Vocational Training (Full Time Education) and Total Population between 18 and 24 Years of Age, 1930-1984

Year	Higher Vocational Training	University	Population 18-24
1930		12,061	1,001,538
1936		12,387	1,037,240
1940		14,825	1,073,306
1947		25,955	1,123,770
1950		29,736	1,133,431
1953		27,987	1,113,304
1955		29,642	1,103,717
1958		35,131	1,131,425
1959	38,300	37,725	1,139,147
1960	40,400	40,727	1,156,564
1961	43,100	43,937	1,191,251
1962	45,300	47,863	1,235,578
1963	46,900	52,400	1,258,429
1964	49,300	58,427	1,354,618
1965	52,500	64,409	1,436,784
1966	55,600	71,260	1,507,024
1967	59,700	77,896	1,554,743
1968	63,800	84,776	1,584,365
1969	68,800	93,594	1,606,684
1970	72,100	103,382	1,647,936
1971	78,300	112,873	1,612,739
1972	83,200	n.a.	1,584,878
1973	89,800	n.a.	1,574,211
1974	97,700	113,707	1,580,140
1975	111,200	120,134	1,602,813
1976	118,400	129,196	1,618,838
1977	123,800	137,426	1,635,711
1978	126,800	143,160	1,654,393
1980	131,800	149,000	1,713,276
1982	140,900	151,000	1,753,142
1983	143,900	157,000	1,765,033
1984	147,000	159,000	1,763,931

Note : Population between 18 and 24 years of age at ultimo current year.

Source : Higher Vocational Training:
    1930-1983 : CBS, *1899-1984. Vijfentachtig jaren statistiek in tijdreeksen*, Staatsuitgeverij, The Hague, 1984, p. 53.
    1984 : CBS, *Zakboek onderwijsstatistieken 1986. Onderwijs cijfergewijs*, Staatsuitgeverij, The Hague, 1986, p. 24.
    University Enrollments:
    1930-1978 : CBS, *Zakboek onderwijsstatistieken 1980. Onderwijs cijfergewijs*, Staatsuitgeverij, The Hague, 1980, pp. 20-21.
    1980-1984 : CBS, *Zakboek onderwijsstatistieken 1986. Onderwijs cijfergewijs*, Staatsuitgeverij, The Hague, 1986, p. 24.
    Population :
    1930-1969 : CBS, *Bevolking van Nederland naar geslacht, leeftijd en bugerlijke staat 1830-1969*, Staatsuitgeverij, The Hague, 1970.
    1970-1984 : CBS, *Maandstatistiek van de bevolking en volksgezondheid. Supplement*, Vol. 19-29 and CBS, *Maandstatistiek van de bevolking*, Vol. 31, no. 6 and Vol. 32, no. 11.

Table A.8.III Number of Enrollments of Universities and Higher Vocational Training per Thousand of the Total Population between 18 and 24 Years of Age, 1930-1984

Year	Higher Vocational Training	University
1930		12
1936		12
1940		14
1947		23
1950		26
1953		25
1955		27
1958		31
1959	34	33
1960	35	35
1961	36	37
1962	37	39
1963	37	42
1964	36	43
1965	37	45
1966	37	47
1967	38	50
1968	40	54
1969	43	58
1970	44	63
1971	49	70
1972	52	n.a.
1973	57	n.a.
1974	62	72
1975	69	75
1976	73	80
1977	76	84
1978	77	87
1980	77	87
1982	80	86
1983	82	89
1984	83	90

Source : table A.8.II

A.9 Newspapers Sent by Inland Mail

Data on the number of newspapers sent by mail were derived from the following sources:

1860-1880	: *Jaarcijfers over 1881 en vorige jaren*. Issued by de Vereeniging voor de Statistiek in Nederland, no. 1., H.L. Smits, The Hague, 1882, p. 103.
1890-1900	: CBS, *Jaarcijfers voor het koninkrijk der Nederlanden. Rijk in Europa (1903)*, Belinfante, The Hague, 1904, p. 241.
1910-1938	: CBS, *Jaarcijfers voor Nederland 1939*, Albani, The Hague, 1940, p. 348.

Midyear population figures were derived from Peter Flora, *State, Economy and Society in Western Europe. 1815-1975. A Data Handbook*. Vol. II, Campus, Frankfurt/London/Chicago, 1987, pp. 65-67.

APPENDIX B: CHAPTER V

B.1 Politics

Annotations to table 5.3

Sources:
- Austria : 1918-1935: Peter Flora, *State, Economy and Society in Western Europe. A Data Handbook in Two Volumes. Volume I*, Campus, Frankfurt/London/Chicago, 1983, p. 156.
  - 1945-1984: K. von Beyme, *Political Parties in Western Democracies*, Gower, Aldershot, 1985, p. 378.
- Belgium : 1918-1940: K. von Beyme, *Die parlementarischen Regierungssysteme in Europa*, Piper Verlag, München, 1973, pp. 904-906.
  - 1945-1984: K. von Beyme, *Op. Cit.*, 1985, p. 380.
- France : 1918-1964: K. von Beyme, *Op. Cit.*, 1973, pp. 927-939.
  - 1965-1984: K. von Beyme, *Op. Cit.*, 1985, p. 386.
- Germany : 1918-1934: K. von Beyme, *Op. Cit.*, 1973, pp. 912-914.
  - 1945-1984: K. von Beyme, *Op. Cit.*, 1985, p. 387.
- Netherlands : 1918-1940: K. von Beyme, *Op. Cit.*, 1973, p. 958.
  - 1945-1984: K. von Beyme, *Op. Cit.*, 1985, p. 399.

Notes : The following notes refer to differences between the sources I used and Peter Flora (1983), pp. 156-157, p. 159, pp. 168-169, pp. 173-174 and p. 182:

- Austria : 1945-1955: excluding the Cabinet Renner (date of investiture April 27th 1945 and the Cabinet Figl (date of investiture October 28th 1952).
  - 1965-1975: excluding the Cabinet Klaus (date of investiture January 19th 1968).
- Belgium : 1918-1925: including the Cabinet Cooreman (date of investiture May 31th 1918).
  - 1935-1940: excluding the Cabinet Pierlot (date of investiture April 1939).
  - 1965-1975: excluding the Cabinet Tindemans (date of investiture June 11th 1974).
- France : 1918-1925: including the Cabinet Clemenceau (date of investiture November 17th 1917).
  - 1945-1955: including the Cabinet De Gaulle (date of investiture November 10th 1944).
  - 1955-1965: excluding the Cabinet De Gaulle (date of investiture June 2th 1958).
  - 1965-1975: excluding the Cabinet Pompidou (date of investiture May 31st 1968 and including the Cabinet Messmer (date of investiture April 5th 1973).
- Germany : 1925-1934: including the Cabinet Muller (date of investiture April 15th 1929 and excluding the Cabinet Hitler (date of investiture January 30th 1933).
  - 1955-1965: excluding the Cabinet Adenauer (date of investiture December 13th 1962).

B.2 Education

Table B.2.I The Number of Elementary School Pupils by Denomination of School, 1930-1984

Year	Total	Public	Denominat. Total	Protestant	Catholic	Others
1930	1,182,528	446,397	736,131	295,507	422,833	17,791
1935	1,141,976	378,920	763,056	293,367	451,788	17,901
1938	1,143,114	357,675	785,439	297,333	469,176	18,930
1945	1,172,337	319,279	853,058	324,586	510,701	17,771
1950	1,215,782	332,628	883,154	334,658	525,920	22,576
1955	1,452,246	417,262	1,034,984	397,049	610,555	27,380
1960	1,415,703	379,837	1,035,866	383,910	625,904	26,052
1964	1,397,795	366,134	1,031,661	386,046	619,797	25,818
1970	1,462,376	401,759	1,060,617	402,093	629,206	29,318
1975	1,453,467	433,820	1,019,647	403,834	583,121	32,692
1980	1,333,342	422,577	910,765	375,434	497,840	37,491
1984	1,094,980	346,288	748,692	314,893	389,168	44,631

Source: 1930-1970 : CBS, *De ontwikkeling van het onderwijs in Nederland, edition 1966, Vol. I: tables*, Staatsuitgeverij, The Hague, 1966, p. 66.
 1970 : CBS, *Statistiek van het gewoon lager onderwijs 1970/'71*, Staatsuitgeverij, The Hague, 1971, p. 19.
 1975 : *Ibidem 1975/'76*, Staatsuitgeverij, The Hague, 1977, p. 17.
 1980 : Ibidem 1980/'81, Staatsuitgeverij, The Hague, 1982, p. 13.
 1984 : CBS, *Statistiek van het Basisonderwijs 1984/'85*, Staatsuitgeverij, The Hague, 1986, p. 5.

Table B.2.II The Number of Pupils of Elementary Schools by Denomination as a Percentage of All Pupils of Elementary Schools, 1850- 1984

Year	Public (1)	Protestant (2)	Catholic (3)	Others (4)	Denom. Quotum (5)	Non. Denom. (6)
1850	n.a.	n.a.	n.a.	n.a.	23.0	77.0
1860	n.a.	n.a.	n.a.	n.a.	21.0	79.0
1870	n.a.	n.a.	n.a.	n.a.	23.0	77.0
1880	n.a.	n.a.	n.a.	n.a.	25.0	75.0
1890	n.a.	n.a.	n.a.	n.a.	29.0	71.0
1900	n.a.	n.a.	n.a.	n.a.	31.0	69.0
1910	n.a.	n.a.	n.a.	n.a.	38.0	62.0
1920	n.a.	n.a.	n.a.	n.a.	45.0	55.0
1930	37.7	25.0	35.8	1.5	60.7	39.3
1935	33.2	25.7	39.6	1.6	65.3	34.7
1938	31.3	26.0	41.0	1.7	67.1	32.9
1945	27.2	27.7	43.6	1.5	71.2	28.8
1950	27.4	27.5	43.3	1.9	70.8	29.2
1955	28.7	27.3	42.0	1.9	69.4	30.6
1960	26.8	27.1	44.2	1.8	71.3	28.7
1964	26.2	27.6	44.3	1.8	72.0	28.0
1970	27.5	27.5	43.0	2.0	70.5	29.5
1975	29.8	27.8	40.1	2.2	67.9	32.1
1980	31.7	28.2	37.3	2.8	65.5	34.5
1984	31.6	28.8	35.5	4.1	64.3	35.7

Note : Till and including 1920 the denominational quotum contains also the category "others". From 1930 onwards this category is counted as non- denominational.

Source : 1850-1930: Ph. J. Idenburg, *Schets van het Nederlandse Schoolwezen*, Wolters, Groningen, 1964, p. 114, staat 1.
1930-1984: table B.2.I.

Table B.2.III Number of University Students by Denomination of University, 1920-1983

Year	Protestant Quotum	Catholic Quotum	Non-Denominational Quotum	Total
1920	225	0	8,327	8,552
1925	348	284	8,806	9,438
1930	494	532	11,035	12,061
1935	619	664	11,345	12,628
1937	611	668	11,226	12,505
1945	933	1,230	19,628	21,791
1950	1,443	1,868	26,425	29,736
1955	2,068	2,382	25,192	29,642
1960	3,116	4,015	33,596	40,727
1963	3,930	5,727	42,743	52,400
1968	7,018	10,511	67,247	84,776
1974	10,097	15,118	83,586	108,801
1979	13,177	19,883	113,991	147,051
1983	13,048	21,696	127,169	161,913

Notes : The years relate to the start of the academic years.
The Protestant quotum refers to the Free University Amsterdam. The Roman-Catholic quotum refers to the universities of Nijmegen and Tilburg. In both denominational quota theological academics are excluded. The non- denominational public quotum includes all public state universities and polytechnics with the exception of the businessschool Nijenrode.

Source : 1920-1963 : CBS, *De ontwikkeling van het onderwijs in Nederland, editie 1966, Vol. I: tabellen*, Staatsuitgeverij, The Hague, 1966, p. 265.
1968 : CBS, *Statistiek van het wetenschappelijk onderwijs 1968/'69*, Staatsuitgeverij, The Hague, 1970, pp. 6-8.
1974 : CBS, *Statistiek van het wetenschappelijk onderwijs 1974/'75*, Staatsuitgeverij, The Hague, 1980, p. 58.
1979 : CBS, *Statistiek van het wetenschappelijk onderwijs 1975/'76- 1979/'80*, Staatsuitgeverij, The Hague, 1982, p. 42.
1983 : CBS, *Statistiek van het wetenschappelijk onderwijs 1983/'84*, Staatsuitgeverij, The Hague, 1986, p. 43.

B.3 Health Service

Table B.3.I Members of Home Nursing Services by Denomination, 1942- 1972

Year	Non-Denomin.	Catholic	Protestant	Total
1942	655,454	335,814	n.a.	n.a.
1947	859,763	453,344	n.a.	n.a.
1951	939,535	504,815	250,000	1,694,350
1958	1,206,095	651,980	264,550	2,122,625
1963	1,350,490	753,701	289,221	2,393,412
1967	1,515,749	846,806	356,000	2,718,555
1970	1,658,656	942,184	355,145	2,955,985
1972	1,726,000	985,000	353,000	3,064,000

Note : In most cases members of home nursing services are heads of households. The figures are measured per January 1st of every year involved. In the province of Limburg the members of the Catholic and the non-denominational home nursing services are counted as members of both organizations.

Source : 1942 and 1947 : CBS, *Jaarcijfers voor Nederland 1947-1950*, De Haan, Utrecht, 1951, p. 47.
           1951   : CBS, *ibidem 1951-1952*, De Haan, Utrecht, 1954, p. 36. For the Protestant home nursing service: *Het Oranje Groene Kruis*, 1951, Vol. II, nr. 2 (Aug), p. 32.
           1958-1968   : CBS, *Statistisch Zakboek '69*, Staatsuitgeverij, The Hague, 1969, p. 30.
           1970   : CBS, *ibidem '72*, Staatsuitgeverij, The Hague, 1972, p. 32.
           1972   : CBS, *ibidem '73*, Staatsuitgeverij, The Hague, 1973, p. 47.

Table B.3.II Number of Hospital Beds by Denomination, 1950-1985 (per December 31st)

Year	Catholic	Protestant	Non-Denom.	Total
1950	19,994	6,733	17,684	44,411
1955	22,322	8,478	20,214	51,014
1960	24,557	10,028	23,433	58,018
1965	27,798	11,972	23,629	63,399
1970	31,108	14,735	25,757	71,600
1975	32,088	12,840	29,339	74,267
1977	30,405	14,882	27,924	73,211
1985	22,902	12,104	28,173	63,179

Note : The category non-denominational includes one Jewish hospital.
Psychiatric hospitals, sanatoria and army hospitals excluded.
The definition of the number of beds differs slightly over time.
The figures for 1985 contain the numbers of recognized beds with the exclusion of the recognized beds in the following hospitals: Leyenburg (The Hague), St. Willibrordus (Deurne), Isselwaerde (Ijsselstein), Diakonessenhuis (Leeuwarden), St. Bonifatius (Leeuwarden), Triotel (Leeuwarden), St. Antoniushove (Leidschendam), Zuiderziekenhuis (Rotterdam), Walcheren Streekziekenhuis (Vlissingen), Diakonessenhuis (Voorburg), Het Nieuwe Spittaal (Warnsveld), De Heel (Zaanstad), Academisch Ziekenhuis (Leiden), Oogziekenhuis (The Hague), Alexander van der Leeuw (Amsterdam), Dijsselhof (Amsterdam), Lyndensteijn (Opsterland), Adriaan Stichting (Rotterdam), Opleidingscentrum Verloskundigen (Rotterdam).
From 1970 onwards the sources present "general Christian hospitals". These beds are allocated to the columns Protestant and Catholic by RK' = RK/(RK+PROT)*C+RK, where:

RK' = number of beds in Roman-Catholic hospitals after allocation of the number of beds in general Christian hospitals;
RK = idem, before allocation of the number of beds in general Christian hospitals;
PROT = number of beds in Protestant hospitals before allocation of the number of beds in general Christian hospitals;
C = number of beds in general Christian hospitals.

Source : 1950   : CBS, *Statistiek der ziekenhuizen 1950*, De Haan, Utrecht, 1953, p. 6.
          1955:   CBS, *Jaarcijfers voor Nederland 1955-1956*, De Haan, Zeist, 1958, p. 34.
          1960-1978 : CBS and Ministerie van Volksgezondheid en Milieuhygiëne, *Compendium Gezondheidsstatistiek Nederland 1979*, Staatsuitgeverij, The Hague, 1980, p. 374.
          1985   : Not published figures from the Bureau Onderzoeksondersteuning en Statistiek of the stafdepartment epidemiology and informatica of the Ministry of Welfare, Health and Culture.

B.4 Broadcasting

Table B.4.I Number of Members of Broadcasting Associations, 1947- 1986 (per December 31st)

Year	NCRV	KRO	AVRO	VARA	VPRO	VOO	TROS	EO
1947	100,172	84,566	86,500	101,146	n.a.	n.a.	n.a.	n.a.
1948	145,452	139,821	n.a.	154,100	n.a.	n.a.	n.a.	n.a.
1949	206,318	196,833	133,400	205,665	85,555	n.a.	n.a.	n.a.
1950	240,114	278,772	195,900	245,886	63,589	n.a.	n.a.	n.a.
1951	270,558	314,894	n.a.	272,606	n.a.	n.a.	n.a.	n.a.
1952	308,495	350,436	251,100	301,620	n.a.	n.a.	n.a.	n.a.
1953	326,967	368,370	n.a.	315,273	n.a.	n.a.	n.a.	n.a.
1954	340,966	397,477	n.a.	345,975	n.a.	n.a.	n.a.	n.a.
1955	361,779	425,000	312,000	378,225	n.a.	n.a.	n.a.	n.a.
1956	384,925	478,504	n.a.	420,810	n.a.	n.a.	n.a.	n.a.
1957	404,793	519,102	360,623	456,146	n.a.	n.a.	n.a.	n.a.
1958	418,462	548,935	378,235	479,581	n.a.	n.a.	n.a.	n.a.
1959	427,173	575,056	400,406	506,337	155,000	n.a.	n.a.	n.a.
1960	435,945	604,243	405,994	533,678	158,000	n.a.	n.a.	n.a.
1961	442,721	607,000	458,986	540,828	145,000	n.a.	n.a.	n.a.
1962	442,894	634,712	459,136	543,350	148,000	n.a.	n.a.	n.a.
1963	450,518	634,000	446,063	535,207	186,000	n.a.	n.a.	n.a.
1964	459,470	627,000	411,901	493,024	185,874	n.a.	n.a.	n.a.
1965	469,535	568,469	384,683	494,028	174,488	n.a.	60,000	n.a.
1966	474,085	567,213	402,016	478,941	135,864	n.a.	100,000	n.a.
1967	477,540	567,765	834,000	468,169	141,096	n.a.	175,000	n.a.
1968	471,545	563,960	836,503	458,421	136,580	n.a.	190,000	n.a.
1969	471,688	542,081	834,691	473,349	117,000	n.a.	240,000	n.a.
1970	476,109	545,262	836,258	485,652	120,425	n.a.	281,000	47,977
1971	479,230	546,785	847,853	488,271	139,215	n.a.	332,000	136,917
1972	480,492	544,179	830,103	492,146	138,260	n.a.	393,000	152,502
1973	477,317	530,119	835,730	492,401	130,787	n.a.	493,000	147,125
1974	488,378	541,987	827,256	493,591	144,396	n.a.	600,000	141,340
1975	488,110	558,530	815,510	480,922	148,532	100,000	650,000	153,311
1976	488,166	584,550	782,409	520,000	154,272	160,000	710,000	157,883
1977	488,704	598,770	783,176	543,818	194,727	180,000	713,000	169,741
1978	497,397	604,129	785,357	544,770	193,647	200,000	730,000	177,001
1979	521,625	609,789	804,494	537,635	188,048	230,000	740,000	205,133
1980	528,896	611,799	809,037	523,678	181,592	300,000	751,000	219,625
1981	527,654	611,254	809,847	527,521	195,688	330,000	763,000	225,814
1982	531,242	607,792	802,186	517,708	351,971	360,000	764,000	270,580
1983	533,271	598,408	801,250	518,121	359,372	460,000	741,000	343,126
1984	532,028	617,490	782,311	527,565	343,791	550,000	745,000	337,505
1985	532,231	632,870	787,227	515,145	318,689	650,000	721,000	326,561
1986	531,482	639,156	795,847	513,290	317,430	726,000	723,056	328,395

Note : According to the law a subscription to a broadcasting associations periodical is also considered as membership with regard to the allocation of broadcasting hours. I therefore included in the figures both membership and subscriptions. The doubling of the members of AVRO in 1967 is caused by the merger of the AVRO's periodical with the periodical Televizier.
As denominational broadcasting associations are considered: NCRV (Protestant), KRO (Roman-Catholic), EO (orthodox Protestant) and VPRO (liberal Protestant) till 1968. As non-denominational are defined: VARA (Social-Democratic), AVRO, TROS, VOO and VPRO in 1968 and after.

Source : Official statistical data are scarce. Besides, they are not consistent and often years are lacking. I therefore had to rely on the figures presented by the broadcasting associations.

Table B.4.II Members of Denominational and Non-Denominational Broadcasting Associations as a Percentage of All Members of Broadcasting Associations, 1949-1986

Year	Non-Denominational	Denominational
1949	41.0	59.0
1950	43.1	56.9
1959	43.9	56.1
1960	44.0	56.0
1961	45.6	54.4
1962	45.0	55.0
1963	43.6	56.4
1964	41.6	58.4
1965	43.6	56.4
1966	45.5	54.5
1967	55.5	44.5
1968	61.0	39.0
1969	62.2	37.8
1970	61.7	38.3
1971	60.8	39.2
1972	61.2	38.8
1973	62.8	37.2
1974	63.8	36.2
1975	64.7	35.3
1976	65.4	34.6
1977	65.8	34.2
1978	65.7	34.3
1979	65.2	34.8
1980	65.3	34.7
1981	65.8	34.2
1982	66.5	33.5
1983	66.1	33.9
1984	66.5	33.5
1985	66.7	33.3
1986	67.2	32.8

Source : Table B.4.I

B.5 Membership Trade-Union Federations

Table B.5.I Numbers of Social-Democratic (NVV), Roman-Catholic (NKV), Protestant (CNV) and Other Trade-Unions, 1914-1980

Year	NVV	NKV	CNV	Others	Total
1914	84,261	29,048	11,023	141,653	265,985
1920	247,748	141,002	66,997	227,721	683,468
1926	190,179	90,475	48,327	164,529	493,510
1933	336,158	192,655	115,006	185,080	828,899
1937	283,382	168,064	108,235	163,905	723,586
1940	319,099	186,137	118,900	174,175	798,311
1947	300,341	224,885	119,051	146,936	791,213
1950	381,554	296,410	155,627	163,435	997,026
1952	420,776	321,478	174,750	176,911	1,093,915
1961	506,964	411,785	223,789	256,237	1,398,775
1964	529,173	420,849	229,068	292,173	1,471,263
1967	556,143	424,012	240,389	313,732	1,534,276
1971	613,000	404,500	237,800	322,700	1,577,900
1975	683,800	360,400	228,100	437,600	1,709,900
1980	751,900	325,900	304,300	407,500	1,789,600

Note : The figures till and including 1971 measured on January 1st; those of 1975 and 1980 on March 31st.
The figures of 1947 till 1967 exclude the members of the EVC.
Before 1945 the NKV had the name "Roomsch-Katholiek Werklieden Verbond; between 1945 and 1964: Katholieke Arbeidersbeweging (KAB).
For 1947 and after the figures include also pensioned members. According to the sources used, figures from 1914 till and including 1940 include probably also pensioners. The figures of 1940 are excluding applicants for membership.

Source : 1914-1941 : CBS, *Statistisch Zakboek 1940*, Albani, The Hague, 1940, p. 28; Column "others" is calculated.
1947-1965 : CBS, *Omvang der vakbeweging in Nederland op 1 januari 1964*, De Haan, Zeist, 1965, p. 6.
1967 : CBS, *Statistisch Zakboek '69*, Staatsuitgeverij, The Hague, 1969, p. 232.
1971 : CBS, *Ibidem 1975*, Staatsuitgeverij, The Hague, 1975, p. 97.
1975 : CBS, *Ibidem 1980*, Staatsuitgeverij, The Hague, 1980, p. 129.
1980 : CBS, *Ibidem 1981*, Staatsuitgeverij, The Hague, 1981, p. 130.

B.6 Strike Activity

Table B.6.I Number of Manyears Lost by Industrial Disputes as a Percentage of the Demand for Labour, Twelve Industrialized Countries, 1920-1979

Year	Austria	Belgium	Denmark	France	Germany	Italy
1920	n.a.	n.a.	0.171	n.a.	0.316	n.a.
1921	n.a.	n.a.	0.349	0.133	0.481	n.a.
1922	n.a.	n.a.	0.590	n.a.	0.509	n.a.
1923	n.a.	n.a.	0.005	n.a.	0.233	n.a.
1924	0.250	n.a.	0.043	n.a.	0.688	n.a.
1925	0.073	n.a.	1.021	n.a.	0.054	n.a.
1926	0.026	n.a.	0.006	0.076	0.023	n.a.
1927	0.052	n.a.	0.030	n.a.	0.112	n.a.
1928	0.060	n.a.	0.003	n.a.	0.366	n.a.
1929	0.030	n.a.	0.010	0.052	0.078	n.a.
1930	0.004	n.a.	0.033	n.a.	0.076	n.a.
1931	0.011	n.a.	0.057	n.a.	0.037	n.a.
1932	0.009	n.a.	0.021	n.a.	0.023	n.a.
1933	0.008	n.a.	0.004	n.a.	0.002	n.a.
1934	0.000	n.a.	0.032	n.a.	n.a.	n.a.
1935	0.000	n.a.	0.003	n.a.	n.a.	n.a.
1936	0.000	n.a.	0.613	n.a.	n.a.	n.a.
1937	0.000	n.a.	0.004	n.a.	n.a.	n.a.
1938	n.a.	0.026	0.018	n.a.	n.a.	n.a.
1950	n.a.	0.324	0.001	0.257	0.007	0.181
1951	0.011	0.069	0.001	0.077	0.029	0.105
1952	0.010	0.101	0.001	0.038	0.008	0.082
1953	0.005	0.048	0.000	0.214	0.026	0.135
1954	0.007	0.052	0.005	0.032	0.028	0.124
1955	0.008	0.115	0.002	0.068	0.014	0.127
1956	0.020	0.107	0.213	0.031	0.026	0.094
1957	0.006	0.426	0.001	0.089	0.017	0.103
1958	0.006	0.033	0.002	0.025	0.013	0.091
1959	0.007	0.112	0.003	0.042	0.001	0.198
1960	0.009	0.038	0.011	0.023	0.001	0.122
1961	0.015	0.010	0.424	0.056	0.001	0.207
1962	0.086	0.030	0.003	0.041	0.007	0.470
1963	0.005	0.027	0.004	0.128	0.028	0.233
1964	0.005	0.048	0.003	0.053	0.000	0.271
1965	0.020	0.008	0.043	0.021	0.001	0.146
1966	0.010	0.057	0.003	0.053	0.000	0.301
1967	0.002	0.019	0.002	0.088	0.006	0.177
1968	0.001	0.039	0.006	n.a.	0.000	0.190
1969	0.003	0.017	0.010	0.046	0.004	0.774
1970	0.004	0.149	0.017	0.036	0.001	0.424
1971	0.001	0.128	0.004	0.090	0.067	0.299
1972	0.002	0.036	0.004	0.076	0.001	0.396
1973	0.022	0.089	0.642	0.079	0.008	0.473
1974	0.001	0.059	0.031	0.067	0.016	0.387
1975	0.001	0.062	0.017	0.077	0.001	0.539
1976	0.000	0.092	0.034	0.099	0.008	0.503
1977	0.000	0.068	0.037	0.072	0.000	0.327
1978	0.001	0.103	0.021	0.043	0.068	0.200
1979	0.000	0.063	0.027	0.071	0.008	0.538

Table B.6.I (continued) Number of Manyears Lost by Industrial Disputes as a Percentage of the Demand for Labour, Twelve Industrialized Countries, 1920-1979

Year	Japan	Netherlands	Norway	Sweden	Switzerland	UK
1920	n.a.	0.318	n.a.	1.118	n.a.	0.486
1921	n.a.	0.172	n.a.	0.347	n.a.	1.723
1922	n.a.	0.162	0.031	0.340	n.a.	0.390
1923	n.a.	0.159	0.261	0.845	n.a.	0.205
1924	n.a.	0.403	1.652	0.145	n.a.	0.160
1925	n.a.	0.098	0.218	0.186	n.a.	0.151
1926	n.a.	0.035	0.770	0.202	n.a.	3.098
1927	n.a.	0.025	0.480	0.047	n.a.	0.022
1928	n.a.	0.076	0.121	0.554	n.a.	0.026
1929	n.a.	0.105	0.063	0.076	0.017	0.154
1930	n.a.	0.027	0.077	0.116	0.046	0.084
1931	n.a.	0.090	2.493	0.303	0.013	0.137
1932	n.a.	0.197	0.127	0.363	0.028	0.127
1933	n.a.	0.058	0.116	0.404	0.012	0.020
1934	n.a.	0.011	0.074	0.088	0.006	0.018
1935	n.a.	0.029	0.052	0.091	0.003	0.035
1936	n.a.	0.009	0.118	0.050	0.007	0.032
1937	n.a.	0.004	0.295	0.098	0.020	0.057
1938	n.a.	0.014	0.162	0.145	0.003	0.023
1950	0.054	0.018	0.013	0.006	0.001	0.027
1951	0.058	0.007	0.011	0.072	0.001	0.033
1952	0.142	0.003	0.037	0.011	0.002	0.035
1953	0.039	0.003	0.012	0.078	0.010	0.042
1954	0.035	0.006	0.032	0.003	0.004	0.047
1955	0.031	0.014	0.032	0.021	0.000	0.072
1956	0.039	0.022	0.291	0.001	0.000	0.039
1957	0.047	0.001	0.008	0.007	0.000	0.158
1958	0.050	0.004	0.018	0.002	0.000	0.065
1959	0.049	0.001	0.015	0.003	0.000	0.098
1960	0.039	0.046	0.001	0.002	0.000	0.055
1961	0.048	0.002	0.126	0.000	0.000	0.055
1962	0.041	0.001	0.024	0.001	0.000	0.106
1963	0.021	0.004	0.067	0.003	0.010	0.032
1964	0.024	0.004	0.000	0.004	0.001	0.041
1965	0.041	0.005	0.003	0.000	0.000	0.053
1966	0.020	0.001	0.001	0.044	0.000	0.043
1967	0.013	0.001	0.001	0.000	0.000	0.051
1968	0.020	0.001	0.004	0.000	0.000	0.085
1969	0.025	0.002	0.006	0.014	0.000	0.123
1970	0.027	0.022	0.013	0.019	0.000	0.197
1971	0.041	0.008	0.002	0.102	0.001	0.244
1972	0.035	0.011	0.003	0.001	0.000	0.430
1973	0.031	0.050	0.003	0.001	0.000	0.128
1974	0.064	0.001	0.081	0.007	0.000	0.260
1975	0.053	0.000	0.003	0.042	0.000	0.107
1976	0.021	0.001	0.033	0.003	0.003	0.059
1977	0.010	0.021	0.006	0.010	0.001	0.183
1978	0.009	0.000	0.015	0.004	0.001	0.168
1979	0.006	0.025	0.002	0.003	0.000	0.520

Note : The figures in these table are computed in the following way:

$$P = \frac{(DL/A)}{(LF - UR * LF)} * 100\%$$

P = Working years lost by industrial disputes as a percentage of the Demand for Labour;
DL = Working days lost by industrial disputes;
A = average number of working days per year per person;
LF = Labour Force;
UR = Unemployment-Ratio.

Sources:

1. Working days lost by industrial disputes (DL)

For the number of working days lost by industrial disputes (DL) I used the following sources:

B.R. Mitchell, *European Historical Statistics 1750-1970*, MacMillan Press Ltd., London, 1975, p. 173-183 for the European countries for the period 1920-1969; B.R. Mitchell, *International Historical Statistics: Africa and Asia*, MacMillan Press Ltd., London/Basingstoke, p.109 for Japan for the period 1950-1975; ILO, *Year Book of Labour Statistics 1979*, Geneva, p. 596-598 and ILO, *Ibidem 1983*, Geneva, p.764-766 for the European countries for the period 1970-1979 and p.763 of the latter source for Japan 1975-1979. In case of difference between these two sources, I used the figure of the most recent source, i.e. ILO Year Book 1983.

Temporal and spatial comparison of the number of days lost by industrial disputes should be done with care, because of differences of definition in the respective sources. The following points should be considered:

Belgium	1970-1977:	excluding workers indirectly affected.
Denmark:	1970-1978:	excluding political strikes and disputes where less than 100 working days were lost.
France:	1970-1978:	excluding agriculture and public administration.
Germany:	1920-1969:	the figures refer to West-Germany after 1945; figure for 1933: only the first quarter of the year; figure for 1950: American and British Occupation Zone only; including West-Berlin from 1960 onwards.
	1970-1978:	excluding disputes lasting less than one day, except if a loss of more than 100 working days is involved.
Italy:	1969-1974:	excluding political strikes.
	1970-1978:	excluding workers indirectly affected.
Japan:	1950-1975:	figures excluding workers indirectly affected and diputes lasting less than 4 hours.
Netherlands:	1920-1969:	until and including 1926: figures refer to disputes started during the year; after 1926 all disputes in being during the year.
Norway:	1970-1978:	excluding workers indirectly affected and disputes lasting less than one day.
UK:	1920-1978:	figures excluding strikes invloving less than ten workers or lasting less than one day, except if the number of days lost exceeded one hundred.
	1970-1978:	excluding disputes not connected with terms of employment or conditions of labour. Disputes involving less than 10 workers or less than one day are not included unless a loss of more than 100 working days is involved.

2. Average working days per year per person (A)

Table B.6.II Average Number of Working Days per Person per Year in Twelve Industrialized Countries, 1920-1938 and 1950-1979

	1920 - 1938	1950 - 1979
Austria	287.06	232.34
Belgium	283.69	255.94
Denmark	284.13	250.41
France	259.06	237.09
Germany	287.50	250.78
Italy	259.69	230.94
Japan	297.19	283.91
Netherlands	281.50	249.19
Norway	275.69	232.69
Sweden	280.44	215.16
Switzerland	287.31	251.50
UK	284.56	225.72

Source : I used A. Maddison, *Phases of Capitalist Development*, Oxford University Press, Oxford/New York, 1986 (1982), p.211, table C9. Maddison gives figures for the annual hours worked per person, but not for all years. Therefore, I calculated two arithmathic means per country, one for the interbellum (arithmathic mean of 1929 and 1938) and one for the period after WO II (arithmathic mean of 1950, 1960, 1970 and 1979). For Italy I took the 1978 figure instead of 1979 because 1979 was not available. In order to get working days in stead of working hours, I divided these means by 8.

3. Labour Force (LF)

Figures about Labour Force were derived from Maddison, 1986, p.209 (table C7). Maddison gives figures for the following years: 1913, 1929, 1938, 1950, 1960, 1970, 1973, 1978 and 1979. His (mid-year) figures refer throughout to 1979 bounderies. Intermediate years I estimated through linear interpolation.

4. Unemployment ratio (UR)

Unemployment ratio's (e.g. the number of unemployed as a percentage of total labour force) were derived from Maddison, 1986, pp. 206-207 (table C6). UR was computed by dividing these figures by hundred.

B.7 Connubium

Explanatory Note

Table B.7.I (column 1 till and including 30) contains the raw data on marriages contracted by denomination from 1936 till 1980. These data were not or only partially published. The data were acquired from the Dutch statistical office, CBS, in The Hague. Data after 1980 are also available. I did not use them because I consider them as less reliable as a result of administrative procedures.

The tables B.7.II till and including B.7.V.b are based on table B.7.I. For each of these tables I will describe how the Catholic-column was computed. The columns of the other denominations were computed likewise. The numbers between brackets refer to the columns of table B.7.I :

Table B.7.II :
{ 2 * (1) / [2 * (1) + (7) + (8) + (9) + (10) + (11) + (16) + (21) + (26)] } * 100.

Table B.7.III.a :
{ (1) / [(1) + (7) + (8) + (9) + (10)] } * 100.

Table B.7.III.b :
{ (1) / [(1) + (11) + (16) + (21) + (26)] } * 100.

Table B.7.IV.a :
{ [(1) + (11) + (16) + (21) + (26)] / (30) } * 100.

Table B.7.IV.b :
{ [(1) + (7) + (8) + (9) + (10)] / (30) } * 100.

Table B.7.V.a. and table B.7.V.b :
The relevant columns of the tables B.7.IV.a and B.7.IV.b were subtracted from the corresponding relevant columns of the tables B.7.III.a and B.7.III.b.

Table B.7.I Marriages Contracted by Denomination of Marriage Partners, Netherlands, 1936-1980

Year:	Denomination of both partners:				
	Catholic	D.-Reformed	Re-Reformed	Others	No
	(1)	(2)	(3)	(4)	(5)
1936	19,282	28,382	n.a.	910	4,330
1937	20,227	29,278	n.a.	901	4,704
1938	20,889	19,849	4,529	1,900	4,878
1939	25,012	23,145	5,064	2,156	6,885
1940	20,951	18,986	4,396	1,917	5,488
1941	22,083	18,175	4,230	1,662	4,892
1942	27,142	24,263	5,471	3,574	6,918
1943	21,967	18,225	4,027	978	5,296
1944	17,439	14,434	3,566	630	3,764
1945	n.a.	n.a.	n.a.	n.a.	n.a.
1946	35,005	29,320	7,177	1,603	9,317
1947	33,338	26,183	6,089	1,273	9,084
1948	30,695	22,634	5,622	1,191	7,697
1949	28,725	21,093	5,250	1,040	7,185
1950	29,437	21,153	5,643	1,043	7,270
1951	31,637	22,996	6,419	1,075	8,519
1952	30,796	21,646	6,201	1,107	8,561
1953	31,139	20,764	6,089	986	9,094
1954	32,466	20,591	6,158	1,021	9,760
1955	33,296	20,283	6,159	1,110	10,120
1956	35,226	20,450	6,533	1,090	10,333
1957	36,646	20,081	6,572	1,049	10,595
1958	36,124	19,445	6,493	1,021	10,150
1959	34,892	17,987	5,956	1,068	10,081
1960	34,735	17,829	6,121	1,114	10,425
1961	36,482	17,459	6,498	1,098	10,678
1962	36,957	17,608	6,517	1,13	10,527
1963	38,002	17,226	6,353	1,136	11,150
1964	41,002	18,395	6,874	1,202	11,971
1965	43,731	18,521	7,176	1,315	13,187
1966	44,673	18,588	7,128	1,255	14,012
1967	44,741	18,592	7,389	1,738	14,357
1968	44,519	17,669	7,231	1,342	15,997
1969	43,837	16,620	7,023	1,406	16,279
1970	45,519	16,656	7,106	1,426	18,162
1971	44,231	15,747	7,115	1,553	18,177
1972	42,715	14,308	6,559	1,413	17,447
1973	36,644	12,416	5,954	1,464	17,456
1974	38,119	11,927	5,907	1,404	17,339
1975	34,406	10,525	5,398	1,405	15,640
1976	33,035	9,692	5,334	1,641	16,047
1977	30,616	9,397	5,177	1,752	15,958
1978	29,412	8,459	4,788	1,778	15,484
1979	27,522	7,952	4,752	1,875	15,466
1980	29,518	8,494	4,951	2,068	15,750

Marriages Contracted by Denomination of Marriage Partners, Netherlands, 1936-1980 (continued)

Year:	Denomination of bridegroom: Roman-Catholic Denomination of bride: Catholic	D.-Reformed	Re-Reformed	Others	No
	(6)	(7)	(8)	(9)	(10)
1936		2,037	n.a.	36	536
1937		2,007	n.a.	28	564
1938		1,674	127	305	598
1939		1,990	107	309	784
1940		1,742	114	254	707
1941		1,708	106	241	676
1942		2,267	147	286	1,038
1943		1,777	123	173	722
1944		1,132	85	168	482
1945		n.a.	n.a.	n.a.	n.a.
1946		2,688	235	328	1,295
1947		2,572	213	301	1,217
1948		2,221	193	270	1,094
1949		2,002	189	269	1,033
1950		1,955	165	221	1,054
1951		1,974	166	223	1,173
1952		1,916	152	214	1,192
1953		1,633	167	222	1,101
1954		1,664	136	171	1,178
1955		1,670	144	215	1,223
1956		1,697	155	227	1,270
1957		1,777	175	212	1,211
1958		1,709	190	169	1,241
1959		1,718	178	244	1,364
1960		1,810	195	258	1,527
1961		1,957	196	259	1,566
1962		2,002	210	260	1,694
1963		2,149	247	302	1,765
1964		2,238	278	321	2,086
1965		2,522	287	347	2,280
1966		2,691	347	374	2,351
1967		2,808	394	439	2,607
1968		3,159	415	411	2,884
1969		3,250	525	492	3,024
1970		3,608	612	499	3,536
1971		3,722	656	518	3,631
1972		3,668	748	607	3,537
1973		3,598	818	506	3,494
1974		3,768	874	575	3,799
1975		3,512	940	543	3,703
1976		3,408	899	580	3,441
1977		3,296	857	601	3,486
1978		3,096	826	528	3,413
1979		2,776	801	560	3,339
1980		3,054	750	649	3,494

Marriages Contracted by Denomination of Marriage Partners, Netherlands, 1936-1980 (continued)

Year:	Denomination of bridegroom: Dutch-Reformed Denomination of bride:				
	Catholic	D.-Reformed	Re-Reformed	Others	No
	(11)	(12)	(13)	(14)	(15)
1936	2,471		n.a.	54	1,918
1937	2,458		n.a.	75	2,010
1938	2,028		1,039	1,048	1,565
1939	2,544		1,210	1,207	2,028
1940	2,099		1,034	870	1,786
1941	1,887		1,038	768	1,689
1942	2,498		1,464	963	2,527
1943	1,962		1,190	709	1,757
1944	1,301		902	530	1,262
1945	n.a.		n.a.	n.a.	n.a.
1946	2,884		1,958	1,193	2,805
1947	2,751		1,758	1,063	2,873
1948	2,552		1,626	916	2,427
1949	2,418		1,405	801	2,313
1950	2,333		1,476	810	2,306
1951	2,241		1,575	778	2,421
1952	2,156		1,521	829	2,448
1953	2,024		1,422	664	2,402
1954	1,905		1,417	713	2,430
1955	1,873		1,448	662	2,297
1956	1,973		1,403	686	2,385
1957	1,962		1,354	684	2,415
1958	1,960		1,289	665	2,317
1959	1,987		1,305	606	2,161
1960	1,918		1,395	598	2,312
1961	2,139		1,404	670	2,469
1962	2,108		1,345	662	2,380
1963	2,212		1,468	685	2,520
1964	2,426		1,615	666	2,662
1965	2,558		1,588	627	2,721
1966	2,830		1,659	646	2,846
1967	3,000		1,895	707	2,884
1968	3,320		1,960	691	3,140
1969	3,617		2,037	628	3,089
1970	4,127		2,161	651	3,090
1971	4,206		2,193	589	3,112
1972	4,271		2,176	533	2,942
1973	4,032		2,193	466	2,713
1974	4,024		2,206	508	2,827
1975	3,871		1,917	419	2,540
1976	3,673		1,866	434	2,486
1977	3,384		1,870	405	2,209
1978	3,286		1,713	395	2,298
1979	3,049		1,731	373	2,104
1980	3,166		1,648	395	2,210

Marriages Contracted by Denomination of Marriage Partners, Netherlands, 1936-1980 (continued)

Year:	Denomination of bridegroom: Re-Reformed Denomination of bride:				
	Catholic	D.-Reformed	Re-Reformed	Others	No
	(16)	(17)	(18)	(19)	(20)
1936	n.a.	n.a.		n.a.	n.a.
1937	n.a.	n.a.		n.a.	n.a.
1938	194	1,143		123	151
1939	170	1,371		119	213
1940	152	1,145		96	199
1941	145	1,180		90	221
1942	209	1,625		117	257
1943	150	1,296		100	213
1944	106	960		73	151
1945	n.a.	n.a.		n.a.	n.a.
1946	245	1,917		133	326
1947	262	1,676		124	282
1948	218	1,511		116	235
1949	212	1,468		101	253
1950	214	1,382		106	267
1951	215	1,589		118	297
1952	228	1,536		105	248
1953	173	1,335		84	292
1954	187	1,392		100	316
1955	196	1,410		68	300
1956	210	1,381		79	290
1957	208	1,359		93	312
1958	195	1,412		102	260
1959	173	1,347		99	267
1960	205	1,358		92	315
1961	223	1,410		114	357
1962	245	1,478		112	332
1963	296	1,488		100	335
1964	290	1,585		88	397
1965	307	1,677		112	422
1966	401	1,733		119	505
1967	406	1,953		132	502
1968	519	2,053		168	595
1969	589	2,103		153	614
1970	712	2,029		132	633
1971	744	2,227		158	654
1972	823	2,152		151	682
1973	848	1,995		167	703
1974	895	2,031		161	757
1975	869	1,969		131	739
1976	822	1,815		156	724
1977	858	1,806		145	714
1978	802	1,728		139	700
1979	757	1,682		129	699
1980	849	1,683		142	680

Marriages Contracted by Denomination of Marriage Partners, Netherlands, 1936-1980 (continued)

Year:	Denomination of bridegroom: Others Denomination of bride: Catholic	D.-Reformed	Re-Reformed	Others	No
	(21)	(22)	(23)	(24)	(25)
1936	46	79	n.a.		48
1937	76	119	n.a.		59
1938	310	788	83		352
1939	386	855	91		483
1940	282	725	86		457
1941	276	622	69		357
1942	340	887	107		492
1943	219	640	72		334
1944	153	450	48		216
1945	n.a.	n.a.	n.a.		n.a.
1946	665	1,336	170		726
1947	321	875	106		530
1948	279	775	88		428
1949	271	769	70		369
1950	237	679	78		399
1951	218	713	88		421
1952	248	624	95		355
1953	190	562	73		359
1954	208	589	84		377
1955	211	574	89		372
1956	242	594	86		394
1957	214	611	87		422
1958	289	571	87		322
1959	277	539	83		389
1960	322	556	85		377
1961	351	608	103		447
1962	383	648	103		392
1963	397	641	106		476
1964	503	703	129		539
1965	569	704	144		555
1966	672	705	183		585
1967	778	820	170		621
1968	829	787	166		641
1969	790	769	196		631
1970	826	690	216		635
1971	846	650	194		705
1972	849	679	192		593
1973	842	570	174		603
1974	799	541	192		585
1975	725	512	188		527
1976	686	440	199		514
1977	776	475	159		587
1978	733	395	182		598
1979	718	441	177		709
1980	854	465	189		796

Marriages Contracted by Denomination of Marriage Partners, Netherlands, 1936-1980 (continued)

Year:	Denomination of bridegroom: No Denomination Denomination of bride:				
	Catholic	D.-Reformed	Re-Reformed	Others	No
	(26)	(27)	(28)	(29)	(30)
1936	746	2,557	n.a.	54	63,486
1937	824	2,637	n.a.	73	66,040
1938	833	1,913	186	535	67,040
1939	1,155	2,372	241	700	80,597
1940	914	2,068	214	538	67,220
1941	894	2,020	223	465	65,717
1942	1,146	2,941	271	609	87,559
1943	882	1,971	219	434	65,436
1944	655	1,454	166	319	50,446
1945	n.a.	n.a.	n.a.	n.a.	n.a.
1946	1,538	3,261	359	737	107,221
1947	1,561	3,239	328	664	98,683
1948	1,292	2,836	253	550	87,719
1949	1,354	2,813	325	533	82,261
1950	1,332	2,752	293	505	83,110
1951	1,533	2,953	345	538	90,225
1952	1,468	2,890	281	585	87,402
1953	1,377	2,789	302	496	85,739
1954	1,502	2,899	304	535	88,103
1955	1,512	2,918	338	549	89,037
1956	1,675	3,033	321	539	92,272
1957	1,699	2,946	352	556	93,592
1958	1,666	2,921	350	560	91,508
1959	1,703	2,732	347	504	88,007
1960	1,781	2,857	365	550	89,100
1961	1,971	3,149	408	567	92,583
1962	2,052	2,981	430	586	93,144
1963	2,160	3,201	423	522	95,360
1964	2,391	3,433	507	612	102,913
1965	2,618	3,426	505	618	108,517
1966	2,830	3,525	607	647	111,912
1967	3,168	3,753	631	630	115,115
1968	3,652	3,992	706	688	117,534
1969	4,194	4,078	799	654	117,397
1970	4,841	4,183	902	679	123,631
1971	4,910	4,161	1,007	689	122,395
1972	4,961	3,917	974	635	117,532
1973	4,823	3,555	996	612	107,642
1974	5,031	3,582	1,088	668	109,607
1975	4,630	3,316	1,076	580	100,081
1976	4,543	3,078	990	538	97,041
1977	4,309	2,972	919	552	93,280
1978	4,213	2,705	868	551	89,090
1979	3,996	2,574	931	535	85,648
1980	4,221	2,629	937	590	90,182

Note: In 1936 and 1937 the total number of Dutch Reformed equals the total number of Protestants; in 1936 and 1937 the column 'others' represents only the Jewish denominations.

Source : CBS, raw data *Huwelijksstatistiek*

Table B.7.II In Current Year Homogeneously Married by Denomination as a Percentage of All in Current Year Married by Denomination, 1936-1980

Year	Catholic	Protestant	Others	No-Denom.	Total
1936	86.8	86.2	85.2	59.6	83.3
1937	87.2	86.3	80.7	60.4	83.4
1938	87.3	84.3	51.7	61.4	80.9
1939	87.0	83.8	51.0	63.3	80.5
1940	87.0	83.4	53.7	61.5	80.2
1941	88.2	83.8	53.5	59.9	81.0
1942	87.3	83.3	65.3	59.9	80.5
1943	88.0	83.6	42.2	61.9	81.0
1944	89.5	85.5	39.2	61.5	82.7
1945	n.a.	n.a.	n.a.	n.a.	n.a.
1946	87.6	83.8	37.7	62.8	80.5
1947	87.9	82.9	39.0	62.9	80.5
1948	88.3	83.0	41.0	62.8	80.9
1949	88.1	82.7	39.5	61.5	80.4
1950	88.7	83.2	40.7	62.0	81.1
1951	89.1	84.1	41.0	63.8	81.8
1952	89.0	83.8	42.0	64.4	81.7
1953	90.0	84.1	42.7	66.6	82.6
1954	90.3	83.9	42.4	67.2	82.6
1955	90.4	84.0	44.8	68.0	82.9
1956	90.4	83.8	43.4	67.6	82.8
1957	90.8	83.5	42.2	68.1	83.0
1958	90.7	83.5	42.5	67.8	83.0
1959	90.1	83.0	43.8	68.0	82.5
1960	89.7	82.5	44.0	67.4	81.9
1961	89.4	81.2	41.3	66.1	81.0
1962	89.2	81.5	41.8	66.0	81.1
1963	88.9	80.4	41.3	66.2	80.6
1964	88.6	80.5	40.3	65.5	80.3
1965	88.4	80.2	41.7	66.7	80.4
1966	87.7	79.1	39.0	66.9	79.6
1967	86.8	78.6	44.7	66.0	78.8
1968	85.4	76.6	38.0	66.3	77.2
1969	84.2	75.2	39.5	65.6	76.1
1970	82.9	74.1	39.7	66.3	75.3
1971	82.1	73.3	41.7	65.8	74.5
1972	81.4	72.0	40.0	65.7	73.8
1973	79.4	70.8	42.6	66.6	72.6
1974	79.4	69.7	41.1	65.4	72.0
1975	78.5	68.6	43.7	64.6	71.2
1976	78.5	68.4	48.1	66.3	71.5
1977	77.7	69.0	48.6	67.0	71.4
1978	77.7	68.0	50.2	66.9	71.1
1979	77.5	68.5	50.7	67.5	71.2
1980	77.6	68.4	50.3	66.9	71.1

Source : table B.7.I

Table B.7.III.a In Current Year Homogeneously Married Men by Denomination as a Percentage of All in Current Year Married Men by Denomination, 1936-1980

Year	Catholic	Protestant	Others	No-Denom.
1936	88.1	86.5	84.0	56.3
1937	88.6	86.6	78.0	57.1
1938	88.5	83.9	55.3	58.5
1939	88.7	83.1	54.3	60.6
1940	88.1	83.1	55.3	59.5
1941	89.0	83.7	55.7	57.6
1942	87.9	83.3	66.2	58.2
1943	88.7	83.5	43.6	60.2
1944	90.3	85.3	42.1	59.2
1945	n.a.	n.a.	n.a.	n.a.
1946	88.5	84.2	35.6	61.2
1947	88.6	82.9	41.0	61.1
1948	89.0	82.9	43.1	61.0
1949	89.2	82.7	41.3	58.8
1950	89.7	83.1	42.8	59.8
1951	89.9	84.3	42.7	61.3
1952	89.9	83.7	45.6	62.1
1953	90.9	84.0	45.4	64.7
1954	91.2	84.0	44.8	65.1
1955	91.1	84.4	47.1	65.6
1956	91.3	84.1	45.3	65.0
1957	91.6	83.8	44.0	65.6
1958	91.6	83.9	44.6	64.9
1959	90.9	83.4	45.3	65.6
1960	90.2	83.1	45.4	65.2
1961	90.2	81.8	42.1	63.7
1962	89.9	82.2	42.6	63.5
1963	89.5	81.2	41.2	63.9
1964	89.3	81.3	39.1	63.3
1965	88.9	81.1	40.0	64.8
1966	88.6	79.8	36.9	64.8
1967	87.7	79.6	42.1	63.7
1968	86.6	77.4	35.6	63.9
1969	85.7	76.2	37.1	62.6
1970	84.6	74.9	37.6	63.1
1971	83.8	74.2	39.3	62.8
1972	83.3	72.8	37.9	62.5
1973	81.3	71.6	40.1	63.6
1974	80.9	70.6	39.9	62.6
1975	79.8	69.8	41.9	62.0
1976	79.9	69.3	47.2	63.7
1977	78.8	70.3	46.7	64.6
1978	78.9	68.7	48.2	65.0
1979	78.6	69.4	47.8	65.8
1980	78.8	69.3	47.3	65.3

Table B.7.III.b In Current Year Homogeneously Married Women by Denomination as a Percentage of All in Current Year Married Women by Denomination, 1936-1980

Year	Catholic	Protestant	Others	No-Denom.
1936	85.5	85.9	86.3	63.4
1937	85.8	86.0	83.7	64.1
1938	86.1	84.8	48.6	64.7
1939	85.5	84.5	48.0	66.2
1940	85.9	83.8	52.2	63.5
1941	87.3	83.8	51.5	62.4
1942	86.6	83.2	64.4	61.6
1943	87.2	83.7	40.9	63.6
1944	88.7	85.6	36.6	64.1
1945	n.a.	n.a.	n.a.	n.a.
1946	86.8	83.4	40.1	64.4
1947	87.2	83.0	37.2	65.0
1948	87.6	83.1	39.1	64.8
1949	87.1	82.6	37.9	64.4
1950	87.7	83.4	38.8	64.4
1951	88.3	83.9	39.3	66.4
1952	88.3	83.8	39.0	66.9
1953	89.2	84.3	40.2	68.6
1954	89.5	83.9	40.2	69.4
1955	89.8	83.6	42.6	70.7
1956	89.6	83.5	41.6	70.4
1957	90.0	83.2	40.4	70.8
1958	89.8	83.1	40.6	71.0
1959	89.4	82.6	42.4	70.7
1960	89.2	82.0	42.6	69.7
1961	88.6	80.7	40.5	68.8
1962	88.5	80.9	41.1	68.7
1963	88.2	79.7	41.4	68.6
1964	88.0	79.6	41.6	67.8
1965	87.8	79.2	43.6	68.8
1966	86.9	78.3	41.3	69.0
1967	85.9	77.7	47.7	68.5
1968	84.3	75.8	40.7	68.8
1969	82.7	74.3	42.2	68.9
1970	81.2	73.2	42.1	69.7
1971	80.5	72.4	44.3	69.2
1972	79.7	71.2	42.3	69.2
1973	77.7	69.9	45.5	69.9
1974	78.0	68.7	42.3	68.5
1975	77.3	67.5	45.6	67.6
1976	77.3	67.5	49.0	69.1
1977	76.6	67.8	50.7	69.5
1978	76.5	67.4	52.4	68.8
1979	76.4	67.7	54.0	69.3
1980	76.5	67.6	53.8	68.7

Table B.7.IV.a Probabilities of in Current Year Married Men by Denomination to Be Married to a Woman of the Same Denomination, 1936-1980

Year	Catholic	Protestant	Others	No-Denom.
1936	35.5	52.1	1.7	10.8
1937	35.7	51.5	1.6	11.1
1938	36.2	46.7	5.8	11.3
1939	36.3	45.2	5.6	12.9
1940	36.3	45.4	5.5	12.8
1941	38.5	44.7	4.9	11.9
1942	35.8	45.0	6.3	12.8
1943	38.5	45.1	3.7	12.7
1944	39.0	46.0	3.4	11.6
1945	n.a.	n.a.	n.a.	n.a.
1946	37.6	45.2	3.7	13.5
1947	38.7	43.6	3.5	14.2
1948	39.9	43.0	3.5	13.5
1949	40.1	43.0	3.3	13.6
1950	40.4	42.8	3.2	13.6
1951	39.7	43.0	3.0	14.2
1952	39.9	42.2	3.2	14.6
1953	40.7	41.0	2.9	15.5
1954	41.2	40.0	2.9	16.0
1955	41.7	39.3	2.9	16.1
1956	42.6	38.6	2.8	15.9
1957	43.5	37.7	2.8	16.0
1958	44.0	37.7	2.8	15.6
1959	44.4	36.6	2.9	16.2
1960	43.7	36.6	2.9	16.8
1961	44.5	35.9	2.9	16.8
1962	44.8	35.8	3.0	16.5
1963	45.2	34.9	2.9	17.0
1964	45.3	34.7	2.8	17.2
1965	45.9	33.7	2.8	17.7
1966	45.9	33.2	2.7	18.1
1967	45.3	33.4	3.2	18.2
1968	45.0	32.4	2.8	19.8
1969	45.2	31.9	2.8	20.1
1970	45.3	30.9	2.7	21.1
1971	44.9	30.8	2.9	21.5
1972	45.6	30.1	2.8	21.4
1973	43.8	30.0	3.0	23.2
1974	44.6	29.3	3.0	23.1
1975	44.5	29.3	3.1	23.1
1976	44.1	28.6	3.5	23.9
1977	42.8	28.9	3.7	24.6
1978	43.2	27.8	3.8	25.2
1979	42.1	27.8	4.1	26.1
1980	42.8	27.5	4.3	25.4

Table B.7.IV.b Probabilities of in Current Year Married Women by Denomination to Be Married to a Man of the Same Denomination, 1936-1980

Year	Catholic	Protestant	Others	No-Denom.
1936	34.5	51.7	1.7	12.1
1937	34.6	51.2	1.7	12.5
1938	35.2	47.2	5.1	12.4
1939	35.0	46.0	4.9	14.1
1940	35.4	45.8	5.2	13.7
1941	37.8	44.8	4.5	12.9
1942	35.3	45.0	6.2	13.6
1943	37.8	45.3	3.4	13.5
1944	38.3	46.2	3.0	12.6
1945	n.a.	n.a.	n.a.	n.a.
1946	36.9	44.7	4.2	14.2
1947	38.1	43.6	3.1	15.1
1948	39.3	43.2	3.1	14.4
1949	39.2	42.9	3.1	14.8
1950	39.5	42.9	2.9	14.6
1951	39.0	42.8	2.8	15.4
1952	39.2	42.2	2.8	15.8
1953	40.0	41.1	2.5	16.4
1954	40.4	40.0	2.6	17.0
1955	41.0	39.0	2.6	17.3
1956	41.8	38.4	2.6	17.2
1957	42.8	37.4	2.5	17.3
1958	43.1	37.3	2.5	17.1
1959	43.6	36.2	2.7	17.5
1960	43.2	36.1	2.8	17.9
1961	43.7	35.4	2.8	18.1
1962	44.1	35.2	2.9	17.8
1963	44.5	34.3	2.9	18.3
1964	44.6	34.0	3.0	18.4
1965	45.3	32.9	3.0	18.8
1966	45.1	32.6	3.0	19.3
1967	44.3	32.5	3.6	19.6
1968	43.7	31.8	3.2	21.3
1969	43.6	31.1	3.2	22.2
1970	43.5	30.2	3.1	23.3
1971	43.1	30.0	3.2	23.6
1972	43.6	29.4	3.2	23.8
1973	41.9	29.3	3.4	25.5
1974	43.0	28.5	3.2	25.3
1975	43.1	28.4	3.4	25.2
1976	42.6	27.8	3.6	26.0
1977	41.7	27.8	4.0	26.5
1978	41.8	27.3	4.1	26.7
1979	40.9	27.1	4.6	27.4
1980	41.5	26.9	4.8	26.8

Table B.7.V.a Differences Between in Current Year Homogeneously Married Men by Denomination as a Percentage of All in Current Year Married Men by Denomination and the Probabilities for Men by Denomination to Be Homogeneously Married, 1936-1980

Year	Catholic	Protestant	Others	No-Denom.
1936	52.6	34.4	82.4	45.6
1937	52.9	35.0	76.4	46.0
1938	52.4	37.1	49.5	47.2
1939	52.4	37.8	48.7	47.7
1940	51.9	37.7	49.8	46.7
1941	50.5	39.0	50.8	45.7
1942	52.1	38.3	59.8	45.4
1943	50.2	38.3	39.9	47.5
1944	51.4	39.3	38.7	47.6
1945	n.a.	n.a.	n.a.	n.a.
1946	50.9	39.0	31.9	47.8
1947	49.8	39.3	37.5	46.9
1948	49.1	39.9	39.7	47.4
1949	49.1	39.7	38.0	45.3
1950	49.3	40.3	39.6	46.2
1951	50.2	41.3	39.7	47.1
1952	49.9	41.5	42.3	47.5
1953	50.2	43.0	42.6	49.2
1954	50.0	44.0	41.9	49.1
1955	49.4	45.1	44.2	49.5
1956	48.7	45.5	42.5	49.1
1957	48.0	46.1	41.2	49.6
1958	47.6	46.2	41.8	49.3
1959	46.5	46.8	42.5	49.4
1960	46.4	46.5	42.5	48.5
1961	45.7	45.9	39.2	46.9
1962	45.1	46.4	39.6	47.1
1963	44.3	46.3	38.3	46.8
1964	44.0	46.6	36.3	46.1
1965	43.1	47.4	37.2	47.1
1966	42.6	46.6	34.2	46.7
1967	42.5	46.3	38.9	45.5
1968	41.7	45.0	32.8	44.1
1969	40.6	44.3	34.2	42.5
1970	39.3	44.1	34.9	42.1
1971	39.0	43.5	36.5	41.3
1972	37.7	42.7	35.1	41.0
1973	37.5	41.7	37.1	40.4
1974	36.3	41.3	36.8	39.5
1975	35.4	40.5	38.8	38.8
1976	35.8	40.7	43.7	39.8
1977	36.0	41.4	43.0	40.0
1978	35.8	40.9	44.4	39.8
1979	36.6	41.6	43.8	39.8
1980	36.0	41.8	43.0	39.9

Table B.7.V.b Differences Between in Current Year Homogeneously Married Women by Denomination as a Percentage of All in Current Year Married Women by Denomination and the Probabilities for Women by Denomination to Be Homogeneously Married, 1936-1980

Year	Catholic	Protestant	Others	No-Denom.
1936	51.0	34.2	84.6	51.3
1937	51.2	34.8	81.9	51.6
1938	50.9	37.5	43.5	52.2
1939	50.5	38.5	43.1	52.2
1940	50.5	38.0	47.0	49.8
1941	49.6	39.1	47.0	49.5
1942	51.4	38.2	58.2	48.0
1943	49.4	38.5	37.4	50.2
1944	50.5	39.5	33.7	51.5
1945	n.a.	n.a.	n.a.	n.a.
1946	49.9	38.6	35.9	50.2
1947	49.1	39.3	34.0	49.9
1948	48.3	40.0	36.0	50.4
1949	47.9	39.6	34.8	49.6
1950	48.2	40.4	35.9	49.7
1951	49.3	41.1	36.6	51.0
1952	49.0	41.6	36.2	51.1
1953	49.3	43.2	37.7	52.2
1954	49.1	43.9	37.6	52.4
1955	48.7	44.7	40.0	53.4
1956	47.8	45.1	39.0	53.2
1957	47.2	45.7	37.9	53.6
1958	46.7	45.8	38.1	53.9
1959	45.8	46.4	39.7	53.2
1960	45.9	45.9	39.9	51.8
1961	44.9	45.3	37.7	50.7
1962	44.4	45.7	38.3	50.9
1963	43.7	45.4	38.5	50.3
1964	43.3	45.6	38.6	49.4
1965	42.5	46.3	40.5	50.1
1966	41.8	45.7	38.2	49.7
1967	41.6	45.1	44.1	48.9
1968	40.5	44.0	37.5	47.5
1969	39.1	43.2	39.0	46.7
1970	37.8	43.1	39.0	46.4
1971	37.4	42.4	41.1	45.5
1972	36.0	41.8	39.1	45.5
1973	35.8	40.7	42.1	44.4
1974	35.0	40.2	39.1	43.2
1975	34.2	39.1	42.3	42.3
1976	34.6	39.7	45.4	43.2
1977	35.0	39.9	46.7	43.0
1978	34.7	40.1	48.3	42.1
1979	35.5	40.5	49.4	41.9
1980	34.9	40.8	49.0	41.9

APPENDIX C: CHAPTER VI

C.1 Marriages Consecrated

Annotations to table C.1.I

Appendix B.7.I. presents the number of marriages contracted by the denomination of the marriage partners from 1936 till and including 1980. Table C.1.I. below is an expansion of table B.7.I. columns (1), (2), (3) and (30). The source of both tables is identical, i.e. the raw data of CBS, *Huwelijksstatistiek*.

Annotations to table C.1.II

The columns (1), (2) and (3) refer to homogeneous marriages consecrated. Column (4) thus includes the columns (1), (2) and (3) as well as the number of mixed marriages consecrated. The source used is CBS, *Huwelijksstatistiek*.

Annotations to table C.1.III

Column (1) equals table C.1.II column (1) divided by the relevant years of table B.7.I. column (1) or table C.1.I. column (1). The columns (2) and (3) are computed likewise. Column (4) equals table C.1.II column (4) divided by the relevant years of table B.7.I. column (30) or table C.1.I. column (4). In 1953 the marriages consecrated include 330 marriages consecrated abroad. Because of that for 1953 the percentage of marriages consecrated deviates from the quotient of the tables C.1.II. and B.7.I.

Table C.1.I Homogeneous Roman-Catholic, Dutch Reformed and Re-Reformed Marriages Contracted and the Total Number of Marriages Contracted, 1981-1986

Year	Roman-Catholic	Dutch Reformed	Re-Reformed	Total
	(1)	(2)	(3)	(4)
1981	27,757	7,854	4,690	85,574
1982	26,569	7,597	4,575	83,516
1983	24,941	7,466	4,479	78,451
1984	23,677	6,802	4,204	81,655
1985	24,061	6,333	4,048	82,747
1986	25,476	6,606	4,197	87,337

Table C.1.II Homogeneous Roman-Catholic, Dutch Reformed and Re-Reformed Marriages Consecrated and the Total Number of Marriages Consecrated, 1951-1986

Year	Roman-Catholic (1)	Dutch Reformed (2)	Re-Reformed (3)	Total (4)
1951	29,928	10,173	5,908	51,608
1952	29,088	9,698	5,773	49,994
1953	29,573	9,736	5,626	50,108
1954	30,840	9,742	5,719	51,737
1955	31,769	9,917	5,785	53,152
1956	33,642	10,189	6,141	55,839
1957	35,094	10,150	6,155	57,491
1958	34,594	9,843	6,089	56,659
1959	33,379	9,210	5,556	54,398
1960	33,158	9,265	5,729	54,741
1961	34,859	9,124	6,119	57,202
1962	35,211	9,452	6,171	58,210
1963	36,217	9,120	5,994	59,250
1964	39,327	9,853	6,547	64,626
1965	41,780	10,105	6,829	68,618
1966	n.a.	n.a.	n.a.	70,688
1967	42,348	10,418	6,998	72,203
1968	41,773	9,888	6,869	72,331
1969	40,870	9,326	6,671	71,343
1970	41,562	9,195	6,655	72,713
1971	40,162	8,715	6,709	71,532
1972	38,391	7,694	6,170	67,758
1973	31,763	6,589	5,456	58,729
1974	33,269	6,454	5,413	62,800
1975	29,692	5,731	4,992	57,061
1976	28,059	5,136	4,910	52,926
1977	25,763	5,096	4,763	48,702
1978	24,529	4,597	4,444	45,911
1979	22,538	4,444	4,384	43,038
1980	23,985	4,685	4,517	45,417
1981	22,210	4,340	4,303	42,403
1982	21,058	4,215	4,205	40,447
1983	18,558	4,063	4,030	36,459
1984	16,892	3,675	3,766	34,110
1985	16,697	3,444	3,614	33,539
1986	16,684	3,506	3,673	33,886

Table C.1.III Homogeneous Marriages Consecrated by Denomination as a Percentage of the Total Number of Homogeneous Marriages Contracted by Denomination and the Total Number of Marriages Consecrated as a Percentage of All Marriages Contracted, 1951-1986

Year	Roman-Catholic	Dutch Reformed	Re-Reformed	Total
	(1)	(2)	(3)	(4)
1951	94.60	44.24	92.04	57.20
1952	94.45	44.80	93.10	57.20
1953	94.74	46.85	92.37	58.22
1954	94.99	47.31	92.87	58.72
1955	95.41	48.89	93.93	59.70
1956	95.50	49.82	94.00	60.52
1957	95.76	50.55	93.65	61.43
1958	95.76	50.62	93.78	61.92
1959	95.66	51.20	93.28	61.81
1960	95.46	51.97	93.60	61.44
1961	95.55	52.26	94.17	61.78
1962	95.28	53.68	94.69	62.49
1963	95.30	52.94	94.35	62.13
1964	95.91	53.56	95.24	62.80
1965	95.54	54.56	95.16	63.23
1966	n.a.	n.a.	n.a.	63.16
1967	94.65	56.03	94.71	62.72
1968	93.83	55.96	94.99	61.54
1969	93.23	56.11	94.99	60.77
1970	91.31	55.21	93.65	58.81
1971	90.80	55.34	94.29	58.44
1972	89.88	53.77	94.07	57.65
1973	86.68	53.07	91.64	54.56
1974	87.28	54.11	91.64	57.30
1975	86.30	54.45	92.48	57.01
1976	84.94	52.99	92.05	54.54
1977	84.15	54.23	92.00	52.21
1978	83.40	54.34	92.82	51.53
1979	81.89	55.89	92.26	50.25
1980	81.26	55.16	91.23	50.36
1981	80.02	55.26	91.75	49.55
1982	79.26	55.48	91.91	48.43
1983	74.41	54.42	89.98	46.47
1984	71.34	54.03	89.58	41.77
1985	69.39	54.38	89.28	40.53
1986	65.49	53.07	87.51	38.80

SUMMARY

Since its sixteenth century origins the Netherlands has been a society of minorities. Pluralism has always been the hallmark of its institutions. However, the content of pluralism changed over time. In the sixteenth and seventeenth century pluralism was primarily manifested in geographical de-centralism. From the late nineteenth century onwards Dutch society became divided along vertical rather than horizontal lines of cleavage. In the literature this process of vertical division is known as pillarization (in Dutch: "verzuiling"), i.e. the conception that Dutch society was built on pillars. They were sovereign in their own domain but shared a national identity.

When pillarization had matured in the first decades of the twentieth century, Dutch society comprised four pillars: a Protestant, a Roman-Catholic, a social-democratic and a liberal pillar. Each had its own political parties and social organizations. With regard to individual behaviour, social distance between the members of different pillars was large. In de 1960's a process of de-pillarization is alleged to have set in.

This study investigates the long term changes in Dutch pluralism and how these are related to interdependent processes of economic and social change, which can be labelled as modernization. In particular two major issues are treated in this book: 1) How is Dutch pillarization as a specific type of institutionalised pluralism related to long term processes of modernization in the Netherlands?; 2) To what extent did Dutch society de-pillarize in the last decades and, if so, how can de-pillarization be related to the progressive modernization of Dutch society?

The study has a long term perspective. It covers the last four hundred years. The approach is multi-disciplinary. It uses studies from the economic, historical, political and social sciences, and tries to connect these so that the sum total will gain.

The first chapter was introductory. The frame of reference is sketched. The problems to be investigated are presented and the research method described.

Chapter II deals with the independent variable, modernization. The problem of this concept is its catch-all character. The sources of scientific thinking on modernization were explored and the major theories were reviewed. Modernization was defined as the process of interdependent changes in the social, cultural, political and attitudinal domains of a society rooted in the transformation of its economy by the purposive development of technology. Modernization has to do with a complex of intertwined variables in which either congruent or discongruent changes may appear as a result of progressive technological development.

The variables involved in the process of modernization relate to each other in a circular causal way. Without adherence to technological determinism, the purposive nature of technological development was considered to be the nucleus of the whole process. In this context "purposive" means the deliberate development and application of technology with the aim to raise productivity. In modern society technology is institutionalised to a high degree in order to raise economic growth.

The concept of compatibility was introduced. Modernization should be conceived as a complex of variables that under specific historical conditions may be mutually more or less compatible. The latter can help to explain acceleration, stagnation and decline in modernization processes. Modernization is the progressive process of the solving of incompatibilities.

The fact that processes of technological and economic development have a logic of their own was not denied. However, it was seriously doubted whether convergence theory is right by stating that modernization is an unilinear process. Instead, the emphasis was laid on social codes as products of historical experience that explain why different societies may choose different solutions to identical problems.

Next to technological development and economic growth chapter II identified six parameters of modernization: 1) the expansion of the division of labour; 2) enlargement of scale (economically and socially); 3) the increase of geographical and social mobility; 4) the bureaucratization and formalization of social and economic relations; 5) the growth of the public sector of the economy; 6) the standardization and greater accessibility of information.

Chapter III monitors Dutch modernization along the lines presented in chapter II. It generally concluded that Dutch society saw three waves of accelerated modernization. The first was concentrated in the sixteenth and seventeenth century. The second wave lasted from 1870 till the 1920's. The third wave started in the 1960's and lasted till 1973.

The period from the last quarter of the seventeenth century till 1820 was one of stagnation and decline, which could be explained by the incompatibilities between the economic, social, cultural and political domains caused by changes in the country's external environment. The period 1820-1870 was considered transitional because while per capita income started to rise the social concomitants of modernization lagged behind.

Chapter IV analyses the relation between modernization of the Netherlands and the associated changes of institutionalised pluralism from the sixteenth to the twentieth century. It described the origins of Dutch pluralism as it was born out of the Revolt against the Spanish.

Till the nineteenth century Dutch pluralism was primarily of a geographic nature. The Dutch Republic had a highly fragmented and de-centralized power structure and was in fact a loosely integrated federation of autonomous provinces and cities. The external geo-political conditions of the time forced it to minimum unity. Gradually institutions arose that made decision making at the national level possible while at the same time safeguarding the autonomy of the parties involved. The most important of these institutions was a culture of "living-apart-together".

In the sixteenth and seventeenth century the Republic's de-centralism was a favourable condition for the economy to flourish. As external economic conditions changed, it became a barrier to adaptation that could not be removed till the French occupation of the country.

When after the French occupation the Netherlands had become a kingdom in 1813, liberalism presented itself as a modernizing force. In the first half of the nineteenth century it promoted technological and social innovation and increasing centralization. It succeeded to delete the power of the traditional elites in the provinces of Holland and Zealand. In the second half of the nineteenth century liberalism came under attack of those Protestants who perceived it as the main representative of the evil of modernity. The first conflicts between liberalism and traditional religious forces took place as early as the beginning of the nineteenth century and centered around the issue of poverty legislation. A second major conflict was the school issue in the last decades of the nineteenth century. Here the Roman-Catholics joined the Protestants.

The Protestant resistance against liberalism - and later rising socialism - was also directed to modern theological ideas in the Dutch Reformed church. Both were regarded as manifestations of modernization. The resistance was led by Abraham

Kuyper, who formed the Protestant Anti-Revolutionary movement. It was his idea that the Protestants should be "sovereign in their own circle" in order to be protected to the evils of modernity. Thus as social differentiation took place as part of the more general process of Dutch modernization, new established organizations were brought under the influence of religious groups, notably the Protestants and the Roman-Catholics. In this way, a triple structure of organizations could arise: Protestant, Roman-Catholic, social-democratic and liberal. For most social functions to be fulfilled, different organizations with different ideological outlooks were founded. This phenomenon of social and political organization around an ideology within the nation state was later on called pillarization. In this way every group could hold an ideological grip on its followers.

Thus at the end of the nineteenth century traditional Dutch geographical pluralism had changed to structural ideological pluralism. Both implied a high degree of de-centralism and autonomy of categories of the Dutch population. In both cases the negative effects of pluralism were overcome by a culture of what I called "living-apart-together".

The first part of chapter V analyses the concept of pillar and the related pillarization process. It concluded that the prevailing pillar concept leads to theoretical and empirical problems because of the implicit emphasis on the religious nature of pillars. A new definition of the pillar concept was given: a subsystem in society that links political power, social organization and individual behaviour and which is aimed to promote, in competition as well as in cooperation with other social and political groups, goals inspired by a common ideology shared by its members for whom the pillar is the main locus of social identification. This definition draws the analysis away from the individual pillar to the relations between pillars as parts of a social system.

A review of the major theories of pillarization led to the conclusion that it should primarily be regarded as a way of social control embedded in Dutch history that resulted from the incompatibi-lities between accelerated modernization and the traditional ways of power allocation in Dutch society.

Part two of chapter V monitored Dutch pillarization as an historical process. The analysis of nineteenth century pillarization showed that the social groups involved pillarized at different moments in time. Further, a difference in the intensity of pillarization between the relevant groups was demonstrated. From the analysis of time series it was learned that pillarization has never been a broad phenomenon that applies to all sectors of Dutch society to the same degree at the same moment in time. One cannot properly speak of the 1950's as the peak of Dutch pillarization. A specification is needed as to what social groups as well as to what kinds of social behaviour the statement is directed.

The final chapter VI monitors and explains de-pillarization. As with pillarization, without specification one cannot properly speak of the de-pillarization of Dutch society. After the 1950's tendencies of de-pillarization set in. However, the process showed several dissimilarities. After the 1950's pillarization lost its effectiveness as a political strategy to guarantee stable constituencies. At the organizational level developments were diverse. Some social activities de-pillarized, others did not. At the level of individual interaction Dutch society can clearly be considered to be de-pillarized since the 1960's. However, with regard to individual behaviour differences between the pillars were presented with regard to moment and degree of pillarization.

The explanation of de-pillarization concentrated on enlargement of scale, secularization and on the rise of the welfare state. All three factors are part of the modernization process. The rise of the welfare state took place after World War II. The first two factors explained the nineteenth century rise of pillarization. I demonstrated that secularization accelerated after World War II and that post-war enlargement of scale differed from that in the second half of the nineteenth century. It is accelerated secularization, the qualitative change of enlargement of scale and the rise of the welfare state that explain why pillarized groups did not tighten up as one would have expected, but, on the contrary lost control over their followers. Instead of further polarization along the orthodox-secular axis, it led to the partial decomposition of the groups involved and therewith contributed to the individualization of Dutch society.

At the end of chapter VI an inverted U-curve relation between modernization and pillarization was presented. It drew the attention to two phenomena. Firstly, in the late nineteenth century and the first half of the twentieth century pillarization was an effective strategy of social control within the conditions of then prevailing modernization. After World War II incompatibilities arose between accelerated modernization and then existing pillarized social control. The very exigencies of post-war modernization eroded pillarization.

Secondly, the question whether or not Dutch society de-pillarized in the last decades, should be specified. It did with regard to individual behaviour. It did only partially on the level of social organization and at the political level the rules of the games were changed.

REFERENCES

A.A.G. Bijdragen, (1965), 13, afd. Agrarische Geschiedenis Landbouwhogeschool, Wageningen.
Adorno, Theodor W. and Walter Dirks (eds) (1955), *Sociologica*, Europaïsche Verlagsanstalt, Frankfurt.
Albeda, W. (1975), *Arbeidsverhoudingen in Nederland. Een inleiding*, Samsom, Alphen a/d Rijn.
Amin, S. (1974), *Accumulation on a World Scale*, Monthly Review Press, New York.
Amin, S. (1976), *Unequal Development*, Harvester Press, Hassocks, Sussex.
Armstrong, W.A. (1974), *Stability and Change in an English County Town. A Social Study of York 1801-1851*, Cambridge.
Aron, Raymond (1955), 'Fin de l'age idéologique?' in: Theodor W. Adorno and Walter Dirks (eds), pp. 219-233.
Bagley, C. (1973), *The Dutch Plural Society. A Comparative Study in Race Relations*, Oxford University Press, London.
Bakker, B.F.M., J. Dronkers and H.B.G. Ganzeboom (eds) (1984), *Social Stratification and Mobility in the Netherlands*, Siswo, Amsterdam.
Baran, P.A. (1957), *The Political Economy of Growth*, Monthly Review Press, New York.
Baran, P.A. and P.M. Sweezy (1966), *Monopoly Capital*, Monthly Review Press, New York.
Bell, D. (1973), *The Coming of Post-Industrial Society*, Basic Books, New York.
Ben-David, Joseph and Terry Nichols Clark (1977), *Culture and Its Creators*, University of Chicago Press, Chicago.
Bendix, R. (1966), 'Tradition and modernity reconsidered' in: *Comparative Studies in Society and History*, pp. 292-346.
Berger, Peter L. (1977), 'Toward a critique of modernity' in: Peter L. Berger, *Facing up to Modernity. Excursions in Society, Politics, and Religion*, Basic Books, New York, p. 75.
Berger Peter L., Brigitte Berger and Hansfried Kellner (1973), *The Homeless Mind. Modernization and Consciousness*, Vintage Books, New York.
Beyme, K. von (1973), *Die parlementarischen Regierungssyteme in Europa*, Piper Verlag, München.
Beyme, K. von (1985), *Political Parties in Western Democracies*, Gower, Aldershot.
Blaas, P.B.M. and J. van Herwaarden (eds) (1986), *Stedelijke naijver. De betekenis van interstedelijke conflicten in de geschiedenis. Enige beschouwingen en case-studies*, Den Haag.
Blom, J.C. (1981), *Verzuiling in Nederland 1850-1925*, Historisch Seminarium Universiteit van Amsterdam, Amsterdam.
Boogman, J.C. (1962), 'Achtergronden en algemene tendenties van het buitenlands beleid van Nederland en België in het midden van de 19e eeuw', in: *Bijdragen en Mededelingen van het Historisch Genootschap*, 76.
Braam, A. van (1957), *Ambtenaren en bureaukratie in Nederland*, De Haan, Zeist.
Braverman, H. (1975), *Labor and Monopoly Capital*, Monthly Review Press, New York.
Bromley, J.S. and E.H. Kossmann (1964), *Britain and the Netherlands II*, Groningen.
Bromley, J.S. and E.H. Kossmann (1971), *Britain and the Netherlands IV: Metropolis, Dominion and Province*, The Hague.

Brugmans, I.J. (1961), *Paardenkracht en mensenmacht*, Den Haag.
Brugmans, I.J. (1977), 'Standen en klassen in Nederland gedurende de negentiende eeuw', in: P.A.M. Geurts and F.A.M. Messing (eds), 1977, pp. 110-129.
Brunt, L. (1972), 'Over gereformeerden en kleine luyden', in: *Sociologische Gids*, 19, pp. 49-58.
Brunt, L. (1973), *Stedeling op het platteland*, Boom, Meppel.
Bryant, C.G.A. (1981), 'Depillarisation in the Netherlands', in: *British Journal of Sociology*, pp. 56-74.
Brzezinski, Zbigniew K. (1967), *The Soviet Bloc: Unity and Conflict*, Harvard University Press, Cambridge, Mass., 2nd ed..
CBS, (1951), *12e Volkstelling 31 mei 1947. Serie A. Deel 4. Statistiek der bestaande huwelijken en van de vruchtbaarheid dezer huwelijken*, Utrecht.
CBS, (1981), *Statistiek van de vakbeweging*, Staatsuitgeverij, The Hague, p. 21.
CBS, (1983), *Statistiek der verkiezingen 1981 Tweede Kamer der Staten-Generaal, 26 mei*, Staatsuitgeverij, The Hague.
CBS, (1984), *1899-1984. Vijfentachtig jaren statistiek in tijdreeksen*, Staatsuitgeverij, The Hague.
CBS, (1987), *Statistiek der verkiezingen 1986 Tweede Kamer der Staten-Generaal, 21 mei*, Staatsuitgeverij, The Hague.
Cohen, Percy S. (1968), *Modern Social Theory*, Heinemann, London.
Constandse, A.K. (1958-1959), 'Acquaintanceships of farmers in a newly colonized area, in: *Sociaal Kompas*, nr. 2.
Daalder, H. (1966), 'The Netherlands, Opposition in a Segmented Society', in: R.A. Dahl (ed.), 1966, pp. 118-236.
Daalder, H. (1974), *Politisering en lijdelijkheid in de Nederlandse politiek*, Van Gorcum, Assen.
Daalder, H. (1984), 'On the origins of the consociational democracy model', in: *Acta Politica*, XIX, pp. 97-117.
Daalder, H. (1985), 'Politicologen, sociologen, historici en de verzuiling', *Bijdragen en mededelingen betreffende de geschiedenis der Nederlanden*, 100(1), pp. 52-65.
Dahl, R.A. (1966), *Political Oppositions in Western Europe*, New Haven.
Dahl, R.A. (1971), *Polyarchy: Participation and Opposition*, New Haven.
Dahl, Robert A. and Charles E. Lindblom (1953), *Politics, Economics and Welfare. Planning and Politico-Economic Systems Resolved into Basic Social Processes*, Harper & Row, New York.
Dalen, P. van (1967), *Wij Nederlanders*, Het Spectrum, Utrecht, Antwerpen.
Damsma, D. and L. Noordegraaf (1977), 'Standen en klassen in een Zuidhollands dorp', in: *Tijdschrift voor Sociale Geschiedenis*, 9 (1977), pp. 243-270.
Daudt, H. (1980), 'De ontwikkelingen van de politieke machtsverhoudingen in Nederland sinds 1945', in: *Nederland na 1945*, Van Loghum Slaterus, Deventer, pp. 178-199.
Daumard, A. (1970), *Les Bourgeois de Paris*, 2nd ed., Paris.
Dekker, G. (1965), *Het kerkelijk gemengde huwelijk in Nederland*, Boom, Meppel.
Dekker, R. (1982), *Holland in beroering. Oproeren in de 17e en 18e eeuw*, Ambo, Baarn.
Deurloo, M.C. and G.A. Hoekveld (1981), 'The Population Growth of the Urban Municipalities in the Netherlands between 1849 and 1970 with Particular Reference to the Period 1899-1930', in: H. Schmal (ed), 1981.

Dijk, H. van (1976), *Rotterdam 1810-1880. Aspecten van een stedelijke samenleving*, Schiedam.
Dijk, H. van (1986), 'Het negentiende-eeuwse stadsbestuur', in: P.B.M. Blaas and J. van Herwaarden (eds), *Stedelijke naijver. De betekenis van interstedelijke conflicten in de geschiedenis. Enige beschouwingen en case-studies*, Den Haag.
Doel, J. van den (1971), *Convergentie en evolutie*, Van Gorcum, Assen.
Doorn, J.A.A. van (1956), 'Verzuiling: Een eigentijds systeem van sociale controle', in: *Sociologische Gids*, vol. 3, pp. 41-49.
Doorn, J.A.A. van (1971), 'Verzuiling als verbinding van organisatie en ideologie', in: J.A.A. van Doorn, *Organisatie en maatschappij*, Stenfert Kroese, Leiden, pp. 62-71.
Durkheim, Emile (1964) (1933), *The Division of Labor in Society*, Macmillan/The Free Press, London/New York.
Eerenbeemdt, H.F.J.M. van de, and J. Hannes (1981), 'Economische en sociale verhoudingen in Noord en Zuid, ca. 1770 - midden 19e eeuw', in: *Algemene geschiedenis der Nederlanden*, (10), Van Dishoeck, Haarlem, 1981, pp. 140-159.
Eijk, C. van der and B. Niemöller (1983), 'Stemmen op godsdienstige partijen sinds 1967' in: *Acta Politica, In het spoor van de kiezer*, Boom, Meppel/Amsterdam, pp. 169-182.
Eisenstadt, S.N. (1973), *Tradition, Change and Modernity*, Wiley, New York.
Eisenstadt, S.N. (1977), 'Dynamics of Civilizations and Development: The Case of European Society' in: *Economic Development and Cultural Change*, Vol. 25, Supplement, Essays on Economic Development and Cultural Change in Honor of Bert F. Hoselitz, pp. 123-145.
Eisenstadt, S.N. (1978), *Revolution and the Transformation of Societies*, The Free Press, New York.
Ellemers, J.E. (1980), 'Ontwikkeling van de samenleving', in: *Nederland na 1945. Beschouwingen over ontwikkeling en beleid* (1980), Van Loghum Slaterus, Deventer, pp. 13-40.
Ellemers, J.E. (1984), 'Pillarization as a process of modernization', in: *Acta Politica*, IXX, pp. 129-143.
Ellman, Michael (1980), 'Against convergence' in: *Cambridge Journal of Economics*, 4, September 1980, pp. 199-210.
Enquête betreffende werking en uitbreiding der wet van 19 september 1874 en naar de toestand van fabrieken en werkplaatsen, 1887.
Etzioni, A. (1961), *Complex Organizations*, The Free Press, New York.
Etzioni, A. and E. Etzioni-Halevy (eds) (1973), *Social Change*, Basic Books, New York.
Etzioni-Halevy, Eva (1981), *Social Change. The Advent and Maturation of Modern Society*, Routledge & Kegan Paul, London.
Faber, J.A. (1979), 'De achttiende eeuw', in: J.H. van Stuijvenberg, 1979, pp. 119-157.
Fennema, M. (1976), 'Professor Lijphart en de Nederlandse politiek', in: *Acta Politica*, vol. 11, nr. 1, pp. 54-77.
Flora, P. (1974), *Modernisierungsforschung. Zur empirischen Analyse der gesellschaftlichen Entwicklung*, Westdeutscher Verlag, Opladen.
Flora, P. (1983), *State, Economy and Society in Western Europe. A Data Handbook in Two Volumes. Volume I. The Growth of Mass Democracies and Welfare States*, Campus Verlag, Macmillan, St. James Press, Frankfurt, London, Chicago.

Foster, R. and Jack P. Greene (eds) (1970), *Preconditions of Revolution in Early Modern Europe*, Baltimore/London.
Frank, A.G. (1970), 'The development of underdevelopment', in: R.I. Rhodes (ed.), 1970, pp. 4-17.
Fusé, Tomoyasa (ed) (1975), *Modernization and Stress in Japan*, Brill, Leiden.
Gadourek, I. (1956), *A Dutch Community*, Leiden.
Gadourek, I. (1980), *Empirische studie van de veranderende maatschappelijke rollen, normen en waarden*, Mededelingen der Koninklijke Nederlandse Academie van Wetenschappen, afdeling Letterkunde, Noord-Hollandsche Uitgevers Maatschappij, Amsterdam, Oxford, New York.
Gadourek, I. (1982), *Social Change as Redefinition of Roles. A Study of Structural and Causal Relationships in the Netherlands of the 'Seventies'*, Van Gorcum, Assen.
Ganzeboom, H. and P. de Graaf (1983), 'Beroepsmobiliteit tussen generaties in Nederland in 1954 en 1977', in: *Mens en Maatschappij*, 1983, nr. 1, pp. 28-52.
Gerschenkron, A. (1962), *Economic Backwardness in Historical Perspective*, Harvard University Press, Cambridge, Mass..
Gerschenkron, A. (1968), *Continuity in History and Other Essays*, Harvard University Press, Cambridge, Mass..
Gerschenkron, A. (1969), 'The early phases of industrialization in Russia: afterthoughts and counterthoughts', in: W.W. Rostow (ed), 1969.
Geurts, P.A.M. and F.A.M. Messing (eds) (1977), *Economische ontwikkeling en sociale emancipatie II*, Martinus Nijhoff, Den Haag.
Goddijn, W. (1957), *Katholieke minderheid en protestantse dominant*, Assen.
Goudsblom, J. (1985), 'Levensbeschouwing en Sociologie', *Sociologisch Tijdschrift*, nr. 1.
Gouré, Leon, Foy D. Kohler, Richard Soll and Annette Stiefbold (1973), *Convergence of Communism and Capitalism: The Sovjet View*, Center for Advanced International Studies, University of Miami, Miami.
Griffiths, Richard T. (1980), 'Achterlijk, achter of anders', Inaugural lecture, Free University of Amsterdam, December 4, 1980, Amsterdam.
Habakkuk, H.J. and P. Deane (1969), , 'The take-off in Brittain', in: W.W. Rostow (ed), 1969.
Harmsen, G. (1985), 'De arbeiders en hun vakorganisaties', in: F.L. van Holthoon (ed), 1985-B, pp. 261-283.
Harts, J.J. and L. Hingstman (1986), *Verhuizingen op een rij; een analyse van individuele verhuizingsgeschiedenissen*, Netherlands Geographical Studies 13, Amsterdam/Utrecht.
Heek, F. van (1954), *Het geboorteniveau der Nederlandse roomskatholieken*, Leiden.
Heek, F. van (ed) (1958), *Sociale stijging en daling in Nederland I*, Brill, Leiden.
Heek, F. van and E.V.W. Vercruysse (1958), 'De Nederlandse beroepsprestige-stratificatie', in: F. van Heek (ed), 1958, pp. 11-48.
Hellemans, S. (1988), 'Verzuiling en ontzuiling van de katholieken in België en Nederland. Een historisch-sociologische vergelijking', in: *Sociologische Gids*, 1988 (1), pp. 43-56.
Hemels, J.M.H.J. (1969), *De Nederlandse pers voor en na de afschaffing van het dagbladzegel in 1869*, Van Gorcum, Assen.
Hendriks, J. (1971), *De emancipatie der gereformeerden*, Samsom, Alphen a/d Rijn.

Hibbs jr., D.A. (1978), 'On the Political Economy of Long-Run Trends in Strike Activity', *British Journal of Political Science*, Vol.8.
Hoffmann, W.G. (1969), 'The take-off in Germany', in: W.W. Rostow (ed), 1969.
Hofstede, B.P. (1964), *Twarted Exodus: Post-war Overseas Migration from the Netherlands*, Martinus Nijhoff, The Hague.
Hollander, A.J.N. den, E.W. Hofstee, J.A.A. van Doorn and E.V.W. Vercruijsse (eds) (1961) (1962), *Drift en koers*, Van Gorcum, Assen.
Holsteyn, J.J.M., G.A. Irwin and C. van der Eijk (1987), *De Nederlandse kiezer '86*, Steinmetzarchief/Swidoc, Amsterdam.
Holthoon, F.L. van (1985-A), 'Verzuiling in Nederland', in: F.L. van Holthoon (ed) (1985-B), pp. 159-175.
Holthoon, F.L. van (ed) (1985-B), *De Nederlandse samenleving sinds 1815*, Van Gorcum, Assen.
Hoselitz, B.F. (1960), 'Theories of stages of economic growth', in: *Theories of Economic Growth*, Free Press, New York.
Hoselitz, B.F. (1964), 'Social stratification and economic development', in: *International Social Science Journal*, 1964, 16, pp. 237-251.
Houtte, J.A. van (1979), 'De zestiende eeuw', in: J.H. van Stuijvenberg (ed), 1979. pp. 49-79.
Idenburg, Ph.J. (1964), *Schets van het Nederlandse Schoolwezen*, Wolters, Groningen.
Inghams, G.K. (1974), *Strikes and Industrial Conflict: Britain and Scandinavia*, London.
Jansen, H.P.H. (1982), *Kalendarium Geschiedenis van de Lage Landen in Jaartallen*, Het Spectrum, Utrecht/Antwerpen.
Jaspers, T. (1980), *Rechtspreken in de maatschappij*, Universitaire Pers, Leiden.
Jonge, J.A. de (1976), *De industrialisatie van Nederland tussen 1850 en 1914*, Sun, Nijmegen, (Scheltema & Holkema, Amsterdam, 1968).
Kaa, D.J. van de (1980), 'Bevolking: a-symmetrische tolerantie of accomodatiepolitiek', in: *Nederland na 1945*, Van Loghum Slaterus, Deventer, pp. 82-102.
Keesing Historisch Archief (1986), Amsterdam/Antwerpen, p. 354.
Kemenade, J.A. van (ed) (1981), *Onderwijs: Bestel en beleid*, Groningen.
Kerr, C. (1983), *The Future of Industrial Societies. Convergence or Continuing Diversity?*, Harvard University Press, Cambridge Mass..
Kerr, C., John T. Dunlop, Frederick H. Harbison and Charles A. Myers (1960), *Industrialism and Industrial Man: The Problems of Labor and Management in Economic Growth*, Harvard University Press, Cambridge Mass..
Kieve, R. (1981), 'Pillars of sand, a marxist critique of consociational democracy in the Netherlands', in: *Comparative Politics*, pp. 313-337.
King, Gregory (1973), 'Natural and Political Observations and Conclusions Upon the State and Condition of England', in: Peter Laslett (ed), 1973.
Knaap, G.A. van der (1978), *A Spatial Analysis of the Evolution of an Urban System: The Case of the Netherlands*, dissertation, Utrecht.
Kooij, P. (1985), 'Stad en Platteland', in: F.L. van Holthoon (ed), 1985, pp. 93-117.
Kruijt, J.P. (1957), 'Levensbeschouwing en groepssolidariteit in Nederland', in: *Sociologisch Jaarboek*, XI, Leiden.
Kruijt, J.P. (1959), *Verzuiling*, Heynis, Zaandijk.
Kruijt, J.P. and W. Goddijn (1962), 'Verzuiling en ontzuiling als sociologisch proces", in: A.J.N. den Hollander a.o., 1962 (1961), pp. 227-263.

Kuiper, D. Th. (1972), *De voormannen*, Meppel.
Kuiper, G. (1954), *Mobiliteit in de sociale en beroepshiërarchie*, Assen.
Kurtz, Paul (ed) (1968), *Sidney Hook and the Contemporary World: Essays on the Pragmatic Intelligence*, John Day, New York.
Kuznets, S. (1969), 'Notes on the take-off', in: W.W. Rostow (ed), 1969.
Kuznets, S. (1979), *Growth, Population and Income Distribution. Selected Essays*, Norton, New York.
Laeyendecker, L. (1984), *Sociale verandering*, Boom, Meppel/Amsterdam.
Laslett, Peter (ed) (1972), *Household and Family in Past Time*, Cambridge University Press, Cambridge, Mass..
Laslett, Peter (ed) (1973), *The Earliest Classics*, London.
Lehmbruch, G. (1967-A), *Proporzdemokratie: Politisches System und politische Kultur in der Schweiz und in Österreich*, Mohr, Tubingen.
Lehmbruch, G. (1967-B), 'A non-competitive pattern of conflict management: The case of Switzerland, Austria and Lebanon', paper reprinted in: McRae (1974)
Lerner, Daniel (1968), 'Modernization - 1. Social Aspects', in: D. Sills (ed), 1968, p. 387.
Levy, Marion (1972), *Latecomers and Survivors*, Basic Books, New York.
Lewis, W. Arthur (1955), *The Theory of Economic Growth*, Allen & Unwin, London.
Lijphart, A. (1968-A, 1975), *The Politics of Accomodation: Pluralism and Democracy in the Netherlands*, University of California Press, Berkeley.
Lijphart, A. (1968-B, 1976, 1979, 1982), *Verzuiling, pacificatie en kentering in de Nederlandse politiek*, De Bussy, Amsterdam.
Lijphart, A. (1977), *Democracy in Plural Societies: A Comparative Exploration*, Yale University Press, New Haven/London.
Lijphart, A. (1987), 'De pacificatietheorie en haar critici', in: *Acta Politica*, vol. 22, nr. 2, pp. 181-225.
Lipset, Seymour Martin (1977), 'The end of ideology and the ideology of the intellectuals' in: Joseph Ben-David and Terry Nichols Clark, 1977, pp. 15-42.
Loo, L.F. van (1981-A), *Den arme gegeven...*, Boom, Meppel.
Loo, L.F. van (1981-B), 'De armenzorg in de Noordelijke Nederlanden 1770-1840', in: *Algemene geschiedenis der Nederlanden*, (10), Van Dishoeck, Haarlem, 1981, pp. 415-435.
Lorwin, V.R. (1971), 'Segmented Pluralism: Ideological Cleavages and Political Cohesion in the Smaller European Democracies', in: *Comparative Politics*, 3/2, pp. 141-175.
Maddison, A. (1982, 1986), *Phases of Capitalist Development*, Oxford University Press, Oxford/New York.
Maddison, A. (1984), 'Origins and Impact of the Welfare State, 1883-1983', *Banca Nazionale del Lavoro Quaterly Review*, 1984, no. 148, pp. 55-87.
Maddison, A. (1987), 'Recent Revisions to British and Dutch Growth, 1700-1870 and Their Implications for Comparative Levels of Performance', in: *Economic Growth in Northwestern Europe: The Last 400 Years*, Research Memorandum nr. 214, Institute of Economic Research, Faculty of Economics, University of Groningen, March 1987, Groningen, pp.19-31.
Marcuse, Herbert (1964), *One Dimensional Man: Studies in the Ideology of Advanced Industrial Society*, Beacon Press, Boston.
Marczewski, J. (1969), 'The take-off hypothesis and French experience', in: W.W. Rostow (ed), 1969.

Matthijssen, M.A.J.M. (1958), *De intellectuele emancipatie der katholieken*, Assen.
McClelland, David (1961), *The Achieving Society*, Van Nostrand, Princeton N.J..
McRae, K. (ed.) (1974), *Consociational Democracy: Political Accomodation in Segmented Societies*, McClelland and Stewart, Toronto.
Middendorp, C.P. (1979), *Ontzuiling, politisering en restauratie in Nederland. Progressiviteit en conservatisme in de jaren 60 en 70*, Boom, Meppel/Amsterdam.
Moore, Wilbert E. (1979), *World Modernization: The Limits of Convergence*, Elsevier, New York.
Mulkay, M.J. (1971), *Functionalism, Exchange and Theoretical Strategy*, Routledge & Kegan Paul, London.
Nakane, Chie (1973), *Japanese Society*, Penguin Books, rev. ed., Harmondsworth.
Nijs, W.F. de (1983), 'Arbeidsconflicten', in: W.H.J. Reynaerts (ed), 1983.
Nilson, S. (1979), 'Toward a theory of cross-cutting cleavages', paper *International Political Science Association*, Moscow.
North, D.C. (1969), 'Industrialization in the United States', in: W.W. Rostow (ed), 1969.
Parsons, T. (1966), *Societies: Evolutionary and Comparative Perspectives*, Prentice-Hall, Englewood Cliffs, New Jersey.
Parsons, T. and E.A. Shills (eds) (1951), *Toward a General Theory of Action*, Harvard University Press, Cambridge.
Pen, J. (1979), 'A Clear Case of Levelling: Income equalization in the Netherlands', in: *Social Research*, 46, 1979, pp. 682-694.
Pen, J. and J. Tinbergen (1976), 'Hoeveel bedraagt de inkomensegalisatie sinds 1938?', in: *Economisch Statistische Berichten*, 1976, pp. 880-884.
Pijnenburg, A.A.G. and J.J.M. Holsteyn (1987), 'Religie en verzuiling', in: J.J.M. Holsteyn, G.A. Irwin and C. van der Eijk (1987), pp. 41-60.
Ponsioen, J.A. (1956), 'Notities voor de sociologische bestudering van de verzuiling', in: *Sociologische Gids*, pp. 50-52.
Pryor, Frederic L. (1973), *Property and Industrial Organization in Communist and Capitalist Nations*, Indiana University Press, Bloomington, Ind..
Reynaerts, W.H.J. (ed) (1983), *Arbeidsverhoudingen. Theorie en praktijk, Deel 2*, Leiden.
Rhodes, R.I. (ed.) (1970), *Imperialism and Underdevelopment*, Random House, New York.
Riesman, D. (1961), *The Lonely Crowd*, Yale University Press, New Haven.
Righart, H. (1986), *De katholieke zuil in Europa*, Meppel, Boom.
Riley, J.C. (1980), *International Government Finance and the Amsterdam Capital Market 1740-1815*, Cambridge.
Rokkan, S. (1968), 'The Structuring of Mass Politics in the Smaller European Democracies', in: *Comparative Studies in Society and History*, pp.173-210.
Rokkan, S. (1970), *Citizens, Elections, Parties*, Oslo.
Rokkan, S. (1975), 'Dimensions of State Formation and National Building: A Possible Paradigm for Research on Variations within Europe', in: Ch. Tilly (ed.), *The Formation of National States in Western Europe*, Princeton.
Rokkan, S. (1977), 'Towards a Generalized Concept of "verzuiling": A Preliminary Note', in: *Political Studies*, vol. 25, pp. 565-565.
Romein, J.N. and A.H.M. Romein-Verschoor (1979), *De Lage Landen bij de Zee. Een Geschiedenis van het Nederlandse Volk*, Querido, Amsterdam.
Roorda, D.J. (1961), *Partij en factie*, J.B. Wolters, Groningen.

Roorda, D.J. (1964), 'The ruling classes in Holland in the Seventeenth century', in: J.S. Bromley and E.H. Kossmann, 1964, pp. 109-132.
Roorda, D.J. (1980), 'De regentenstand in Holland', in: C.B. Wels, 1980-A, pp. 221-241.
Rose, R. (ed.) (1980), *Electoral Participation*, Sage, London/Beverly Hills.
Ross, A.M. and P.T. Hartman (1960), *Changing Patterns of Industrial Conflict*, New York.
Rostow, W.W. (1966), *The Stages of Economic Growth*, Cambridge University Press, Cambridge.
Rostow, W.W. (ed) (1969), *The Economics of Take-off into Sustained Growth*, MacMillan, London.
Rostow, W.W. (1971), *Politics and the Stages of Growth*, Cambridge University Press, Cambridge.
Ruse, Michael (1979), *Sociobiology: Sense or Nonsense?*, Reidel, Dordrecht.
Schendelen, M.P.C.M. van (1978), 'Verzuiling en restauratie in de Nederlandse politiek', *Beleid en Maatschappij*, V, pp. 42-64.
Schendelen, M.P.C.M. van (1984), 'The views of Arend Lijphart and collected criticisms', in: *Acta Politica*, pp. 19-57.
Schmal, H. (ed) (1981), *Patterns of European Urbanisation Since 1500*, London.
Schöffer, I. (1956), 'Verzuiling, een specifiek Nederlands probleem', in: *Sociologische Gids*, vol. 4, pp. 121-127.
Schöffer, I. (1968), 'De Nederlandse confessionele partijen, 1918-1938, in: *De confessionelen: Ontstaan en ontwikkeling van hun politieke partijen*, Utrecht, pp. 41-61.
Schöffer, I. (1973), 'Het politiek bestel van Nederland en maatschappelijke verandering, *Kleio*, vol. 14, pp. 518-534.
Scholten, I. (1980), 'Does Consociationalism Exist? A Critique of the Dutch Experience', in: R. Rose (ed), 1980, pp. 329-355.
Schumpeter, Joseph A. (1942), *Capitalism, Socialism and Democracy*, Harper, New York.
Sills, D. (ed) (1968), *International Encyclopedia of the Social Sciences*, Vol. 10, MacMillan and Free Press, New York.
Sixma, H. and W.C. Ultee (1983), 'Trouwpatronen en de openheid van een samenleving; de samenhang tussen de opleidingsniveaus van (huwelijks) partners in Nederland tussen 1959 en 1977', in: *Mens en Maatschappij*, 1983, pp. 360-382.
Smelser, N.J. (1973), 'Toward a theory of modernization' in: A. Etzioni and E. Etzioni-Halevy (eds), 1973, pp. 268-284.
Smit, J.W. (1970), 'The Netherlands Revolution', in: Foster, R. and Jack P. Greene (eds) (1970), pp. 19-54.
Sociaal en Cultureel Rapport 1986 (1986), Staatsuitgeverij, The Hague.
Spoormans, H. (1988), *'Met uitsluiting van voorregt'. Het ontstaan van de liberale democratie in Nederland*, SUA, Amsterdam.
Steiner, J. (1974), *Amicable Agreement Versus Majority Rule*, North Carolina University Press, Chapell Hill.
Steiner, J. (1981), 'The consociational theory and beyond', in: *Comparative Politics*, pp. 339-354.
Steininger, R. (1975), *Polarisierung und Integration*, Meisenheim am Glan.

Steininger, R. (1977), 'Pillarization and political parties', in: *Sociologische Gids*, pp. 242-257.
Stuijvenberg, J.H. van (ed) (1979), *De economische geschiedenis van Nederland*, Wolters-Noordhoff, Groningen.
Stuijvenberg, J. H. van (1981), 'De economie in de Noordelijke Nederlanden, 1770-1970', in: *Algemene geschiedenis der Nederlanden*, Vol. 10, Van Dishoeck, Haarlem.
Stuurman, S. (1983), *Verzuiling, kapitalisme en patriarchaat*, Sun, Nijmegen.
Sweezy, Paul M. and Charles Bettelheim (1971), *On the Transition to Socialism*, Monthly Review Press, New York.
Szirmai, A. (1988), *Inequality Observed. A Study of Attitudes Towards Income Inequality*, Gower, Aldershot.
Teijl, J. (1971), 'Nationaal inkomen van Nederland in de periode 1850-1900. Tasten en testen.', in: *Economisch en Sociaal-Historisch Jaarboek*, 34, 1971, pp. 232-263.
Thane, P. (1982), *The Foundations of the Welfare State*, Longman, London.
Thurlings, J.M.G. (1971), *De wankele zuil. Nederlandse katholieken tussen assimilatie en pluralisme*, Nijmegen/Amersfoort.
Tijn, Th. van (1977), 'Voorlopige notities over het ontstaan van het moderne klassebewustzijn in Nederland', in: P.A.M. Geurts and F.A.M. Messing (eds), 1977, pp. 124-144.
Tijn, Th. van (1979), 'De negentiende eeuw. De periode 1814-1914', in: J.H. van Stuijvenberg (ed), 1979, pp. 218-259.
Tijn, Th. van (1980), 'The Party Structure of Holland and the Outer Provinces in the Nineteenth Century', in: C.B. Wels (ed), 1980-B, pp. 100-129.
Tinbergen, J. (1959), *Selected Papers*, North Holland, Amsterdam.
Tinbergen, J. (1964), *Central Planning*, Yale University Press, New Haven.
Toffler, Alvin (1970), *Future Shock*, Random House, New York.
Touraine, A. (1971), *The Post Industrial Society*, Random House, New York.
Tsuru, S. (1969), 'The take-off in Japan', in: W.W. Rostow (ed), 1969.
Tulder, J.J.M. van (1962), *De beroepsmobiliteit in Nederland van 1919 tot 1954. Een sociaal-statistische studie*, Stenfert Kroese, Leiden.
Ultee, W.C. (1984), 'The Ups and Downs of the Mobility Problem in Dutch and International Sociology', in: B.F.M. Bakker, J. Dronkers and H.B.G. Ganzeboom (eds), 1984, pp. 11-29.
Verweij-Jonker, H. (1957), 'De psychologie van de verzuiling', in: *Socialisme en Democratie*, pp. 30-39.
Verwey-Jonker, H. (1961), 'De emancipatie-bewegingen', in: A.N.J. den Hollander a.o., 1961 (1962), pp. 105-125.
Vries, Jan. de (1981), *Barges and Capitalism: Passenger Transportation in the Dutch Economy. 1632-1839*, Utrecht.
Vries, Jan de (1984), 'The Decline and Rise of the Dutch Economy, 1675-1900', in: Technique, Spirit and Form in the Making of the Modern Economies: Essays in Honour of William N. Parker, *Research in Economic History*, 1984, Suppl. 3, pp. 149-189.
Vries, Joh. de (1978), *The Netherlands Economy in the Twentieth Century. An Examination of the Most Characteristic Features in the Period 1900-1970*, Van Gorcum, Assen.

Vries, Joh. de (1983, 1977), *De Nederlandse economie tijdens de 20ste eeuw. Een verkenning van het meest kenmerkende*, Fibula-Van Dishoeck, Bussum.
Wallerstein, I. (1974-A), *The Modern World System*, Academic Press, New York.
Wallerstein, I. (1974-B), 'The rise and future demise of the world capitalist system', in: *Comparative Studies in Society and History*, 1974, 16, pp. 387-415.
Wansink, H. (1980), 'Holland en zes bondgenoten: de Republiek der Zeven Verenigde Provinciën', in: C.B. Wels (ed), 1980-A, pp. 199-221.
Weisglas, M. (1969), 'Het Nederlandse bedrijfsleven in de jaren zestig', *Economisch Statistische Berichten*, 1969, p. 1280.
Wels, C.B. (ed) (1980-A), *Vaderlands verleden in veelvoud I*, Martinus Nijhoff, The Hague.
Wels, C.B. (ed) (1980-B), *Vaderlands verleden in Veelvoud II*, Martinus Nijhoff, The Hague.
Wilson, E.O. (1975), *Sociobiology: The New Synthesis*, Belknap, Cambridge, Mass..
Windmuller, J.P. (1969), *Labor Relations in the Netherlands*, Ithaca, New York.
Windmuller, J.P. (1970), *Arbeidsverhoudingen in Nederland*, Het Spectrum, Utrecht/Antwerpen.
Windmuller, J.P. and C. de Galan (1979), *Arbeidsverhoudingen in Nederland*, third revised edition, Het Spectrum, Utrecht/Antwerpen.
Wittfogel, K.A. (1978), *Oriental Despotism. A Comparative Study of Total Power*, New Haven/London.
Wolfe, Bertram D. (1968), 'A Historian Looks at the Convergence Theory' in: Paul Kurtz (ed), 1968, p. 70.
Zappey, W.M. (1979), 'De negentiende eeuw I', in: Van Stuijvenberg (ed), 1979, pp. 201-218.

INDEX

A
action group, 146, 147
Afscheiding, 88
agriculture, 47, 49
   See also Republic
Albeda, W., 129
Algemeen Nederlandsch Werklieden
   Verbond, 94
Amsterdam, 77
anomia, 21, 169, 170
ANWV
   See Algemeen Nederlandsch Werklieden
   Verbond
Armstrong, W.A., 44
ARP
   See Protestant
Asia, 19
Austria, 83
AVRO, 122

B
Bagley, C., 73
Belgium, 74, 84, 157, 158
Bell, D., 25
Berger, P.L., 21, 33
Berlin, I., 25
besogne, 81
Boerenpartij, 145
Boerhaave, 1
Braam, A. van, 59
Brabant, 78, 86, 90
break-through, 166
Britain, 79
Brzezinski, Z.K., 26
Bücher, 15
bureaucracy, 32, 73, 161
   and social control, 52
   and social relations, 31, 58
   See also modernization
   See also pillarization
bureaucratization, 31

C
CDA
   See Christian-Democrats
census system, 85
Charles V, 74
Christian-Democrats
   and church membership, 142
   electoral support, 142, 143
   political party CDA, 114, 141, 146, 156
CHU
   See Protestant
church organization, 92
cleavage, 109
   ideological, 147, 153
   main lines of, 100
   See also pillarization

   See also pluralism
CNV
   See Protestant
Communists
   Communist Party CPN, 145
compatibility, 29, 30, 31, 32
   See incompatibility
Comte, A., 8
connubium
   defined, 131
consociational democracy, 101, 107, 108
Constandse, A.K., 100
constitution
   of 1814, 85, 86
   of 1815, 86
   of 1848, 84, 85, 86
   of 1917, 91
constitutional monarchy, 85
Continental System, 36
convergence hypothesis, 14, 23, 24, 25, 26, 28, 30
corporatism, 126
CPN
   See Communists
cultural change
   See culture
culture
   cultural change, 156, 160, 162
   cultural heterogeneity, 51, 52, 55
   diffusion of, 12
   of "living-apart-together", 81, 82, 104, 111, 158
   political, 156

D
D'66, 145
Daalder, H., 2, 73, 87, 105, 106, 109, 110, 111, 139
Dahl, R.A., 25, 108, 109
Dahrendorf, R., 25
Darwin, 19
Daudt, H., 109
Daumard, A., 44
de-colonization, 8, 19, 20
de-pillarization, 3
   and broadcasting, 149
   and educational system, 156
   and floating vote, 146
   and health services, 149
   and incompatibilities, 156
   and individualization, 167, 168, 169
   and industrial relations, 149, 152, 158
   and modernization, 3, 7, 141, 156, 157
   and political polarization, 146
   and secularization, 163
   and social control, 167, 170
   and social identification, 170
   and social inertia, 155
   and strike activity, 152
   and the welfare state, 157, 158
   Belgian Roman-Catholic, 157

political, 142, 143, 144
Dekker, G., 100, 131, 138
Dekker, R., 78
democratization
   and industrial relations, 150
   and political parties, 144
   and the company, 150
   of society, 144
Denmark, 74
denominational
   content of school curriculum, 148
   education, 90, 147, 148
Dijk, H. van, 44, 46, 83
division of labour, 8, 19, 20, 31, 32, 43, 46, 47, 49, 59
   See also modernization
Doel, J. van den, 25
Doleantie, 88
Domicilie van onderstand, 86
Doorbraak
   See break-through
DS'70, 146
Durkheim, E., 8, 9, 20, 21, 51, 59, 169
Duverger, 25

E

economic growth, 13, 17, 30, 32, 41, 43, 45
   and industrial relations, 151
   and liberalism, 84
   and population, 40, 51
   decline, 38
   leading sector, 17
   modern, 17, 35, 40, 51
   paths, 36, 38
   stagnation of, 37
   See also modernization
economic reform, 45
education, 30, 41, 50, 66, 83, 84, 92
   and legitimation, 89
Eenheidsvakcentrale EVC, 125, 126, 150
Eijk, C. van der and B. Niemöller, 144
Eisenstadt, S.N., 27, 28, 111
Elizabeth of England, 79
Ellemers, J.E., 3, 73, 156, 172
Ellman, M., 25
emancipation hypothesis, 105, 106
England, 1, 36, 37, 74
enlargement of scale, 7, 31, 41, 43, 47, 49, 51, 52, 58, 156, 157
   and education, 162
   and rationalization, 57, 160
   and role expectations, 157
   and social control, 160, 161
   defined, 160
   of agriculture, 57
   of enterprises, 55, 56
   See also modernization
entrepôt trade, 35
EO
   See Protestant
estates, 83
Etzioni, A., 27, 159

EVC
   See Eenheidsvakcentrale EVC

F

Federatie Nederlandse Vakbeweging FNV
   See FNV
Flora, P., 115
FNV, 152, 166
France, 1, 17, 35, 36, 74, 79, 82
franchise
   and social control, 112
   extension of, 111, 112
Frank, A.G., 13
Fusé, T., 9

G

Gadourek, I., 100, 118, 168, 169, 170
Galbraith, J.K., 25
Ganzeboom, H., 50
German Empire, 74
Germany, 17, 45, 83
Gerschenkron, A., 26
Goddijn, W., 92, 101, 117, 120
Golden Age, 36, 37, 43
Goudsblom, J., 170
government
   stability of, 116
GPV, 153
Graaf, P. de, 50
Griffiths, R.T., 37
Groot, Hugo de, 1

H

Habsburg
   House of, 74
   monarchs, 75
Hall, van, 84
Hartman, P.T., 129
Heek, F. van, 49, 92
Hellemans, S., 157, 158
Hemels, J.M.H.J., 69
Hendriks, J., 93
heterogamy, 50
Hildebrand, 15
Historic School, 15
historical determinism, 27
Holland, 35, 38, 76, 77, 79, 80, 81, 84, 85
Holthoon, F.L. van, 86
Hoselitz, B.F., 16
Huygens, 1
hydraulic management, 79

I

Idenburg, Ph.J., 66
income inequality, 51
incompatibility, 22, 23
   and modernization, 23
individualism

248

See modernization
individualization, 21
    defined, 168
    See also de-pillarization of society
Indonesia, 45
industry
    sector of, 17, 48
intellectuals, 145
interest group, 146, 147

J
Japan, 16, 17, 19, 21, 22, 27, 42
Jonge, J.A. de, 45, 46

K
Keesing Historisch Archief, 144
Kerr, C., 24, 25, 30, 31, 32
King, Gregory, 37
Knaap, G.A. van der, 52
knowledge
    content of, 30
Kooij, P., 54
Korea, 19
KRO
    See Roman-Catholic
Kruijt, J.P., 100, 101, 117, 120, 155, 166
Kuiper, D.Th., 89
Kuiper, G., 100
Kulturkampf, 87
Kuyper, Abraham, 88, 89, 90, 101
Kuznets, S., 17, 35
KVP
    See Roman-Catholic

L
Leopold III, 157
Levy, M., 9
Lewis, W.A., 9, 13, 20
liberalism, 83, 84, 85, 88
Liberals
    political party VVD, 145, 146
Lijphart, A., 2, 91, 101, 102, 107, 108, 109, 110
Limburg, 78, 86, 90, 95
Lindblom, Ch.E., 25
Lipset, S.M., 25
List, 15
living
    patterns of, 31
Local Government Act, 87
Luxembourg, 74, 84

M
Maddison, A., 35, 36, 37, 38, 41, 42, 43, 44, 70, 151
Maine, 9
Marx, K., 8, 14, 19, 20, 21
McClelland, D., 9, 21
mercantilist policy, 35
merchant class, 35, 77, 85
Middendorp, C.P., 156, 168, 169

mixed marriage
    attitude to, 154
model
    of concentric circles, 101, 155, 156, 166
modernization, 1, 7, 18, 20, 21, 24, 27, 28, 29, 30, 41, 43, 47, 49, 111
    and authority, 59
    and bipolarity, 14
    and bureaucratization, 18
    and division of labour, 18
    and economic development, 9, 16
    and economic growth, 10, 11
    and enlargement of scale, 160, 172
    and evolutionism, 8, 19
    and individualism, 20, 22
    and industrialization, 7, 14
    and instrumental rationality, 18
    and secularization, 18, 87, 103, 160, 172
    and social consciousness, 33
    and social control, 161
    and social differentiation, 73
    and social theory, 7
    and strain, 15, 23
    and technological development, 10, 18
    and technology, 10
    and the Third World, 13
    and the welfare state, 172
    concept, 8, 9, 18, 29
    defined, 11
    degree of, 9
    dimensions of, 8, 22
    stages of, 15, 16, 17
    See also pillarization
Moore, W.E., 26
Munster
    Bishop of, 79
    peace of, 75

N
NCRV
    See Protestant
newspaper tax, 69
Niemöller, B.
    See Eijk, C. van der
NKV
    See Roman-Catholic
nobility, 74, 75, 76
NVV
    See Social-Democratic

O
occupational prestige, 49, 50
occupational structure, 52
oligarchy, 89
optimum regime, 24, 30
Orange
    House of, 75, 76, 77, 78, 80
Orangists, 77, 78

P

Pacification Democracy, 91, 108
Parsons, T., 8, 14, 16, 17, 20, 21, 25
patrimonialism, 158
Patriots, 82
pattern variables, 14, 16, 21
PBO
    See Publiekrechtelijke Bedrijfsorganisatie
Pen, J., 51
Philip II, 74
pillar, 2, 100, 104, 138
    and ideology, 2, 104
    and religion, 103
    and social functions, 100
    and social identification, 104
    and social organization, 104
    defined, 104
    the idea of a general, 103
    See also pillarization
pillarization, 2, 86, 108
    and accomodation, 108, 109, 110, 146, 151
    and broadcasting, 122, 139
    and bureaucracy, 110
    and class society, 102
    and cleavage, 105, 107, 108, 138
    and ideology, 162
    and industrial relations, 124, 129, 130, 149
    and modernization, 3, 7, 73, 103, 112, 141, 156, 157, 172
    and political restoration, 142
    and political stability, 91, 108, 114, 117
    and politics, 101
    and role expectations, 160
    and school system, 117, 148, 149
    and social class, 107, 127
    and social control, 107, 112, 114, 134, 141, 157, 159, 173
    and social identification, 100, 160
    and social organization, 101, 117
    and strike activity, 127, 128, 130, 149
    and the economy, 123
    and the inverted U-curve, 173
    and the press, 69
    and the welfare state, 157
    defined, 104
    index of, 147, 148
    indicator of, 101
    model of, 100, 103, 105, 113
    See also pillar
pluralism, 1, 2, 3
    and cleavage, 2, 73, 74, 78, 89, 109
    and the Republic, 78
    change of, 85, 110, 158
    degrees of, 109
poor relief, 44, 83, 84, 86, 87
poor relief act, 87
Post-Industrialism, 14
principle of proportionality, 89, 91, 108, 158
production
    organization of, 31
    resources of, 31
productivity, 41, 42, 43
professionalization, 157
protection hypothesis, 106
Protestant
    broadcasting association EO, 123
    broadcasting association NCRV, 122, 123
    Christelijk Nationale Werkmansbond, 94
    modernism, 88, 89, 94
    orthodoxy, 87, 88, 89
    Patrimonium, 94, 96
    political party ARP, 88, 166
    political party CHU, 88, 114, 166
    trade-union federation CNV, 125, 152, 166
Provo, 144
Pryor, F.L., 26
public sector
    and pillarized organization, 158
    growth, 31, 151
    growth and modernization, 63
Publiekrechtelijke Bedrijfsorganisatie PBO, 126
PvdA
    See Social-Democratic

Q

Quanta Cura, 92

R

Raad van Vakcentrales, 124
raadspensionaris, 76
rationality, 12, 157, 160
rationalization, 49
    and production, 47
    and the judicial system, 61
    See also enlargement of scale
regenten
    See regents
regents, 76, 79, 80
religion
    intensity of, 141
repartition system, 80
representation
    proportional, 91, 142, 163
Republic, 38, 84, 89, 91
    agriculture, 38, 44
    and education, 65
    and external enemies, 79
    and factions, 77
    and integration, 79
    and national unity, 75
    and oligarchy, 77
    and political parties, 77
    and social class, 77, 78
    industrial structure, 38
    investment rate, 38
    labour force, 38
    legal system, 82
    political structure, 76
    power structure, 73, 76, 79, 83
    ruling class, 76
    service sector, 38
    social stratification, 44
    trade monopoly, 35

transportation system, 40
unemployment, 44
revolution
   and class consciousness, 83
   of 1848, 83
Ricardo, 8
Riesman, D., 9, 25
Righart, H., 105, 106
Riley, J.C., 38, 40
riots, 78, 85
Rokkan, S., 109
Roman Empire, 74
Roman-Catholic
   and social organization, 91, 92, 93
   broadcasting association KRO, 122
   church crisis, 157
   leadership, 86
   Mandement, 124, 125, 126
   Nederlandsch Roomsch-Katholieke Volksbond, 95
   orthodoxy, 87
   political party KVP, 166
   position in the Republic, 78
   radicalization, 87
   restoration of episcopal hierarchy, 86, 87
   trade-union federation NKV, 125, 152, 166
Roorda, D.J., 77
Ross, A.M., 129
Rostow, W.W., 9, 15, 16, 17, 18, 22
Rotterdam, 44, 46
Russia, 17, 27, 74

S
Saint-Simon, 8
Schendelen, M.P.C.M. van, 107
Schlesinger Jr., A.M., 25
Schmoller, 15
school issue, 67, 89, 90, 91, 92, 108, 119
Schumpeter, J.A., 27
secularization, 157, 170
   and church membership, 163
   and education, 148
   and marriage consecration, 164
   defined, 163
   of Dutch society, 166
   See also de-pillarization
   See also modernization
SER, 126, 151
service sector, 47
SGP, 153
Shill, 25
Singapore, 19
Sixma, H., 50
Smelser, N.J., 20
Smit, J.W., 75
Sociaal en Cultureel Planbureau, 147, 168
Sociaal en Cultureel Rapport 1986, 148, 154
social class, 44, 46, 47, 49
   segmentation, 93, 95, 96
social codes, 28, 30, 32, 73, 89, 111
   defined, 28
social control, 2, 20, 83, 86, 162, 168

social differentiation, 3, 8, 20, 59
social distance, 32, 44, 83, 100, 130, 132, 133, 153, 170
Social Economic Council
   See SER
social ethics
   See social codes
social mobility, 22, 31, 33, 43, 44, 46, 47, 49, 50, 51, 83, 89, 162
   See also modernization
social question, 108
Social-Democratic
   broadcasting association VARA, 122, 123, 167
   newspaper Het Vrije Volk, 167
   organization, 167
   political party PvdA, 145, 146, 167
   trade-union federation NVV, 94, 125, 152, 166
   youth movement AJC, 167
Social-Democrats
   and state regulation, 159
socialism, 88
socialization, 89
Sombart, 15
Spain, 74
Spanish Empire, 74
Spencer, H., 8
Spinoza, 1
Staatsgezinden, 77, 78
Stadtholder, 76, 77, 80
standardization, 47
Staten, 76, 81
Staten-Generaal, 75, 76, 77, 81, 87
Stichting van de Arbeid, 125
strike activity, 127, 128, 129, 130
   See also pillarization
strike volume, 127, 128
Stuijvenberg, J.H. van, 37
Stuurman, S., 91, 96, 102, 107
suburbanization, 55
suffrage, 85, 108, 111
   See also franchise
Sweden, 74
Switzerland, 74
Syllabus Errorum, 90
Szirmai, A., 51

T
Taiwan, 19
take-off, 9, 17, 18
technological determinism, 12
technological development, 12, 13, 19, 29, 30, 32, 42, 43, 45, 47, 49, 50, 156
   and values, 12
   defined, 10
   logic of, 18, 30
   purposive, 11
   See also modernization
technological innovation
   See technological development
technology
   defined, 10
   See also modernization

Thorbecke, 85, 86, 87
tiende penning, 75
Tijn, Th. van, 44, 45, 47, 93, 94
Tinbergen, J., 24, 25, 30, 51
Tönnies, 9
Touraine, A., 14
trade union, 94
    and class struggle, 94
    development, 94
    inderdenominational, 95
    See also Protestant
    See also Roman-Catholic
    See also Social-Democratic
TROS, 123
Tulder, J.J.M. van, 49, 50

U

UK, 17, 35, 38, 41
Ultee, W.C., 50
urbanization, 52
USA, 17, 25, 26, 27, 35, 42

V

VARA
    See Social-Democratic
vassalage, 35
Vercruysse, E.V.W., 49
vertrossing
    in broadcasting, 123, 149
    in Dutch politics, 142
verzuiling
    See pillarization
Vienna
    Treaty of, 84
Vietnam, 162
VOO, 123
Vries, Jan de, 36, 37, 38, 39, 40, 41, 43, 44, 47, 70
Vries, Johan de, 38, 41
vroedschap, 75
VVD
    See Liberals

W

wage policy, 150, 151
Wansink, H., 76, 80
Weber, M., 8, 9, 12, 21, 33, 59
Weber, Max, 160
welfare state, 157
William I, 45, 84
William II, 84
Windmuller, J.P., 95, 96, 129
Witt, Johan de, 1
Wolfe, B.D., 27
work
    patterns of, 31

Z

Zealand, 76, 77, 79, 80
zuil(en)
    See pillar

DATE DUE

MAR 2 6 1996			
MAR 1 5 1998			
ILL: 2193559			

Demco, Inc. 38-293